MEDITERRANEAN MARINE ECOSYSTEMS

NATO CONFERENCE SERIES

I	Ecology
II	Systems Science
III	Human Factors
IV	Marine Sciences
V	Air–Sea Interactions
VI	Materials Science

I ECOLOGY

Volume 1 Conservation of Threatened Plants
edited by J. B. Simmons, R. I. Beyer, P. E. Brandham, G. Ll. Lucas, and V. T. H. Parry

Volume 2 Environmental Data Management
edited by Carl H. Oppenheimer, Dorothy Oppenheimer, and William B. Brogden

Volume 3 The Breakdown and Restoration of Ecosystems
edited by M. W. Holdgate and M. J. Woodman

Volume 4 Effects of Acid Precipitation on Terrestrial Ecosystems
edited by T. C. Hutchinson and M. Havas

Volume 5 In Vitro Toxicity Testing of Environmental Agents: Current and Future Possibilities (Parts A and B)
edited by Alan R. Kolber, Thomas K. Wong, Lester D. Grant, Robert S. DeWoskin, and Thomas J. Hughes

Volume 6 Trace Element Speciation in Surface Waters and Its Ecological Implications
edited by Gary G. Leppard

Volume 7 Trends in Ecological Research for the 1980s
edited by June H. Cooley and Frank B. Golley

Volume 8 Mediterranean Marine Ecosystems
edited by Maria Moraitou-Apostolopoulou and Vassili Kiortsis

MEDITERRANEAN MARINE ECOSYSTEMS

Edited by

Maria Moraitou-Apostolopoulou

and

Vassili Kiortsis

University of Athens
Athens, Greece

Published in cooperation with NATO Scientific Affairs Division

PLENUM PRESS · NEW YORK AND LONDON

Library of Congress Cataloging in Publication Data

Main entry under title:

Mediterranean marine ecosystems.

(NATO conference series. I, Ecology; v. 8)

NATO Advanced Research Institute on "Mediterranean Marine Ecosystems": a workshop held in Heraklion, Crete, 9/23–27/83; sponsored by NATO Eco-Sciences Special Programme Panel and Marine Science Panel.

"Published in cooperation with NATO Scientific Affairs Division."

Bibliography: p.

Includes index.

1. Marine ecology—Mediterranean Sea—Congresses. I. Moraitou-Apostolopoulou, Maria. II. Kiortsis, V. (Vassili), 1925– . III. NATO Advanced Research Institute on "Mediterranean Marine Ecosystems" (1983: Heraklion, Crete). IV. North Atlantic Treaty Organization. Special Program Panel on Eco-Sciences. V. NATO Special Program Panel on Marine Science. VI. Series.

QH93.M43 1985 574.5′2636′091638 85-525

ISBN 0-306-41910-6

Proceedings of a NATO ARW on Mediterranean Marine Ecosystems,
held September 23–27, 1983, in Heraklion, Crete, Greece

A Division of Plenum Publishing Corporation
233 Spring Street, New York, N.Y. 10013

Printed in the United States of America

PREFACE

This book contains the papers presented at a NATO Advanced Research Institute on "Mediterranean Marine Ecosystems", held at Heraklion-Crete, Greece, from September 23-27, 1983. A workshop rather than a conference, it was sponsored by the Eco-Sciences Special Programme Panel, in cooperation with the Marine Science Panel. The third of its kind, it was scheduled in the framework of a project on a multidisciplinary integrated approach to the study of the Mediterranean.

This Sea and the surrounding land was not only the cradle of many civilizations but is, up to the present time, one of the major world areas of marine traffic, communication and exchanges, fisheries and aquacultures, inshore human activities and... pollution. To a certain degree it constitutes a gigantic natural laboratory, where the fate of threatened aquatic and terrestrial ecosystems including the human one, is tested.

The Mediterranean Sea, with its geological history and present-day geographic, hydrological and climatic conditions is believed to form an ecological entity. Important exchanges and mutual influences take place with the surrounding land area and the water masses, naturally (Atlantic, Black Sea) or artificially (Red Sea), connected to the Mediterranean. Therefore, a better and in-depth knowledge of the various ecosystems, benthic, planktonic and nektonic, neritic or pelagic, in the Western or the Eastern Basin seems to be a prerequisite to any action in preserving, upgrading and managing the natural resources of the area.

In a time, when so many agencies and organizations show an increasing concern for this sea and its problems, amidst conflicting priorities, a rigorously scientific dialogue of specialists from Mediterranean and other countries on the present and future state of marine ecosystems there, appears of interest.

The Eco-sciences and Marine Sciences Panels, the organizing committee of the A.R.I. and the editors of these proceedings, thought that this would be a useful contribution. Time will show if their expectations would be fulfilled.

As editors and organizers we should like to express our thanks to N.A.T.O. Eco-sciences Panel who supported this meeting; to the other members of the organizing committee, Prof. B. Battaglia (co-director) and Prof. C. Almaça; to all participants for their stimulating contributions and the small staff for logistic support. Our gratitude goes mainly to Dr. Trevor Platt (Canada), whose invaluable advice, help and scrutiny, down to details, made our editorial task an easy one. Last but not least thanks are due to Plenum Press for their willingness to publish this book.

Maria Moraitou-Apostolopoulou

Vassili Kiortsis

CONTENTS

ECOLOGICAL FACTORS AND THEIR BIOGEOGRAPHIC CONSEQUENCES IN THE MEDITERRANEAN ECOSYSTEMS

Michele Sara

Istituto di Zoologia, Universita di Genova
Via Balbi 5, Genova, Italy

INTRODUCTION

The biogeographical research of today is mainly concerned with the explaining of the pattern of species and ecosystem distribution from a causal point of view. Unfortunately this effort involves difficult questions to which the present development of biogeography does not give precise answers. Recently three hypotheses have been advanced concerning terrestrial and freshwater biogeography; they are known as the vicariance, the dispersal and the ecological models, according to the set of factors that is considered more important in determining the distribution of plants and animals (Endler, 1982). The vicariance and dispersal hypotheses tend to emphasize especially the historical factors, whereas the ecological hypothesis reflects the present ones. However it seems clear that a synthesis is needed which may incorporate these competing models and that in every case ecology plays a major role together with dispersal and vicariance (Davis, 1982). Consequently it appears very useful also for marine biogeographical research to adopt a plural approach which should take into consideration the different sets of factors and their interrelationships. As other branches of biology, biogeography is directly concerned with evolutionary problems. When we analyze the role of ecological factors we cannot consider only the present one, but also that which they played in geological time, influencing the immigration and extinction of species and the evolutionary rates of taxa and ecosystems.

Marine biogeography is, both from a causal and descriptive point of view, less developed than terrestrial biogeography. Progress in marine biogeography is hampered by some general features of the seas such as their continuity, ecological complexity and diversity of inhabiting organisms, and by the fact that the groups more important in terrestrial biogeography, such as mammals, are, in the sea, scarcely represented or lacking. Thus a deeper insight into the connection between ecological and historical factors should be considered as an essential task of future marine biogeography, to-day generally restricted to descriptive data on the geographic distribution of species and ecosystems in relation to single environmental variables.

The Mediterranean Sea, for which it is possible to utilize considerable knowledge from taxonomical, ecological and geopaleontological points of view, shows a high number of endemic species and a considerable diversity of environmental conditions, which lead to the occurrence, in different areas, of both temperate and subtropical benthic populations (Pérès, 1967). It represents, therefore, a favourable place for testing, also from a general standpoint, the role of the ecological factors - abiotic and biotic, operating to-day or in the past, and in their connection with the historical ones - in determining the present day composition and distribution of its fauna and flora.

ECOLOGICAL FACTORS AND THE HISTORY OF MEDITERRANEAN BIOTA

It is interesting to consider first the role of ecological factors during the geological events which occurred in the Mediterranean basin from its formation to the present time. The impact of paleoecology on the Mediterranean biogeography is in fact very significant.

It is well known that the present configuration of the Mediterranean within the wider Tethyan Sea has been shaped by tectonic events during upper Miocene, Pliocene and Quaternary times (Ruggieri, 1967). During the Miocene, communication with the Atlantic ocean was restricted and that with the Indian ocean, which previously existed across Syria, was closed. An irregular communication was left across the Balkan peninsula and Turkey with the large northern Paratethys Sea, inhabited by a distinctive Sarmatic fauna. However the ancient, and in a large measure tropical-like tethyan population was destroyed, with the exception of some possible relics, when the communication between the Mediterranean basin and the Atlantic ocean was interrupted, and a salinity crisis arose at the end of Miocene, in the Messinian. The Mediterranean was transformed into a series of lagoons, but some of them were gradually desalinified and developed, in the western area, an oligohaline warm-water fauna with gastropods of the genus Melanopsis. The Melanopsis fauna was also destroyed with the subsequent reestablishment, in the Pliocene, of a communication between Atlantic and Mediterranean, in the place of the present Strait of Gibraltar, and the establishment of normal marine conditions. The Mediterranean was repopulated by the reintroduction of the Atlantic fauna from the area off the Strait of Gibralter;

Ruggieri (1967) suggests that this area may have represented an asylum for Indopacific relicts during the salinity crisis of the Messinian. It is an open question whether the presence of the tropical forms now inhabiting the Mediterranean, apart the Lessepsian immigrants, is due to the ancient Tethyan population or depends on the subsequent pliocenic and quaternary, Senegalese immigrations. Probably, the great majority of these forms should be attributed to immigration (Kosswig, 1956; Pérès, 1967; Por, 1978). Already in the early Pliocene, according to Klausewitz (1973), a Senegalian fauna of the tropical-subtropical West African coasts repopulated the Mediterranean basin. However, during the Pliocene, as is shown by data on extinction of molluscs (Ruggieri, 1967), a gradual transformation of biota occurred following the changing of the climate towards cooler conditions, and a consequent substitution of temperate for tropical forms.

The most important events which have shaped the actual biogeographic pattern of the Mediterranean are bound to the climatic fluctuations related to the glacial and interglacial periods which characterize the course of the Quaternary. During glacial maxima, surface water temperatures were some degrees lower than now and salinities were variable: this is connected also with the opening and closure of the Bosphorus - through which flow the less-saline waters of the Black Sea due to the fluctuations in the sea level.

The climatic fluctuations were accompanied not only by changes in the temperature and salinity of the waters but also by changes during transitions from glacial to interglacial periods in density stratification of water layers, due to the additional effect of melting of glaciers, reestablishment of communication with the Black Sea and increase in precipitation and river runoff. These changes have been related to a variation in the current direction at the Strait of Gibraltar, and consequently with the type of immigrants from the Atlantic.

As we know, in the present time (comparable with an interglacial period), the Mediterranean Sea has a deficient hydrobiological balance, with evaporation exceeding the supply of fresh water from streams and precipitation, this deficiency being mainly compensated by the flow of Atlantic waters through the Strait of Gibraltar. This flow enters with a surface current less salty and less dense while another deeper current flows from Mediterranean to Atlantic. The boundary between the two currents varies, according to the season, between 100 and 300 m of depth. According to Mars (1963) and his theory of current inversion in the Strait of Gibraltar, at the end of the glacial periods, for the various reasons mentioned, the Mediterranean waters become more diluted than the Atlantic ones. This is followed by the penetration of cold, deep Atlantic waters into the Mediterranean with a further decrease of salinity. In the course of the interglacial periods a deficient hydrobiological balance is gradually established in the Mediterranean, with the consequence that a warm water current flows from the Atlantic and is accompanied by an increase of salinity. Between these periods and the related current inversions are brief period of deficient circulation with a consequent relative impoverishment of the benthos. In fact, the quaternary Mediterranean fauna is characterized by an alternate appearance of deep-water northern immigrants favoured by the low temperatures and by the low salinities, and of littoral Senegalese immigrants favoured by the opposite conditions. According to Ruggieri (1967) the overall effect of this alternation of climatic conditions has been the reduction of the number of subtropical species of Pliocene origin with an impoverishment of the Mediterranean fauna only partially balanced by the acquisition of new species through Pleistocene immigrations. On the contrary Valentine (1967) observes that fluctuating climates act as a sort of diversity pump, enhancing diversity during climatic deterioration and returning some fraction of the new lineages during climatic ameliorations. The Mediterranean fauna was enriched by this process and also by recruitment of species from the extra-Mediterranean provincial margins, especially during ameliorating climatic phases.

Thus pleistocenic events show the remarkable role of fluctuations of the environmental parameters, and also that of the different types of immigration waves in shaping the biogeographic pattern of the Mediterranean. It is also noteworthy to underline the interlacing of the ecological and historical factors because the environmental changes determined the conditions of species dispersal (Fig. 1). The quaternary history of the Mediterranean gives to this sea its peculiar biogeographic physionomy in which the relationship between the Mediterranean and Atlantic (between the Channel and Cape Bianco in Mauritania) fauna and flora is fundamental. The Mediterranean is therefore generally considered (Mars, 1963; Briggs, 1974, etc.) as a subprovince of the Atlanto-mediterranean province in the Atlanto-mediterranean region which includes also the Black Sea province (according to Briggs, 1974, two other provinces of this region are the Caspian Sea and the Aral Sea; according to Por, 1981, a Lessepsian province could be established for the eastern Mediterranean not including the Aegean and the Adriatic seas). Anyway, the Strait of Gibraltar did not function as an important obstacle to the intermingling during the Pliocene and Quaternary times of Atlantic and Mediterranean species. More important, also in relation to the original adaptive characters of their biota, of Tethyan and Sarmatian origin respectively, was the obstacle represented by the Bosphorus between the Mediterranean and the Black Sea.

The Mediterranean Sea, however, is a special part of the Atlanto-mediterranean province because its fauna is considerably richer than that of the Atlantic coasts and is characterized by many endemic species. The percentage of endemism is very high for the sessile or sedentary groups such as ascidians with 50.4% (Pérès

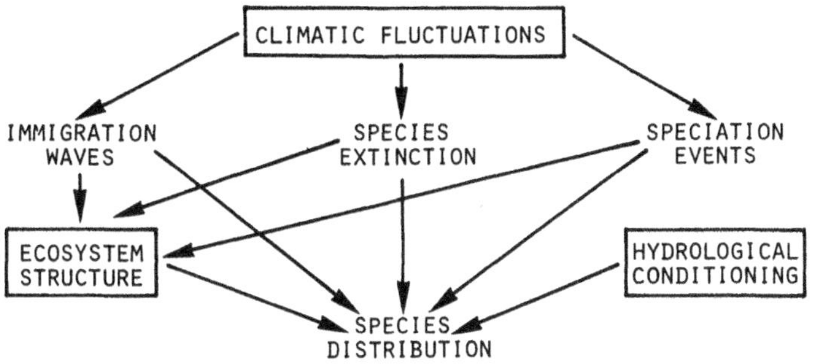

Fig. 1 Interdependence of ecological (in rectangles) and historical processes in the Mediterranean biogeography.

and Picard, 1964), sponges with 42.4% (Vacelet, 1981), hydroids with 27.1% (Pérès and Picard, 1964), echinoderms with 24.3%(Tortonese, this book) but it is also considerable for the other groups such as decapod crustaceans with 13.2%(Pérès and Picard, 1964; Pérès, 1967), and fishes with 10.9% according to recent data of Tortonese (this book). The great richness of the Mediterranean fauna with its many endemic species compared to the Atlantic coasts, which have extremely low endemism, leads to the conclusion (Briggs, 1974) that the Mediterranean basin has probably functioned as a primary center for the evolution and radiation of the eastern Atlantic warm temperate fauna. According to this hypothesis a large contingent of the Atlantic coast species, possibly the majority, must have come from the Mediterranean. Two mechanisms may be taken in consideration for these evolutionary facts. On one side a high rate of speciation may have been favoured by the climatic fluctuations and the consequent niche variations: recent evolutionary literature (Schopf, 1981) admits as possible a very rapid speciation rate, also in the order of few thousands of years. On the other hand the Mediterranean may have played a role of refuge for many Atlantic species which have found suitable conditions for their survival only in this sea. These two mechanisms, related to the vicariant and dispersal hypotheses, respectively, have probably cooperated under the general conditioning of the ecological factors.

Another aspect of Mediterranean biogeography in which a historical event has had a considerable impact thanks to suitable ecological factors is very recent and regards the unidirectional migration of Red Sea species to the Mediterranean after the opening of the Suez Canal, the so-called Lessepsian migration (Por, 1971, 1978). According to Por (1978) the Lessepsian migrants can be numbered at about 500 species of which 200 listed, concentrated along the Levant coasts. Some fishes, decapod crustaceans and molluscs have advanced beyond the limits of the Levant basin. The Lessepsian migration followed the historical event of the opening of the Suez canal but was fortuitous in the ecological situation, the Levant basin being a depauperated area. Also in this case it is possible to observe a concurrence of dispersal and ecological factors.

The biogeographical situation of the Eastern Mediterranean, characterized by the scarcity of species, arose as a consequence both of the history and the ecology of the basin. As shown by Udvardy (1969) the Pleistocene climatic fluctuations, accompanied by distance and time barriers, were a severe obstacle for many species to reach the eastern extremity of the Mediterranean. Therefore the eastern Mediterranean and especially the Levant basin may be considered as just a periphery of the western Mediterranean. Differences between the western and eastern zones arose because many western species never, or only episodically, reached the eastern basin; in fact, there is no indication of species peculiar to the eastern basin during the Pleistocene. Eastern Mediterranean climatic and sea-level fluctuations cannot be easily correlated (Por, 1978) to the fluctuations which characterize the Pleistocene period in the western basin. This fact is reflected in faunistic differences. For example the cold water and low salinity fauna of the Arctica islandica association which characterizes the Calabrian, Sicilian and Würm periods is not represented in the eastern Mediterranean, with few exceptions (Zaccaria, 1968; Mars and Picard, 1958). On the contrary, the Tyrrhenian warm water complex with Strombus bubonius is well represented. A significant role in determining floro-faunistic differences with the western zone has probably been played by the recurrent semi-isolation of the eastern basin at the Sicily channel, which during the Riss was only 300 m deep and few kilometers wide (Pfannenstiel, 1960). This barrier was not only a hindrance to the diffusion of species but also an obstacle to water exchange, with a deficient oxygenation at bottom levels due both to the lack of water change and to the increase of saline stratification. Thus, during the Pleistocene, environmental instability, fluctuating salinity and oxygenation, higher temperatures made the eastern Mediterranean an impoverished appendix of the western Mediterranean. The last occasion for a large faunal exchange between the two basins occurred during the Tyrrhenian. After this period there is an increased diversification. In the last two glacial periods (Wurm II and III) which followed one another closely there occurred a massive extinction of the warm water fauna of the Tyrrhenian phase with the disappearance of the Senegalian molluscs. Fissurella nucleata, Cypraea lurida and Purpura haemastoma, now living on the west African subtropical and tropical coasts and in the southern and eastern Mediterranean, but not represented in the western and northern Mediterranean, are probably relics of this fauna (Pérès, 1967).

The floro-faunistic impoverishment of the eastern Mediterranean compared with the western Mediterranean richness in species is well documented. The species diversity of all the environments decreases sharply from the western Mediterranean to the Aegean basin and from this to the Levant one. As results from the data summarised by Por (1978), this impoverishment is true of the plankton as well as the benthos. Planktonic Hyperiidea (Amphipoda) are represented in the Atlantic by 118 species, in the western Mediterranean by 46 species and in the eastern part only by ten species. Not one of the eight species of Pteropoda of the western Mediterranean reaches the eastern zone. For the benthos a richer documentation is available. The impoverishment is impressive among the decapod crustaceans, the echinoderms, the ascidians. For example the Molgulidae are represented by eight species in the western Mediterranean, by four in the Sicily-Tunisian area, by one in the Aegean Sea, none is found along the Israel coasts. Facies-building species such as Posidonia oceanica and the gorgonians are less represented in comparison with the western Mediterranean. Fredj (1974) calculates that the eastern Mediterranean as a whole contains only 48.4% of the 1244 species globally analyzed, while the western Mediterranean contains the 92% of them. Furthermore, according to Fredj, the eastern Mediterranean has only the 4.2% of species strictly limited to this basin against the 35.4% of species found only in the western Mediterranean. Of the 247 Mediterranean endemic species 49% are exclusive to the western Mediterranean and only 4.5% to the eastern basin with the exclusion of the Aegean Sea, which may be considered as a part of the eastern Mediterranean clearly separated from the Levant basin. For the Levant basin, practically the only contingent of characteristic species are the Lessepsian migrants, but this contingent, with a percentage of 10-15% compared to the autochthonous species, represent such important biogeographic novelties for the Mediterranean, at the level of genera and families, that Por (1981) suggests the establishing of a new subtropical province for the eastern Mediterranean, not including the Aegean Sea and the Adriatic Sea: a Lessepsian province, interposed between the warm temperate province of the western Mediterranean and the tropical province of the Red Sea.

The drop in diversity, which is to-day only partially compensated by the Lessepsian migrants and is accompanied occasionally by a quantitative increase of species well adapted to warm and saline waters, such as the green algae Caulerpa racemosa and C. scalpelliformis, should be attributed to an interlacing of historical and environmental factors. Neither the historical factors, for example the isolating barriers, nor the ecological ones, such as the high temperatures and salinities, the scarcity of nutrients and lack of oxygenation are to be considered separately. Rather it is right to consider that the actual ecological conditions are not suitable for many western Mediterranean species. When historical events have allowed the immigration of other species of tropical origin such as the Lessepsian ones, the ecological conditions have not represented a significant obstacle to their successful implantation.

The interlacing between isolating barriers and ecological factors has operated also in the past in determining the extraordinary paucity of nutrients which characterizes the Mediterranean as a whole but especially the eastern basin. The values of nutrient concentrations in the Levant basin are extremely low (a mean value of total phosphorus of 0.8 mg/l). Thus the Levant basin is not only a subtropical, saline "cul de sac" of the warm temperate Atlantic Mediterranean biogeographic area (Por, 1978), but is also characterized by the lowest values of primary production recorded in oceanic waters. Following data by Por (1978), the mean gross yearly production of the coastal waters of Israel is around 60% less than that of the western Mediterranean. Benthic level-bottom biomass between 200 and 1000 m (Tchukhtchin, 1964) is of 0.089 g/m^2, against 4.12 g/m^2 on the Sicily-Tunisian sill, 266 g/m^2 in the Atlantic, and 388 g/m^2 in the Black Sea. In circalittoral areas Gilat (1964) finds 8.16 g/m^2 for the eastern Mediterranean and only 0.059-0.075 g/m^2 for the coasts of Israel against 48-126 g/m^2 for the Adriatic. It is indeed not surprising to find high biomass values also in the eastern Mediterranean where, exceptionally, special local conditions, such as near the delta of the Nile, are present. Thus, in the fishery grounds of Alexandria, Vatova (1935) has found the high biomass value of 551 g/m^2. This is of course related to the local influence of Nile waters rich in nutrients, but this enrichment is strictly localized and seasonal in character and has little importance for the Levant basin. A peculiar effect of the scarcity in nutrients of this basin is the so-called "Levantine nanism" which is recorded for Sipunculida (Stephen, 1958), Porifera (Levi, 1957) for fishes and also as a general phenomenon (Tortonese, 1951).

It is remarkable that this shortage of nutrients has not represented an important obstacle to the immigration of species from the Red Sea. Evidently it has been counterbalanced by the number of free niches, well-suited for tropical species adapted to warm and saline waters (Por, 1978). This recent and presently occurring biotic modification of the Levant basin is a clear example of the role of immigration in biogeography, but the immigration would not be possible if ecological conditions were not suitable and if there were not wide ecological niches available. The new immigrants can modify the composition of the communities either after competition and substitution of the preexisting species or simply by occupying wide niches. According to Por (1978) this last hypothesis occurred for the Lessepsian migration as confirmed by the fact that there are few cases of congeneric replacement (vicariance) but rather appearance of new ecological types of animals (Por, 1981). Immigrant establishment in the communities succeeded also in relation to the vacuum left by the cold water complex of species after the end of the glacial period, with the subtropical warming of the Levant Sea. Indeed a complex series of interrelationships between ecological and historical factors has favoured the success of the Lessepsian migration (Fig. 2). Also, the further biogeographic subdivision of the western and eastern mediterranean basins is generally dependent both on historical events and on paleo- and modern ecological

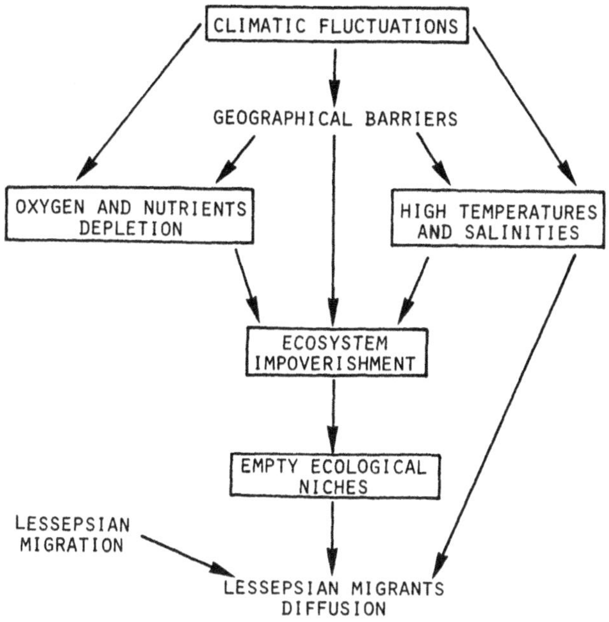

Fig. 2 Interdependence of ecological (in rectangles) and historical processes leading to the successful diffusion of Lessepsian migrants in the eastern Mediterranean basin.

factors. An environmental conditioning, operating also to-day, clearly allows a separation of the western Mediterranean into a northern part along the coasts of Spain, France and Italy and a southern part along the North-African coast. The special subdivision of the Alboran Sea near the Strait of Gibraltar, on the other hand, is due largely to the historical events of the Atlantic immigration. The eastern basin is usually considered as a very heterogeneous area. The Adriatic Sea and the Aegean Sea are characterized by a particular history, but also by a particular hydrography. In these two basins the temperature and salinities are lower than those of other parts of the Mediterranean. Along the African coast there is an inshore area between the two opposing gulfs of Sidra and Gabès confined between the west coast of Cyrenaica and the east coast of Tunisia, where the salinities are of the west-Mediterranean type, i.e. below 39 o/oo, but temperatures are high and subtropical, and the tides very pronounced, so that this area is somewhat transitional between the western and the eastern basins. Also the Ionian Sea, north of the Sidra Sea and south of the Adriatic, bordered to the east by Crete and to the west by Sicily, is in fact a transitional body of water without any peculiar characteristics. Finally there is the Sea of Levant or Levant basin, delimitated by the south-Anatolian coast, the Levant shores and the Egyptian and north-Sinai coast with its peculiar, previously-analysed, biotic characteristics.

The western limits of the Levant basin should, according to Por, be defined by hydrographical parameters, i.e., the annual surface isotherm of 20 °C and the minimum surface isohaline of 39 o/oo. These boundaries, fluctuating of course, form an arc extending from the Nile delta to the eastern top of Crete and the Dodecanese. With its low salinity waters, the Atlantic current flowing in anti-clockwise direction along the African coasts, determines the deep horizontal deflection of the isohalines along the Egyptian coasts.

THE BIOGEOGRAPHIC ROLE OF ABIOTIC AND BIOTIC ECOLOGICAL FACTORS

Analyzing single ecological factors, and firstly the abiotic ones, it would appear that the most relevant biogeographic role in the Mediterranean has been probably played by temperature. In fact, temperature, directly or indirectly, has been of outstanding importance, as a paleoecological factor in relation to the pliocenic and pleistocenic climatic fluctuations, in shaping the present-day composition of the Mediterranean flora and fauna. But also in the present-day climatic situation temperature plays an important role on a biogeographic scale. With average annual values ranging from the 15.7°C of Banyuls surface waters to the 18.8° of Naples waters and to 22.8° of Alexandria ones, temperature helps to maintain the differentiation of population of the northern and southern part of the Mediterranean. In special areas of the northern Mediterranean such as along the Ligurian coast thermophilic oases can arise. A peculiar situation, with the lowest temperatures of the Mediterranean, occurs in the northern Adriatic Sea where Fucus virsoides is present, the only mediterranean species of this Atlantic genus, and where Zostera marina is well developed. On the other hand, a tropical species such as Halophila stipulacea, a Lessepsian migrant found favourable conditions in the Levant basin (Gessner, 1970) and now has reached Malta. Macrophytes are very sensitive to temperature and it has been observed that the distributional pattern of the mediterranean marine vegetation reflects not only the mean annual temperature but also the seasonal temperature fluctuations. In a temperate sea such as the Mediterranean, the differences between the highest and lowest local water temperatures are generally very great. At Banyuls the annual temperature amplitude is 12°C, at Alexandria 8°C. Furthermore, in the Mediterranean, the distributional limits of species may often be determined by occasional temperature extremes occurring at intervals of several years rather than by the normal intermediate thermal patterns. Concerning the biogeography of the deep sea fauna, it is very important to remember that the fact that in the deepest water also, Mediterranean temperature cannot exceed the minimal value of 13°C represents a serious obstacle to the penetration of many deep-water Atlantic species.

As a biogeographical factor, temperature acts, of course, together with other abiotic factors such as salinity and light. It can be also realized that it affects species distribution not only with its most obvious effect of excluding certain species from areas with unsuitable thermal regimes but also because of its impact on life histories, population dynamics and synecological relations (Kinne, 1970). Of special importance appear to be the temperature-induced changes in maturity and reproduction efficiency and time (length of breeding period, number, survival and development of offspring). Data on the effect of temperature on Mediterranean whole ecosystems are unfortunately lacking. The importance of these studies is indicated by the work of Golikov and Scarlato (1967) on japanese ecosystems (biocoenoses), in which these authors have observed that average temperature intensities and temperature fluctuations are the most effective force in determining the dynamics of these shallow-water biocoenoses of semi-enclosed coastal areas, comparable to many Mediterranean ecosystems. Temperature-related modifications refer to the distribution of communities, their species composition and their relative degree of complexity and stability. Certainly the temperature effect must be seen in the framework of other environmental factors such as the nature of substrate and the quantity of food available, and of what is called by these authors the biocenotic background, i.e. the occurrence and aggregation of certain species which can modify the general conditions and create additional niches. It must be also remembered that seasonal temperature changes may affect the species composition of the ecosystem, causing migration, periodic appearance and disappearance of species with formation of resting or active-life stages and may also influence the predator-prey relations.

Salinity, whose relevant role in the paleoecology of the Mediterranean has been previously indicated, performs also to-day in connection with temperature, but also by itself, a significant role. It is one of the essential hydrological factors differentiating the warm and high saline waters of the eastern and southern Mediterranean from the less warm and less saline waters of northern and western Mediterranean. Its influence on Mediterranean biota is particularly relevant in local conditions such as those near the deltas of rivers and in coastal brackish-water lagoons, such as the Venice lagoon. In these habitats, with a pronounced salinity gradient and reduced salt concentration, salinity becomes a major factor influencing sharply the distribution and composition of populations which may include a variety of marine, brackish and also limnic species. The general biogeographic significance of the brackish water habitats, with the wide fluctuations of hydrological parameters, may be also found in their functioning as active centers of speciation and species radiation, as indicated in Mediterranean by researches on the crustacean Tisbe (Battaglia, 1970).

The salinity factor is very important in distinguishing the Black Sea province from the Mediterranean one: the surface salinity of the Black Sea surface waters is only 17-18 o/oo against the 38-39 o/oo of

Mediterranean waters. The Mediterranean waters of the Aegean Sea also seem not significantly influenced by the Black Sea waters flowing through the Bosphorus. The water exchange pattern between the Mediterranean and the Black Sea is, however, a matter of discussion (Gessner and Schramm, 1971).

Water movement influences many phenomena affecting the distributional pattern of marine animals and plants (Riedl, 1971). These influences on marine animal distribution vary according to the requirements of organisms and the intensity of water-movement patterns. Fluctuations in water level, due to tides and similar factors, and to wave action, are of outstanding importance in determining the local horizontal and vertical distribution of organisms in the littoral zone. This action is in close relationship with the character of the substrate, owing to its different influence on the different types of rocky and soft bottoms. Water-movement due to tides and wave action can also play, beyond the local effects, a biogeographic role in relation to its general pattern which is different, also according to the type of coasts, in different Mediterranean zones. It is also important to evaluate, in Mediterranean as in another sea, the biogeographic role of currents which regulate the dispersal of larvae and propagula.

Light is the other most relevant physical factor besides water-movement, influencing the local community pattern. It affects the ecological distribution of plants and animals, vertically, but also, in shallow waters, horizontally in relation to the orientation of the coastlines and their topographic and micro-topographic features. From a biogeographic point of view its relevance is connected to the general pattern of Mediterranean coasts. However, the turbidity of water which affects the penetration of light diminishes from north to south, and we observe a major expansion in depth of infralittoral communities in the southern basins of the Mediterranean.

Oxygen requirements by faunal assemblages have been utilized in support of theories on the origin of faunas occupying a given biogeographical region (Vernberg, 1971). For example, according to Caspers (1957), habitats in the Black Sea with low oxygen contents are colonized entirely by Mediterranean forms, which have lower oxygen requirements compared to the Ponto-Caspian biogeographical element.

The scarcity of nutrients and their gradual impoverishment from the western to the eastern part of the Mediterranean is, as already discussed, an important ecological feature of this sea, which has had, and has also to-day, a very important biogeographical effect. The biogeographic role in Mediterranean of organic substances, with their subtle informational value, such as allomones and pheromones, and their connection with the principle of animal exclusion (Hardy, 1956) should be also evaluated.

Hydrological factors can be taken in consideration also when in some areas a species changes its ecological valence from eurytopic to stenotopic. This is the case for the sponge Crambe crambe, very frequent in the western Mediterranean infralittoral, but restricted mainly to bivalve shells in the Adriatic Sea (Sarà, 1967). It ought to be an essential task of biogeography, also of Mediterranean biogeography, even if very complex, to analyze the relationships between the large-scale and the small-scale distribution of biota, i.e. between their geographical and ecological distribution. This study could give a deep insight into the causality of biogeographical patterns. There exist on one side environmental factors such as, in the Mediterranean, temperature, salinity and currents which have an overall macrodistributional, biogeographical relevance. Other factors such as water turbulence, light, the characters of substrate, have chiefly a local ecological and micro-distributional effect. Yet it may be realized that local factors can also play a general biogeographic role when their action influences a large area. This is particularly true for the edaphic factor of substrate. Here the degree of patchiness of types of substrate is very important. From the occurrence of a certain type of substrate depends, indeed, the settling of benthic species and consequently the direction and rate of their dispersal.

The difficulty to evaluate the real biogeographical impact of the environmental factors stands out when we consider the abiotic factors in their relation to the biotic ones at the community level. Here the stronger differential gradients, and then the most effective differentiating action, are given by the local factors such as water-movements, light, substrate, certain local situations of temperature, salinities and nutrients (as in the lagoons). All this influences the biotic factors and processes which in turn regulate the structure and dynamics of the ecosystems and consequently the recruitment, expansion and restriction of species. Among these processes may be mentioned food webs, competitive and cooperative networks, reproductive and dispersal strategies. As Por (1978) says: "the success of a migrational advance is bound to the dispersal capacity but also to the ability to find a niche in the new ecosystem and to maintain it through successful reproduction, inside the interspecific web of relations and against the abiotic ecologic valences of the competing species".

The biogeographic processes should be closely connected with the community processes. Considering then the modifications in time of communities and species, there exists also a link with the micro-evolutionary processes. In this way the ecological and historical points of view can strictly cooperate.

CONCLUSIONS

The biogeographic approach, which refers to whole ecosystems and not to separate species or taxa is very promising for understanding the causal factors of biogeographic patterns, even if accompanied by a greater complexity of analysis. General questions arising, also for the Mediterranean, at this level, are those of biotic diversity and stability. For the Mediterranean, the general tenet may be challenged that climatic conditions suitable for many forms of life and rather uniform for long periods of time are favourable for the formation of ecosystems with a high degree of species diversity, integration and stability, whereas intensively fluctuating climatic conditions lead to ecosystems poor in species, little integrated and unstable. And we may also take into consideration the additional hypothesis that, while the members of stable ecosystems are largely regulated by biotic processes such as interspecific relations, the members of unstable ecosystems are governed primarily by the intensities of abiotic factors. If we consider the impressive climatic fluctuations that have affected the Mediterranean during all its geological history, and on the other side the richness in species and ecosystems of this sea, it seems difficult to apply schematically the foregoing general statements to this sea. In fact Valentine (1967) considered for the Mediterranean that species diversity has been enhanced by the climatic fluctuations which have functioned as a pump of novelties, and Briggs (1974) emphasizes the role of the Mediterranean as a center of species radiation towards the Atlantic coasts and thus its function as a species reservoir. On the other side, the observations of Pérès (1967) on the non-equilibrium position of the Mediterranean biocoenoses, because not all the resources are exploited and not all the niches are covered, can be explained by the scarcity of nutrients. However, in spite of this, the Mediterranean can support a larger number of species, as the success of the Lessepsian migration demonstrates. This success is probably gradually increasing as the addition of new species creates additional niches and increases the biomass and indirectly perhaps also the food resources.

In conclusion, whereas paleogeographic and paleoecological factors, together with the history of immigrations and extinctions, explain the basic structure of the Mediterranean ecosystems, the study of existing ecological factors gives the key to understanding the present-day pattern and dynamics of these systems and their evolutionary perspectives. However, historical factors are also operating to-day and this may be taken in account for these perspectives. One such is the impressive Lessepsian migration from the Red Sea and the subsequent expansion of the immigrants in the Mediterranean; another, with a smaller impact, is the penetration of Atlantic species through the Strait of Gibraltar and the Sea of Alboran. Finally, of outstanding importance among the historical factors now operating is the impact of man with his increasing pollution and management of the Mediterranean Sea.

In order to control the changes in Mediterranean flora and fauna which may be accelerated by these historical events it would be very useful to have an efficient monitoring method at the level of communities. This method may consist in a comparative study for a period of years of the same ecosystem in different Mediterranean stations in order to analyze changes in its distribution, structure and dynamics. This ecosystem should be largely diffuse, easily recognizable and delimited, as for example, the Posidonia ecosystem on which a considerable effort of research is presently made in Mediterranean. The biogeographical extension of the Posidonia ecosystem covers nearly all the Mediterranean with exception of some areas mainly in the Levant basin. Its structure, with a single facies-building organism, Posidonia oceanica, and a large set of epibionts, makes easy its comparative distributional and structural study in different areas and in course of time. Furthermore, it seems that the Posidonia ecosystem is a good indicator of ecological variations and is severely affected by the anthropogenic impact.

SUMMARY

For a causal analysis of Mediterranean biogeography, the ecological factors acting to-day, and in the past history of the Mediterranean Sea have been taken in consideration in this work, as well as their connection with historical factors. Historical processes, such as the immigration waves of Boreal and Senegalese elements and the speciation events in the Mediterranean Sea, are strictly bound to the pleistocenic climatic fluctuations. The present-day Lessepsian migration through the Suez Canal has been favoured, in spite of a scarcity of nutrients in the Levant Basin, by the occurrence of empty ecological niches, related to its paleoecological and geological history.

The Mediterranean Sea is a well-characterized subprovince of the Atlanto-Mediterranean province, with a high number of endemic species and a considerable diversity of environmental conditions leading to the occurrence in different areas of this sea of both temperate and subtropical benthic populations. The biogeographic role of temperature, salinity and other environmental variables is analyzed. Moreover, the role of the biotic factors, regulating the structure and dynamics of the communities, with the reproductive and colonizing strategies of species and interspecific relationships such as sociality, competition, predation, parasitism and symbiosis, is emphasized: these factors represent the essential ecological cues for the evolutionary processes of taxa and ecosystems. Among the historical events in the Mediterranean biogeography the present-day impact of anthropogenic activities should be considered. It would be useful, in this regard, to analyze the changes that may occur in different areas and in the course of time, in the distribution and structure of a well-recognizable and delimited Mediterranean ecosystem such as, for example, the Posidonia one.

REFERENCES

Battaglia, B., 1970, Cultivation of marine copepods for genetic and evolutionary research, Helgol. wiss. Meer.,20:385.

Briggs, J.C., 1974, "Marine zoogeography," McGraw-Hill, New York.

Caspers, H., 1957, Black Sea and Sea of Azoy, in: "Treatise on marine ecology and paleoecology," Vol. I Ecology, J.W. Hedgpeth, ed., Geol. Soc. America, New York.

Davis, G.M., 1982, Historical and ecological factors in the evolution adaptive radiation and biogeography of freshwater Mollusks, Am. Zool., 22:375.

Endler, J.A., 1982, Problems in distinguishing historical from ecological factors in biogeography, Am. Zool., 22:441.

Fredj, G., 1974, Stockage et exploitation des données en écologie marine. C. Considérations biogéographiques sur le peuplement benthique de la Méditerranée, Mem. Inst. océan. Monaco, 7:1.

Gessner, F., 1970, Temperature: plants. in: "Marine Ecology," Vol. I, Environmental factors, O. Kinne, ed., Wiley, London.

Gessner, F., and Schramm, W., 1971, Salinity - Plants. in: "Marine Ecology," Vol. I, Environmental Factors, O. Kinne, ed., Wiley, London.

Gilat, E., 1964, The macrobenthonic invertebrate communities on the Mediterranean continental shelf of Israel, Bull. Inst. Océanogr. Monaco, 62:1.

Golikov, A.N. and Scarlato, O.A., 1967, Ecology of bottom biocoenoses in the Possyet Bay (the Sea of Japan) and the peculiarities of their distribution in connection with physical and chemical conditions of the habitat, Helgol. wiss. Meer.,15:193.

Hardy, A.C., 1956, Plankton ecology and the hypothesis of animal exclusion, Proc. Linn. Soc. Lond., 148:64.

Kinne, O., 1970, Temperature: animals. Invertebrates. in: "Marine Ecology," Vol. I.Environmental factors, O. Kinne, ed., Wiley, London.

Klausewitz, W., 1973, New aspects of the palaeogeography of the Mediterranean ichthyofauna, Icthyologia (Belgrad), 5:74.

Kosswig, C., 1956, Beitrag zur Faunengeschichte des Mittelmeers, Pubbl. Staz. Zool. Napoli, 28:78.

Levi, C., 1957, Spongiaires des côtes d'Israel, Bull. Res. Coun. Israel, 6B:201.

Mars, P., 1963, Les faunes et la stratigraphie du Quaternaire méditerranéen, Rec. Trav. St. Mar. Endoume, 28:61.

Mars, P. and Picard, J., 1958, Note sur les gisements sous-marins a faune celtique en Méditerranée, Rapp. Comm. Int. Mer Médit., 15:325.

Pérès, J.M., 1967, The mediterranean benthos, Oceanogr. mar. biol. Ann. Rev., 5:449.
Peres, J.M. and Picard, J., 1964, Nouveau manuel de bionomie benthique de la mer Méditerranée, Rec. Trav. St. Mar. Endoume, 47:1.
Pfannenstiel, M., 1960, Erläuterungen zu den bathymetrischen Karten des östlichen Mittelmeeres, Bull. Inst. Oceanogr. Monaco, 1192:15.
Por, F.D., 1971, One hundred years of Suez Canal. A century of Lessepsian migration: retrospect and viewpoints. Syst. Zool., 20:138.
Por, F.D., 1978, "Lessepsian migration," Springer-Verlag, Berlin.
Por, F.D., 1981, The Lessepsian biogeographic province of the eastern Mediterranean, in: "Journ. Etud. syst. évol.," CIESM, Cagliari.
Riedl, R., 1971, Water movement: general introduction. in: "Marine Ecology," Vol. I, Environmental factors, O. Kine, ed., Wiley, London.
Ruggieri, G., 1967, The miocene and later evolution of the Mediterranean Sea, in: "Aspects of Tethyan biogeography," C.G. Adams, and D.V. Ages, eds., Syst. Ass. Publ., n. 7, London.
Sarà, M., 1967, La zoogeografia marina e il litorale pugliese, Arch. Bot. Biogeogr. It., 43:327.
Schopf, T.J.M., 1981, Evidence from findings of molecular biology with regard to the rapidity of genomic change: implications for species durations, in: "Paleobotany, Paleoecology and Evolution," Vol. I, K.J. Niklas, ed., Praeger Publ., New York.
Stephen, A.C., 1958, The sipunculids of Haifa Bay and neighbourhood, Bull. Res. Coun. Israel, 7B:129.
Tchukhtchin, V.D., 1964, Trudy Sevastopol Biol. Stn, 16:215.
Tortonese, E., 1951, I caratteri biologici del Mediterraneo orientale e i problemi relativi, Arch. Zool. Ital., Suppl.,7:205.
Tortonese, E., This book, Distribution and ecology of endemic elements in the Mediterranean fauna (fishes and echinoderms).
Udvardy, M.D.F., 1969, "Dynamic zoogeography with special reference to land animals," Van Nostrand Reinhold, New York.
Vacelet, J., 1981, Les affinités du peuplement de spongiaires de la Méditerrannée, Journ. Etud. syst. evol., CIESM, Cagliari: 29.
Valentine, J.W., 1967, The influence of climatic fluctuations on species diversity within the Tethyan provincial system, in: "Aspects of Tethyan biogeography," C.G. Adams, and D.V. Ager, eds., Syst. Ass. Publ. n. 7, London.
Vatova, A., 1935, The fishery grounds near Alexandria. II. A bottom sample taken at Alexandria, Notes Mem. Fish. Res. Dir. Cairo, 9:1.
Vernberg, F.J., 1971, Dissolved gas. Animals. in: "Marine Ecology," Vol. I, Environmental factors, O. Kinne, ed., Wiley, London.
Zaccaria, M., 1968, Studi geologici sulle isole del Dodecaneso (Mare Egeo); VIII. Fauna Calabrian dell'isola di Rodi, Riv.Ital. Paleont., 74:275.

THE MEDITERRANEAN BENTHOS: REFLECTIONS AND PROBLEMS RAISED BY A CLASSIFICATION OF THE BENTHIC ASSEMBLAGES

D. Bellan-Santini

Station marine d'Endoume, 13007, Marseille, France

Investigations on Mediterranean macrobenthos have led to a classification of the assemblages which, although varying a little depending on the research workers involved, complies with a certain number of principles. The first of these principles is that it is the ambient factors that mainly influence the distribution of the assemblages.

THE ENVIRONMENTAL FACTORS

The environmental factors are of two types:

- Climatic or general factors
- Edaphic or local factors

Climatic Factors

Three factors are responsible for the presence of the different zones: humidity, light and hydrostatic pressure.

In the shallowest levels, the humidity gradient is the dominant factor. Humidity is the result of sea spray and sometimes brief immersions in waves during storms.

In the submerged region, light is the principal factor; its variations control primary productivity and thus directly influence the existence of a phytal system. The gradient of light decreases with depth.

The influence of hydrostatic pressure is felt above all, at great depths: it is less understood and its action on living organisms is the subject of numerous investigations.

Edaphic Factors

If climatic factors acted alone, there would be only a single assemblage per zone with a gradient corresponding to the variation of the dominant environmental factor. In reality, however, many edaphic (local) factors act in a synergistic manner either with themselves or with the climatic factors. Among these edaphic factors are (considering only those of natural origin and not those caused by human activity):

- hydrodynamism, which deals with the agitation of surface waters or bottom currents;
- local variations of salinity, due to influx of freshwater (rivers, springs) or to evaporation;
- the nature of the substrate: solid or soft bottomed, with all possible variations;
- temperature should be considered separately for it is partly the result of the factor light, and has a general action on the water mass as a whole. However, it is also an edaphic factor for it can vary strongly from one region to another and from one day to another.

The distribution and the interaction of the different factors - influencing the marine benthos result in a very complex distribution of the assemblages, which thus do not correspond to a linear type of organisation.

CLASSIFICATION OF THE MEDITERRANEAN BENTHOS

The system of classification of benthic assemblages utilised in Mediterranean waters comprises two large groups classified as a function of the action of light:

- phytal system
- aphytal system

We will see later that although the separation of these two systems is not difficult in theory, its application in the environment presents certain minor difficulties.

Each system is bathymetrically divided into sub-zones or stages in which ecological conditions, as a function of depth with respect to sea level, are fairly constant or vary regularly between the two critical levels marking the limits of this zone (Pérès and Picard, 1958).

Subsequently, the assemblages are most often separated depending on the quality of the substrate and the dominant influence of a specific edaphic factor. Depending on the manner in which the study has been conducted and particularly on the specialization of the scientist involved and the geographical area studied, a certain factor or a group of species could dominate. This could lead to a biased classification of the assemblages, which would not fit into an overall pattern.

In the Mediterranean, the system of classification of the benthic communities is highly elaborate, having been studied for a long period. The characteristics of the Mediterranean benthos are many:

- the fauna and the flora are rich in species;
- the conditions of the environment are diversified;
- the benthic populations of the European seas, considered as a whole, are the best known in the world;
- the Mediterranean is a warm temperate sea and is thus unique in permitting the study of homologies with well-known temperate seas and less-known tropical regions;
- the Mediterranean fauna and flora are composed partly of endemic species partly of North and South Atlantic species and partly of an increasing number of immigrant species from the Red Sea.

Apart from furnishing general data on certain characteristics of the system of classification and characteristics of the Mediterranean benthos, the most-utilised classification of the benthic communities in the Mediterranean will not be discussed in detail, having been already the subject of an in-depth study by Pérès and Picard (1958), subsequently reviewed and extended (Pérès and Picard, 1964; Pérès 1967,1982). I will cite, among the recent local investigations, the work of Gamulin-Brida (1967) in the Adriatic in which local biogeographic peculiarities of this sea are emphasized, and the sum total of the publications of the 2nd Iberian Symposium (1981-1982) in which numerous benthic communities of the Spanish coast have been studied, both by classical methods and by factorial analysis, confirming and improving our knowledge of this region.

I should like to present, in a very simplified manner, the problems still remaining at present and the work under way and that being initiated.

For the Mediterranean:

The phytal system consists of 4 zones (in French "etages)

- Supralittoral
- Mediolittoral
- Infralittoral
- Circalittoral

The aphytal system consists of 2 zones:

- Bathyal
- Abyssal(?)

Supralittoral Zone

The supralittoral is populated by organisms which require a high degree of humidity but which are never submerged. The dominant factor in this zone is humectation, with a gradient increasing from high to low water.

Hard substrata. Supralittoral rock biocenosis. The main species are the lichen *Verrucaria symbalana*, the prosobranch *Littorina neritoides* and the isopod *Ligia italica*. This biocenosis is rich in Cyanophyceae and Chlorophyceae, epilithic and endolithic on calcareous rocks. The Cyanophyceae are responsible for the blackish-brown coloration of this zone.

Soft substrata. two biocenoses may be recognized: Biocenosis of quickly-drying wracks. The main species are *Talitrus saltator mediterraneus* or var. *briani* for the Adriatic, *Tylos europeus* and diverse *Bledius*.

Biocenosis of slowly-drying wracks. Diverse *Orchestia* are present; *Tylos sardous* is replaced in the Adriatic by *Tylos ponticus* spp. *adriaticus*, diverse pulmonate gastropods and insects. These wracks sometimes correspond to the "banquettes" of *Posidonia* in which *Orchestia* abound. A facies of *Salicornia* and Cyanophyceae could be attached to this biocenosis.

The supralittoral zone is poor in characteristically-sessile species. It is populated by very motile species which, on the one hand, are influenced by humidity differences and on the other hand, perform numerous trophic migrations. Predators coming from other zones are also found in this zone. Little is known about the zone with the exception of the Cyanophyceae and a recent study (Bigot, Picard, Roman, 1982) demonstrating four different biochores: two of sand and two of wracks and tree trunks, the motile fauna of which is essentially of terrestrial origin. Faunistically, the whole is extremely poor.

Mediolittoral Zone

Except for the northern part of the Adriatic and the gulf of Gabès (Tunisia) where the tides are of some importance, the Mediterranean Sea is practically lacking in tides. The fluctuations in water level are restricted to wave movements and to variations due to wind and atmospheric pressure.

The Mediolittoral zone comprises communities consisting of species which tolerate more-or-less regular immersions, but, in no case, continuous immersion.

The amplitude of the Mediolittoral is a subject of discussion by different workers, but in the Mediterranean it could generally be considered that the upper limit corresponds to the highest level normally attained by the waves, while the lower limit corresponds to the lowest level of normal waves.

Hard substrata. Two subzones linked to humectation differences, occupied by two distinct biocenoses, can be recognized. Depending on transient local conditions, the motile species are able to descend or ascend and live momentarily in enclaves corresponding to particular conditions, limited in time and space. They do not intervene, in any case, in the average composition of the biocenosis.

Although this is the case in all the zones, it is particularly so in the Mediolittoral, partly due to the narrowness of the zone and partly due to the significant differences in conditions of life in the neighbouring zones.

Biocenosis of the Upper Mediolittoral Rock (R.M.S.). Characterized by Chthamalus stellatus, Patella lusitanica, Patella ferruginea (very localised species in Corsica and Sardinia) and numerous epilithic and endolithic Cyanophyceae. Three algal facies are present, as belts, in this region:

- Bangia fuscopurpurea
- Porphyra leucosticta
- Rissoella verruculosa (calcifugous species)

Certain motile Hyale as Hyale perieri may enter this facies which is, from the point of view of altitude, the lowest.

Biocenosis of the Lower Mediolittoral Rock (R.M.I.). This biocenosis is characterised above all by the presence of Melobesieae: Lithophyllum tortuosum, L. papillosum and two molluscs: Patella aspera and Middendorfia caprearum. Owing to the presence of Melobesieae, numerous cavities and crevices forming infralittoral enclaves are created, in which a minute fauna abounds. This biocenosis is represented above all as much by its numerous geographic facies - Fucus virsoides in the Adriatic and Hydroclathrus clathratus in Greece - as by its edaphic facies: Nemoderma tingitanum on very exposed rocks and the facies Lithophyllum tortuosum constituting (in the western Basin) "trottoir" or "corniche" formations of great importance. These formations do not appear to have been found in the living state in the Eastern Mediterranean. They have been noted in the fossil state in Rhodes by Laborel (1981). The facies of Neogoniolithon notarisii is frequent in the Mediterranean, as a whole, and sometimes, as in Sicily or Tunisia, is very well-developed.

About the hard substrata of the Mediolittoral zone, two remarks should be made:

1. Considering the entablure of Lithophyllum tortuosum, this original formation has not, up to the present, been sufficiently well-studied. Laborel considered it an excellent indicator of sea level fluctuations. The maximal growth of algae is at average sea level. Certain fossil or subfossil entablatures have been studied.

2. A third biocenosis that of the Mediolittoral caves of the hard substrate has always been delineated in the Mediolittoral zone, but almost nothing is known of the biocenosis, which appears to be characterized by the rhodophycean alga Hildenbrandtia prototypus.

Concerning the hard substrata of the two zones, Supra- and Mediolittoral, three remarks are necessary:

1. A recent study of the Chthamalus of the Mediterranean (Relini 1981, 1983) showed, mixed with two known species Euraphia depressa (=Chthamalus depressus) and Chthamalus stellatus, a third species, Chthamalus montagui, confused up to the present with the others. The study of its distribution in space and time as well as its possible replacement is under way. The discovery of this third species, demonstrated up to the present only in the Western Basin, could permit the resolution of certain enigmas with respect to the distribution of the Chthamalus in certain special conditions.

2. The pools of variable salinity have been most often presented as either Supra or Mediolittoral biocenoses depending on the area in which they are situated. It appears at present that the species present must be considered as Infralittoral, but as particularly tolerant of environmental modifications, owing to the feeble part played by the macrobenthos in these biochores, which are however, of great interest.

3. The endolithic Cyanophyceae have been specially studied in the course of the last decade (Le Campion, 1978). Their importance in the processes of biodestruction of calcareous rocks and their

action in the formation of biokarst has been demonstrated. On the other hand, the systematics of the group having been entirely revised utilising modern methods of investigation, it has been shown that the forms of the galleries are specific. The advantage of the simultaneous discovery of the specificity of the galleries and of the microzonation of the species is considerably great. It has even made possible the ecological interpretation of the fossil biokarst.

Soft substrata. Although the conditions of humectation are the same for the fauna of both hard and soft substrates, the problem is complicated in the latter case by absorption of porewater, a factor particularly difficult to evaluate.

Three biocenoses can be distinguished:

Biocenosis of the Mediolittoral coarse detritic bottoms (D.M.): between the pebbles exist a mass of organic detritic matter deposited by the sea, in which lives a population composed essentially of detritivorous species: *Echinogammarus olivii* (=*Gammarus olivii*), *Sphaeroma serratum*, *Allorchestes aquilinus*. A part of the Mediolittoral, the "banquettes" of dead leaves of *Posidonia*, could be included in this biocenosis.

Biocenosis of Mediolittoral sand (S.M.): it is no longer considered at present that *Ophelia bicornis radiata* and *Nerine cirratulus* form a single sandy biocenosis - the two species have different behaviour with respect to the modification of their biochore by water movements. To these species can also be added *Eurydice affinis* and *Mesodesma corneum*.

Biocenosis of the mud or sandy mud bottom in brackish waters: this deals with a little-studied biocenosis in which the Cyanopyceae could form a sort of mat. Two facies, *Salicornia herbacea* and *Juncus maritimus* have been pointed out, which retain the sediments, leading to a shallowing of the bottom and the installation of a population which could be categorized as belonging to the superior zone.

Apart from mentioning that only a small number of detailed studies have been conducted on Mediolittoral soft substrata populations, no particular remarks appear to be called for. A general problem concerning the soft-bottomed biotopes exists and the limits of the marine and terrestrial domain will be treated below.

Infralittoral Zone

The Infralittoral corresponds to the zone compatible with the presence of the phanerogams or the photophilic algae. It begins beneath the median level, i.e., at the beginning of the zone where immersions are no longer only accidental. Its amplitude is directly linked to the turbidity of the water and could vary from a few meters to 40 or 50 m in particularly clear waters.

Hard substrata. Biocenosis of photophilic algae (A.P.): It is one of the richest communities in the Mediterranean benthos, considered as a whole. The algae are abundant there and, considering the multitude of microenvironments created by the action of different factors-light; hydrodynamism, nature, orientation and form of the substrate - it could be stated that a very rich fauna and flora are found in optimal living conditions.

Taking into account the richness of the biochores, the biocenosis (characterised by very high biomass) is most often present in the form of facies at one or many strata, with a considerable development of sciaphilous species.

The biocenosis of the photophilous algae is diversified into a great number of facies, the most important of which are:

- facies of Cystoseira stricta: these facies can also be termed a community of "Cystoseires clydonophiles". Many varieties of C. stricta are present in the Mediterranean. This facies, which is found on many strata, and often on concrete substrata, lives in places very exposed to waves and sun.
- Facies of Cystoseira crinita: this facies equally rich, lives in more protected zones than the preceding one and depends on conditions most often linked to a gradient of environmental factors.

Numerous other facies succeed each other, the most important of which are:

- facies of Cystoseira abrotanifolia
- facies of Corallina mediterranea
- facies of Halopteris scoparia
- facies of Cladostephus verticillatus
- facies of Acetabularia mediterranea
- facies of Lithophyllum incrustans, which perhaps can be considered as a degradation facies, notably on account of sea urchin grazing.
- facies of Vermetus cristatus in zones not subject to very cold winters (Sicily, Algeria, Lebanon).
- facies of Mytilus galloprovincialis, absent in the southern part of the Eastern Mediterranean, in zones rich in organic matter.

The algae, being particularly abundant on the hard substrata of the Infralittoral zone, have been the subject of considerable research (Feldmann, 1937; Molinier, 1960; Boudouresque 1973a,b; Boudouresque and Passelaigue, 1972; Giaccone, 1977). This has led to a hierarchical classification based on algal associations corresponding to the facies of the biocenosis of photophilic algae. On the basis of recent statistical work, Romero-Martinengo (1981) regrouped the most abundant species in a group with affinities to the biocenosis of photophilic algae.

Soft substrata. The biocenosis of the soft substrata of the Infralittoral zone, although well-studied over a long period, was the subject of a recent revision by Picard (1983). This author, taking into account the various constraints to which the populations of the high level are subject, and the biocenotic dynamics that they provoke, considers that five of the six biocenoses of the soft substrata could be ordered along a hydrodynamic gradient (Fig. 1). The average level is occupied by the meadows of *Posidonia* (H.P.), a climax biocenosis.

On progressing towards the higher energy areas, the Biocenosis of fine well-sorted sands (S.F.B.C.) is found. The Biocenosis of the fine sand of the high level (S.F.H.N.) follows. In extremely high energy conditions, substitution facies of *Donax trunculus* and *Nerine mesnili* are found.

As one progresses towards calmer areas, the Biocenosis of muddy sands in calm waters (S.V.M.C.), with its facies of epiflora of *Cymodocea* and *Zostera* is found, followed in the calmest areas by Euryhaline and Eurythermal Biocenosis in brackish waters (L.E.E.) which gives way to the substitution facies of *Cerastoderma glaucum* and *Hydrobia* cf. *ulvae*. The Biocenosis of sand and gravel under bottom currents (S.G.C.F.) must be considered apart, since, although well developed in the Infralittoral, it is also found in the Circalittoral. This biocenosis, directly affected by the presence of a strong current, and also called "*Amphioxus* sand", has, in the high levels, a substitution facies of *Saccocirrus papillocercus* and *Lineus lacteus*.

The sixth biocenosis identified in the soft substrata is the impoverished Biocenosis of Infralittoral gravel and pebbles (G.I.), which has been little studied, but of which the fish *Gouania wildonewi* is a good indicator.

Two important problems have to be noted in the Infralittoral zone:

1. *Posidonia* meadows: this climax biocenosis, poses above all, a problem pertaining to substrate since it exists (rarely, it is true) on hard as well as on soft substrates.

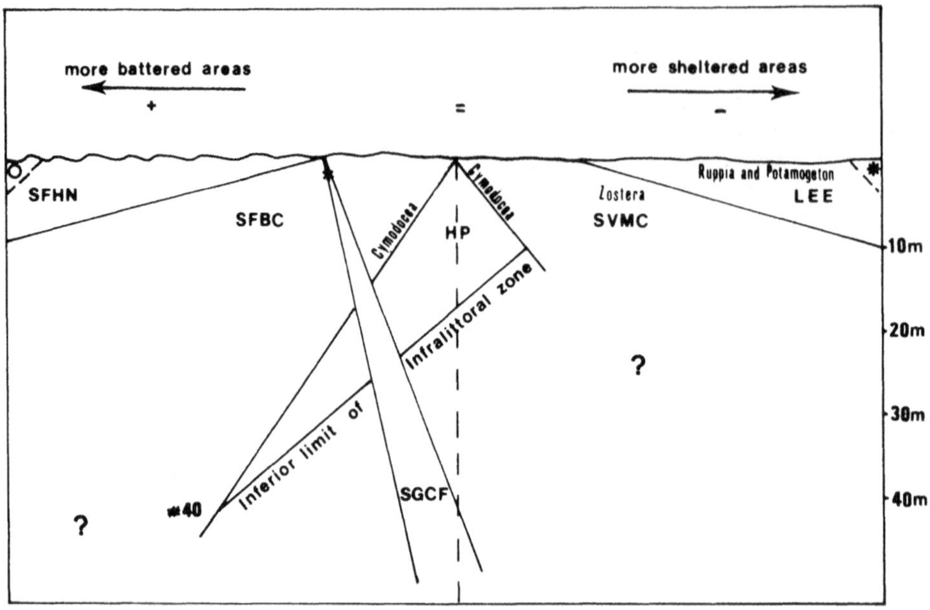

1 Substitution facies with Donax trunculus and Nerine mesnili

2 Substitution facies with Cerastoderma glaucum and Hydrobia cf. ulvae

3 Substitution facies with Saccocirrus papillocercus and Linneus lacteus

Fig. 1. Succession of soft substrata biocenoses along a hydrodynamic gradient in Picard (1983).

Similarly, a complex mixture of the two substrates is present, partly because of its biogenic origins and partly to the significant biogenic and organogenic sedimentation that it causes.

Dwelling on this biocenosis, its important and complexity can be seen by the observation of Giaccone (1983) who notes that for the period 1980-82, 14.2% of the bibliography consecrated to the Mediterranean phytobenthos concerns the phanerogam meadows.

2. Transition from superficial formations to biocenoses of the Circalittoral zone.

Although the superficial populations, especially those of the hard substrata, have been much studied, this is not the case between 5 meters depth and the lower limit of the Infralittoral. The most sciaphilous facies of the biocenosis of the photophilic algae are found there, but a biocenotic assessment of the zone, on which the photophilic deep algae could be identified, is not available.

Circalittoral Zone

This zone extends just up to the survival limit of the multicellular autotrophic algae.

Hard substrata. Coralligenous biocenosis (C). Laubier (1966) considered this biocenosis as an eco-ethological junction within the populations of the hard substrate of the littoral system. The definition demonstrates the complexity of this climax biocenosis with its multitude of microenvironments and extreme diversity of appearance.

Two types of coralligenous biocenoses, based on the origin of formation, can be mentioned:

- Coralligenous biocenosis of the Lower Side of the Littoral Rock, specifically that formed from rocky surfaces;
- Coralligenous banks, when the biocenosis results from a biological concretionary process originating from the coarse elements of the soft substrata. Evidently, it could be very difficult to distinguish these two types.

The characteristic coralligenous species are numerous and consist, above all, of Algae, serpulid polychaetes, bryozoans, sponges which participate in the concretionary process and the associated motile species: polychaetes, crustaceans, pelecypods.

The following facies are particularly notable:

- Pre-coralligenous facies
- Gorgonian facies
- Facies of *Astroides calycularis*

The coralligenous biocenosis has been particularly well studied in the North Western Mediterranean and the Adriatic, but is little or poorly-known otherwise.

Biocenosis of semi-dark caves (G.S.O.)

This biocenosis is purely animal and is dominated by sponges, anthozoans and bryozoans; among the facies that could be described, three are dominant:

- *Parazoanthus axinellae* facies
- *Corallium rubrum* facies
- impoverished facies, which could take various forms.

Biocenosis of dark caves (G.O.)

This biocenosis in total darkness is characterised by three typical species, two of which are pharetronid sponges: *Petrobiona* massiliana and Plectoroninia hindei. At times, water circulation may be imposed for long periods, causing massive kills of organisms. The distribution of certain species shows the extreme affinity of this biocenosis to the bathyal zone, and the hypothesis of a possible enclave of this zone cannot be rejected.

Biocenosis of offshore rocky bottoms (R.L.)

Submarine dives have permitted the study of this biocenosis based on sponges, anthozoans, bryozoans and brachiopods. It is situated on the edge of the continental shelf.

By studying the hard substrates, it has been possible to observe that the major problem in the Circalittoral is the quantity of light which arrives at the substrate level. It is evident that they key biocenosis is the Coralligenous one and that the three other biocenoses, in spite of their climatic affinities with the Bathyal, belong to the Circalittoral.

Soft substrata. Four biocenoses are very well-characterised on the basis of the origin and the granulometry of the sediments, and the relative proportions of the different sedimentary fractions.

Biocenosis of the coastal detritic (D.C.).

This corresponds to detritic sediments from the Infralittoral or Circalittoral. Small areas of solid substrate are very often found, and some of the organisms are from the Coralligenous biocenosis. This biocenosis presents numerous facies, the majority of which have one or many algae as pilot species which form nodules or are fixed on a small solid substrate. Among these could be cited:

- "Prâline" facies
- facies of Halarachnion spatulatum
- facies of free Squamatiaceae
- facies of "Maërl"
- facies of Vidalia volubilis

Biocenosis of muddy detritic bottoms (D.E.)

Installed on detritic sediment, this contains a dominant fraction of mud. Picard (1965) has described 12 species as characteristic of this biocenosis, which presents two facies:

- facies of Ophiotrix quinquemaculata
- facies of Alcyonium

Biocenosis of terrigenous mud (V.T.C.)

Depending on the mechanism and the rapidity of the deposition of the mud, the biocenosis presents more-or-less soft muddy facies, where certain species develop preferentially.

Biocenosis of the shelf-edge detritic (D.L.)

This occurs in the depths between 90-95 m and 120-150 m. The sediment is a Quaternary fossil detritic with which is mixed sand and mud. It presents various facies, the most important of which is the facies of the large hydroids Lytocarpia and Nemertesia.

For a long time the limits between these four biocenoses and more particularly the ordinance of the D.E. posed a problem. A recent important work comprising both classic analysis and numerical analysis (Salen-Picard, 1982) has delimited these biocenoses, and the transitions between them, mark very well. However, Falconetti (1980) showed that Neolampas rostellata facies, otherwise considered as D.L. facies, is in fact an impoverished facies corresponding to a mixture of the two biocenoses D.C. and SCGF.

Bathyal Zone

The aphytal system commences with the Bathyal zone at the limit of growth of the multicellular algae and extends over the continental slope. Its lower limit coincides with that of the distribution of the eurybathyal species.

Hard substrata. Biocenosis of the deep-sea corals: Very little is known about this biocenosis; it consists essentially of two characteristic species, Lophelia prolifera and Madrepora oculata. The associated fauna is poor. This biocenosis exists on rocks below 300 m, but it is evident that, at this depth, rocky peaks emerging from the mud are rare. It is currently believed that the living portion of these reefs are implanted on bigger reefs, the deeper portions of which could be subfossil.

Soft substrata. Biocenosis of the bathyal mud (V.P.): The deep Mediterranean sediment is usually a greyish-yellow, relatively-compact mud, the fauna of which is very sparse although rich in characteristic species.

Several facies can be distinguished, depending on the quality and the compactness of the sediment. The most important are:

- fluid mud facies of Brissopsis lyrifera
- soft mud facies of Funiculina quadrangularis
- compact mud facies of Isidella elongata
- sandy mud facies of Thenea muricata

• sandy mud mixed with gravel facies of Terebratula

• decandation area facies of Pheronema grayi

Up to the present, owing to limited knowledge of these environments, it was supposed that in the Mediterranean, the plains and trenches which in other seas of the world correspond to the Abyssal Zone were occupied by the Biocenosis of the Bathyal Mud. This general problem will be dealt with later in the light of recent investigations.

GENERAL PROBLEMS POSED BY THE CLASSIFICATION OF THE BENTHIC ASSEMBLAGES IN THE MEDITERRANEAN SEA

Subsequent to this concise review of one type of classification of the Mediterranean benthos as a whole, a certain number of questions of a general order remain. It is difficult, owing to the actual extent of our knowledge, to provide complete answers.

Knowledge of the Mediterranean Fauna and the Problem of its Richness

Specific richness. The Mediterranean benthic biocenoses have a higher species richness than coresponding biocenoses in the Atlantic.

The reasons for this phenomenon are various and are related to:

- the history and diverse origins of the Mediterranean fauna;
- the diversity of the environments existing in the Mediterranean which has permitted the installation and development of species of diverse origins;
- the importance of endemic species which could attain, depending on the groups, a percentage as high as 44.6% (in the demosponges, (Vacelet, 1981) and even 50.4% (in the ascidians (Pérès, 1967).

Nevertheless, I believe that this view still understates reality. For the majority of the systematic groups, the studies have essentially been conducted on the west coasts of Europe and it is usually on the basis of similar characteristics that Mediterranean fauna have been identified. On the other hand, investigations on certain groups in the Mediterranean are still lacking and a more complete modern study could lead to the appearance of an unsuspected species richness. A recent investigation on crustacean Amphipoda conducted in 1976 indicated the presence of about 170 species. Following a systematic revision of the group, taking into account the known collections and based on the redescription of all species, the number of species is now estimated to be over 400.

Quantitative richness. Quantitative biomass studies conducted in the Mediterranean are not very numerous, and are usually very localised in the northern part of the Western Basin and in the Adriatic. The other scattered results available do not permit the calculation of mean values and variances for the Mediterranean as a whole.

The available values in the north of the Western Basin can be summarized in the following manner (expressed in dry decalcified weight).

• Hard substrata (Bellan-Santini, 1966; True *in* Pérès, 1967).

- Photophilic algae (A.P.): 234 to 2217 g/m^2
- Coralligenous (C): 503 g/m^2
- Semi dark caves (GSO): 489 to 967 g/m^2

• Soft substrata (Reys *in* Pérès, 1967)

- Fine well-sorted sand (S.F.B.C.): 1.82 to 4.39 g/m^2
- Sand and gravel under bottom currents (S.G.C.F.): 6.89 g/m^2
- Coastal detritic (D.C.): 1.53 to 16.55 g/m^2
- Muddy detritic (D.E.): 2.34 to 10.07 g/m^2

Similar values have been obtained in the Adriatic. With respect to the Aegean Sea, Kisseleva (1961, 1964) has recorded at different depths a biomass base level not exceeding (on soft substrate) 5 g/m^2 in the most fertile regions.

Considering the bathyal zone, Vinogradova et al. (1982) recorded, at depths between 1000 and 3000 m, values inferior to 1 g/m^2

Table 1

	g/m^2		
	1000-2000 m	2000-3000 m	+3000 m
Alboran Sea	0.32	.	.
Tyrrhenian Sea	0.2-0.22	0.08	0.02
Ionian Sea	0.34	.	0.01
Adriatic Sea	0.49	.	.

Recent studies have not resulted in substantial changes. The conclusions of Pérès (1967) could thus be recapitulated.

1. On Infralittoral hard substrates, biomass values are generally lower than those of analogous or homologous populations in North European seas.
2. On hard substrata of the Circalittoral, the biomass found in the Mediterranean is relatively high.
3. On soft substrata, the biomasses obtained in the Marseille region are, in general, greater than those found by other workers in Mediterranean. Above all, in the Oriental Basin, they are much smaller than those noted in the Northern Sea.

Considering benthic productivity, very little is known. Nevertheless, the life cycles of a good number of species, being considerably shorter and the seasonal variations very strong the benthic productivity could be much higher than believed.

Work is in course but it must be intensified, partly to study actual bottom productivity, especially on the Infra and Circalittoral, and also to investigate the different diets of the organisms and paths of circulation of matter and of energy.

Heterogeneity of Studies in the Mediterranean

Attempts to review studies of the Mediterranean benthos are based on data obtained in the northern part of the Western Basin and in the Adriatic, augmented by scattered local data, certain of which date back more than 20 years, and which have never been updated utilising new methods.

To demonstrate and evaluate the importance of lacunae currently existing in our knowledge of one of the better known seas of the world, I present as a first example, the chart (Fig. 2) depicting the sum total of known stations at which the gammarid amphipods have been studied in the Mediterranean.

A second example is a census of work on the zoobenthos of the Mediterranean and adjacent seas from 1974 to 1982 (Bellan-Santini, 1983a).

Table 2

	Number of titles	Western Mediterranean %	Adriatic %	Eastern Mediterranean %	Black Sea and Red Sea %
1974-76	572	43.8	18.5	11.7	23.1
1976-78	552	39.4	22.9	14.8	12.6
1978-80	557	46.8	14.4	17.4	20.7
1980-82	633	49.5	17.7	14.9	17.9

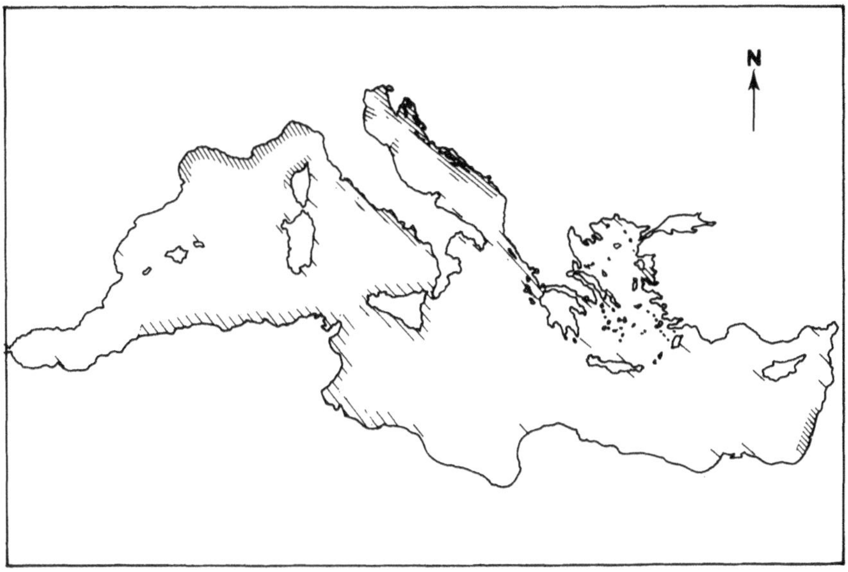

Fig. 2. Showing the distribution of the stations in which amphipods have been sampled and studied in "The Amphipods of the Mediterranean". The shading is in proportion to the intensity of investigations in the zone.

More than half the literature concerns only the Western Basin and again, in this basin, it deals essentially with the northern part.

It is thus imperative that work concerning the benthic communities be extended to little-known regions, where climatic as well as biogeographic conditions are such that research in these regions might provide complementary data, or, even new concepts such as the importance of new species endemic to the deep mud found in the Eastern Mediterranean. This might permit the description of a characteristic abyssal stock.

Importance of Bioconstruction in the Benthic Domain

Very few authors have emphasized the importance of biological construction in the Mediterranean. Without attaining the same degree of development as in tropical seas, constructive as well as destructive organisms are essential in the creation of many of the substrates.

Considering hard substrata, the most important bioconstructive formations are: the entablature of Lithophyllum tortuosum, of which the importance as a tracer in the palaeoenvironment has been pointed out, and the Coralligenous biocenosis, whether on the plateau or on the lower side of the littoral rock. This extremely rich "ecological junction", climax biocenosis is, as already pointed out, installed in and on the entirely bioconstructive massif. Its central portion, which is fossilised (Hong, 1980), is the result of a fragile equilibrium between constructive and clogging organisms on the one hand and destructive species on the other. The essential basis of its richness is the result of a tremendous colonisation of its surface in the multiplication of cavities and channels, allowing a very strong aggregation of organisms which find nourishment and shelter in abundance. In the hope of creating such environments (absent on certain flat sandy coasts or in regions destroyed by man) experiments with artificial reefs patterned on the bioconstructed masses (Duval et al., 1983) have been attempted.

Lastly we take into account the concretions of the base of algal Infralittoral populations, the facies of calcified algae: "prâlines" of Melobesiae, Peyssonnelia, the populations of madreporarian and the white coral of the Bathyal Zone.

The antagonistic action of biodestruction is very strong at two levels. In the Supralittoral and the Mediolittoral zones, as already pointed out, it is the combined action of cyanophycean and chlorophycean algae and grazing organisms which is responsible for the biokarst of the calcareous coasts. In the Infralittoral and Circalittoral zones, a non-negligible portion of the detritic bottoms results from the destruction of calcareous or non-calcareous constructive organisms.

The Diverse Concepts or Organisation of the Different Assemblages

Marine ecologists have achieved close to unanimity in describing the distinct assemblages of the macrobenthos. This is not the case in descriptions of the details of their structure.

Organismic assemblages. The most-utilised assemblages involve, at one and the same time, both the fauna and flora. The disputes concerning the relative value of quantitative or qualitative factors in delimiting the assemblages in the Mediterranean are well known. Generally, the notion of the "biocenosis" as determined qualitatively is utilised, and it is considered that the "community" (sensu Petersen) differs very little, especially if we consider also the quantitative biomass. In 1982, with a view to making the concept universal, Pérès introduced the term "organismic assemblage".

Disputes about terminology, although underlining the importance of an adequate definition of the terms utilised, are of little interest to me, compared to the importance of agreement on the concept itself: that a qualitative or quantitative approach is utilized according to local richness. This concept consists of viewing, in a global way, the "group of living beings which are often linked by a reciprocal dependence and which by reproduction maintain themselves permanently in a certain place".

Living organisms, owing to their own particular characteristics of motility, of longevity, and of reproduction, do not all respond in the same manner to environmental conditions. It follows therefore that if assemblages are preferentially based on a certain group, the separation will not coincide.

The most characteristic example is that furnished by vegetal associations. The algae, being fixed organisms, are subject to variations of light and to restricted microenvironments. They are separable into numerous distinct and remarkably-hierarchical assemblages (Molinier, 1960; Boudouresque, 1973a).

The fauna is less strictly dependent on microenvironments and constitute assemblages corresponding to vegetal associations highly placed in the phytosociological hierarchy. The marine environment can be conceived as comprising, at one and the same time, the soft substrata, generally deprived of vegetation, and hard substrata which carry the phytal and aphytal systems; in each case the concept of an organization is unchanged and the "biocenosis" alone appears to respond to the need.

Zones of transition. Another aspect of the problem is the transition from one biocenosis to a neighbouring one. In the marine environment, the majority of factors vary progressively. Abrupt transitions are rare although it is possible to find, at the extremity of a rocky point, a very exposed area on one side,and on the other a relatively protected one; or a very strong gradient of light at the entry to a cave. Generally one passes thus from one biocenosis to another by a more-or-less wide transition zone called an "ecotone" where most often the number of species and the number of individuals are higher than in the neighbouring zones themselves. Salen-Picard (1982) demonstrated that in the case of two soft substrata biocenoses DC/DE or DE/VTC, five zones can be distinguished, with, for each biocenosis, the creation of a facies of an accompanying species. In the true central zone of transition, an increase in the number of species or individuals is not necessarily produced but a specific diversity results owing to the regression of lateral facies.

Certain authors consider that the distribution of organisms depends on a "continuum" in relation to the gradient of environmental factors. The zones where several species are most abundant constitute the "noda" (Boudouresque, 1971). But in fact, if the continuum is taken in its strictest sense, the following considerations are overlooked:

1. The threshold of sensitivity of the species, which often strictly limits their distribution;

2. The notion of synergy between the factors which further reduces the limits of distribution of the species in an essentially complex environment where even a single factor does not act according to a strictly regular gradient;

3. If, on the other hand, the notion of continuity is taken in a broader sense and if the "nodum" is related not only to species themselves, but also to a certain number of ecological factors, the notion of the sum total returns to that of the biocenosis separated by the ecotones.

Recent statistical studies have confirmed this notion of homogenous groups separated by transition zones.

Notion of stability and instability of the assemblages. For a long time the notion of Instability of an assemblage has attracted the attention of benthologists. The definition of an"unstable soft bottom biocenosis" has refined the problem, to evaluate precisely the phenomenon of Instability and to study it. This phenomenon of Instability is more-or-less strongly manifested in different conditions: when a process of periodic silting and dredging affects a bottom; when a significant decantation provokes the formation of temporary "muck heaps"; when active hypersedimentation leads to a transition in time and in space, from one biocenosis to another in an irreversible manner. In all these cases - and in others that it would be too many to enumerate here - the delimination of a geographic or temporal space is achieved. Within this, an evolutionary stage develops, populated with species indicative of instability, whereas species characteristic of this biocenosis disappear. The population is then partially disorganised. If this phenomenon of Instability is provoked by an ecological catastrophe or by the extreme action of a natural factor, a real "astructuration" of the population and the development of an oligospecific population is seen (for example facies of infralittoral sand of the high level in a very exposed area consists of *Donax trunculus* and *Nerine mesnili*, population of *Capitella capitata* in very polluted zones).

In fact it appears completely logical in a system consisting of certain biocenoses (groups of living beings in equilibrium in biotopes),to conceive of unstable, non-equilibrated zones where environmental conditions vary strongly or are extreme and the populations of which destroy themselves or do not succeed in restructuring themselves.

Limit of the terrestrial and maritime domain. Problem of Supralittoral and Mediolittoral biocenoses based on motile species

In a practically tideless sea, the biotopes of the soft substrata of beaches pose a zonation problem. Little work has been conducted for the fauna are rare, essentially motile, variable in time and space and, finally, the component groups are not well-known.

On the exposed coasts of Camargue (delta of the Rhône) a recent investigation (Bigot et al., 1982) tried to fill this gap by conducting an intensive study for more than one year.

The beaches above the average sea level represent zones subject to strong redistribution of sand by air and water movements to the action of waves and to wide variations in temperature. In the case of strong perturbation, the organisms can make active displacement to reach areas where the factors acting on their distribution are more optimal.

A certain number of consequences result:

1. The motile species of the sand are independent of their location and are distributed only as a function of their own humidity requirements;
2. The motile species can thus escape consequences of climatic extremes which could kill sessile species. They can therefore reconstitute their assemblage rapidly when average conditions have returned.
3. Short-term or long-term variations in the limits between sandy populations are thus possible;
4. There is no stable transition zone corresponding to a progressive change.

The zones studied comprise:

Adlittoral zone	:	Living dune
Supralittoral zone	:	High beach
Mediolittoral zone	:	Medium level beach
Infralittoral zone	:	Fine sand of the high level biocenosis (S.F.H.N.)

Beyond these profoundly unstable zones are situated much more stable populations: at the summit of the Adlittoral: the dead dune fixed by vegetation; in the Infralittoral: the fine, well-sorted sand (S.F.B.C.) not subject to direct wave action.

The Adlittoral-Supralittoral was considered to correspond to the interface between the limits of the terrestrial and maritime domains. However, these authors suggest that, in the case of the sandy littoral of a tideless sea, the interface Supralittoral-Mediolittoral better corresponds to this limit for three reasons:

1. There is practically no faunal exchange and no interpenetration between the populations situated on the two sides of the limit while the exchanges are active in the case of the other limits (trophic exchanges).
2. The displacement of the organisms are always made in the different fluids on the two sides of the limit Supra - Mediolittoral, air above, water below.
3. Above, all the species are referable to terrestrial species which can leave the water and migrate to land (Talitridae); below, all the species could be referred to their own genus in the marine domain.

In fact the absence of true marine species in the high Supralittoral beach could be linked to extreme conditions of dryness and temperature.

It would be interesting to assess if, on beaches where the conditions are still more rigorous (Algeria, Lybia, Israel), this retreat of the marine domain, to the benefit of the terrestrial domain, is confirmed.

Considering hard substrata, practically no research is available separating the terrestrial fauna from the purely marine fauna. Picard (pers. comm.) considers that the permanence of the solid substrate is a primordial factor. It seems advisable, in this case, to take into consideration the physiology of terrestrial and marine organisms. Treating this problem in a broader manner, an ecobiological limit at the level of the minimum of faunal exchanges could be denoted, as in the case of the beaches.

Problem of the Bathylittoral Zone

The zone of contact between phytal and aphytal systems is little known. Its limit corresponds theoretically to that of the existence of unicellular algae. In reality the question is more complex, for the survival of the algae in the quasi-totàl absence of light is not impossible and the problem posed by their actual distribution beyond the limit of the Circalittoral has not been resolved.

The problem has recently been tackled by Falconetti (1980) who found, in addition to the disappearance of the alga *Cryptonoemia tunaeformis*, an alga symbiotic on *Gryphus vitreus*. The facies of *Gryphus vitreus* (Branchiopoda) and *Cidaris cidaris* (echinoderm) living on heterogenous bottoms presumably situated at the summit of the Bathyal led to a belief in the possible existence of a Bathylittoral zone the definition of which, and its coherence with the other zones were difficult to conceive. Falconetti (1980) studied more particularly the different groups of the species; on the one hand that linked to coarse sediments: "mixticoles", "gravellicoles" and on the other hand that linked to fine sediments: muddy species exclusively VTC and VP. It is considered that this hinge population assures a transition between the detritic biocenoses of the Circalittoral and the Bathyal mud. It could perhaps be considered as the "skeleton" of a "larval" assemblage of the detritic Bathyal which could no longer inhabit deeper regions because of the generalised silting which characterizes the Bathyal zone.

There no longer seems any argument for maintaining a Bathylittoral zone. However, Reyss (1974) emphasizes the bathymetric importance of this border: 125-150 m to 250-300 m. This is also a critical zone for the distribution of the species with respect to characterizing a border of bathylittoral contact between the 2 systems: the phytal and the aphytal. The status of this border is thus different from that of the borders of contact between the superior zones. The latter present only minor problems: a few centimeters of transition, small, easily characterisable enclaves.

This solution which demonstrates above all our ignorance, has the advantage of underlining the complexity which exists in the bathymetric distribution of certain species, the area width of the border and finally the difficulty, owing to lack of documentation, of understanding the aphytal system in the Mediterranean.

Problem of the Abyssal Zone in the Mediterranean

Deep sea expeditions in the Mediterranean are not numerous and the extent of our knowledge declines strongly from the littoral towards the great depths; it becomes insignificant beyond a depth of 300 m. This is probably due to our relatively better knowledge of the summit of the aphytal system. The bathyal zone was almost "unknown soil" just up to the last decade.

The existence of the biocenosis of the deep corals and of the deep mud is of course well known. Deep sea diving, submersibles and photos permit a better knowledge of the morphology of the bottoms. Nevertheless the fauna remains very poorly known; almost all new expeditions furnish numerous new species (to the extent of 20% for the amphipods: Bellan-Santini (in press) and cumaceans (Reyss, 1973b). These two biocenoses are comparable to their Atlantic homologues but in general are poorer, although some reservations must still be made.

Two important facts have to be remembered about the distribution of fauna in the depths of the Mediterranean: this sea is homothermic at 13°C which, of course, eliminates the majority of cold stenothermic species, but which, on the contrary, favours the eurythermic and the warm stenothermic species; on the other hand, certain zones are very oligotrophic, others clearly less so. For a benthic community, limited in general in food supply, this relative heterogeneity is significant.

Up to recently it was considered that in the Mediterranean only a rather uniform bathyal population existed on the soft substrate and that this population covered the whole of the slope, the plains and the pits. Improved technology over the last decade has permitted the intensification of collection of small species; more data permitting a restudy of this question are thus available.

- It has been shown (Reyss, 1973b) that the species of the deep Mediterranean fauna are inferior in size to those found generally in the same Atlantic biotopes, but correspond to deeper species living in equally precarious trophic conditions. This could explain, in part, the lack of knowledge that one has of certain groups.
- This same author has shown that the number of species and specimens is much smaller in the Mediterranean than in the Atlantic, especially up to 2000 m. Beyond this depth, it is much less clear where the stations of comparable richness could be found.

Recent investigations (Chardy et al., 1973a,b,c; Laubier, 1972; Reyss, 1971, 1972, 1973a,b; Bellan-Santini, in press) reveal that: the pelecypods conform to the pattern described; their diversity is weak. However, certain dense populations could be collected. No species seems to be endemic or abyssal.

The polychaetes are well represented and could be classified into 2 groups:

a) Atlanto-mediterranean bathyal or abyssal species;

b) Bathyal or abyssal endemic species.

The Cumacea have been well studied. Reyss (1973b) distinguished 3 groups among the species which have never been collected above 200 m:

- one group restricted to the Aegean Sea, the depth of which does not exceed 1500 m;
- one group living between 500 and 2500 m consisting of endemic species;
- one Atlanto-Mediterranean group which extends up to 3000 m.

The amphipods have been the subject of a recent study (Bellan-Santini, in press); they could be classified into four groups:

- one group which does not appear beyond the depth of the continental slope;
- one group collected from the slope to the plain and comprising muddy species, generally speaking, and Atlanto-Mediterranean bathyal species;
- one group of strictly bathyal species comprising 9 endemics;
- one group of five new species, collected at 2500 m and beyond.

Although presenting variations due to their occurrence in certain confined basins (North Aegean Sea) and zones of extreme poverty (Ionian Sea), it could be concluded that:

- the fauna of the depths seems more rich, at least insofar as certain groups (polychaetes, cumaceans, amphipods, isopods, tanaidaceans) are concerned than one would have believed at first. This is due to the small size of the individuals which have thus never been collected before;
- this fauna comprises different groups of species, the most characteristic of which are:
    - the Atlanto-Mediterranean species (which are bathyal for the most part);
    - the eurybathyal species, probably eurythermal, exhibiting a preference for muddy areas;
    - one group of endemic bathyal species, not having been collected beyond 2500 m, but which could perhaps be found again in a more thorough survey;
    - one group of deep endemics consisting of polychaetes (Laubier, 1972), amphipods (Bellan-Santini, in press), Tanaidacea (Kudinova-Pasternak, 1982), isopods (Pasternak, 1982) which could constitute the characteristic abyssal nucleus.

It is generally considered that the Abyssal fauna is characterised by a faunal renewal, but, taking into account particular conditions of the history and the temperature of the Mediterranean, nothing opposes the descent of the eurybathyal species to the greatest depths; the renewal could take place only between purely bathyal and purely abyssal forms.

Pressure of Human Action on the Benthos and the Transformations Induced

In this review the action of man and the transformation induced by him in the benthic domain has not been taken into account for it deals with a subject too important and too vast to be treated here. It is the subject of another synthesis. Nevertheless it is timely to here point out its importance and to make certain remarks.

Human pressure is exerted, above all on the phytal system, and is of two types: pollution and modification of the littoral.

Considering pollution, a certain number of pollution-related biocenoses have been categorized. They have been shown to be degradation stages of natural biocenoses submitted to anthropogenic action.

Pollution proceeds by successive stages, to biocenotic destruction of the natural population. This process is now well known. Although the phytal system seems the most affected, it must not be forgotten that pollutants reach the greatest depths (Arnoux et al., 1981) and that the Mediterranean, being an enclosed sea, it is not improbable that they could accumulate.

Considering that littoral modifications, harbour developments, marinas and diverse rough stone breakwaters affect directly the upper zones, in particular the Infralittoral, by reducing its surface, by destroying bottoms such as those of lagoons, sea grass meadows, by reducing, owing to the influx of suspended matter, the penetration of light and thus primary production and the maintenance of populations. The Circalittoral is itself strongly perturbed by the growth of quantities of suspended material and the hypersedimentation which follows. It is necessary to remember that in certain sectors: north of the Western Basin, High Adriatic, the zones subject to anthropogenic perturbations are already bigger than those which are not. In these sectors at least, the problem of degradation by pollution must be taken in to account during all biocenotic studies.

ACKNOWLEDGMENTS

I am deeply grateful to Dr. Jacques Picard for his comments and suggestions and Dr. Sheila Jacob for translating the manuscript.

SUMMARY

A homogenous classification of the Mediterranean benthic biocenosis has been established progressively as research work has been enlarged in scope and depth by investigations in little known zones and by utilisation of new methods of investigation (diving, apparatus permitting quantitative collection) and data treatment (development of indices and of numerical analyses).

In numerous recent research works a more integrated treatment has been made of the various environmental factors, which have sometimes been dealt with as independent units. A coherent system has nevertheless not been formulated and certain problems have been raised.

The questions that could be posed are of two broad types: basic questions about the biocenosis concern the lack of knowledge of certain biocenoses, and of their links with neighbouring biocenoses. Questions of a more general order deal with the theoretical concepts themselves, or with complex population units which access may be particularly difficult.

Among the more general problems raised, two groups can be distinguished: on the one hand, problems inherent to the sea studied: its richness and the uneven knowledge of it; and, on the other hand, problems linked to interpretation that could be made: the various concepts utilised for the study of the benthos, the uncertainty that one has of different boundaries, as those for the terrestrial and marine domain, or of the phytal and aphytal systems and, finally, of the existence of an abyssal zone. Interposed between the two groups is the problem created by the pressure of human action on the benthos as a whole, and the transformations subsequently induced.

The author hopes that she could thus contribute to a deepening of the thinking of Mediterranean benthologists and to the development of research work capable of proposing solutions to the problems raised.

REFERENCES

Arnoux, A., Bellan-Santini, D., Monod, J.L., Tatossian, J., 1981, Polluants minéraux et organiques dans les sédiments prélevés entre la Provence et la Corse (Mission BIOMEDE 1), Ves Journées Etud. Pollutions, Cagliari, CIESM: 421.

Auteurs Divers, 1981-82, Actas del II Simposio Iberico de Estudios del Bentos marino, (I): Oecologia aquatica, Barcelona, (5) 1981:1; (II) Oecologia aquatica, (6) 1982: 1; (III) Cuadernos de Biologia marine, 1982:1, J. Ros and F.X. Niell, eds.

Bellan-Santini, D., 1966, Elément de bionomie quantitative des peuplements de l'étage infralittoral sur substrat rocheux en Méditerranée nord occidentale, Second Int. oceanog.Congr., Moskva, (Abstr. Papers) 2:29.

Bellan-Santini, D., 1983a, Rapport bibliographique sur le zoobenthos de la Méditerranée, la mer Noire et la mer Rouge (1980-82), Rapp. Comm. int. Mer Médit., 28:29.

Bellan-Santini, (in press), Amphipodes profonds de Méditerranée, Boll. Mus. liv. Storia nat. Verona.

Bellan-Santini, D., Karaman, G., Krapp-Schickel, G., Ledoyer, M., Myers, A., Ruffo, S., and Schiecke, U., 1982, The Amphipoda of the Mediterranean, Mem. Inst. oceano. Monaco, 13:1.

Bigot, L., Picard, J., et Roman, M.L., 1982, Contribution à l'étude de peuplements des invertébrés des milieux extrêmes. 1. La plage et les dunes vives de l'Espiguette (Grau du Roi, Gard), Ecologia Mediterranea, 8:3.

Boudouresque, C.F., 1971, Le concept de nodum en bionomie et sa généralisation, C.R. Acad. Sci. Paris, 272:1260.

Boudouresque, C.F., 1973a, Recherches de bionomie analytique, structurale et expérimentale sur les peuplements benthiques, sciaphiles de Méditerranée occidentale (fraction algale). Les peuplements sciaphiles de mode relativement calme sur substrats durs, Bull. Mus. Hist. nat. Marseille, 33:147.

Boudouresque, C.F., 1973b, La sous strate sciaphile des peuplements de grandes Cystoseires clydonophiles en Méditerranée occidentale (fraction algale), Rapp. Comm. int. Mer Médit., 21: 637.

Boudouresque, C.F., Passelaigue, F., 1972, Note préliminaire sur le peuplement algal des biotopes sciaphiles superficiels battus de substrat rocheux dans la région de Marseille, Bull. Mus. Hist. nat. Marseille, 32:125.

Chardy, P., Laubier, L., Reyss, D., et Sibuet, M., 1973a, Données préliminaires sur les résultats biologiques de la campagne Polymède I. Dragages profonds, Rapp. Comm. int. Mer Médit., 21:621.

Chardy, P., Laubier, L., Reyss, D., et Sibuet, M., 1973b, Données préliminaires, Rapp. Comm. int. Mer Médit., 22:107.

Chardy, P., Laubier, L., Reyss, D., et Sibuet, M., 1973c, Dragages profonds en Mer Ionienne. Données préliminaires. Rapp. Comm. int. Mer Médit., 22:103.

Duval, C., Bellan-Santini, D., et Harmelin, J.G., 1983, Habitats artificiels benthiques immergés en Méditerranée Nord-occidentale. 1. Mise au point d'un module cavitaire expérimental, Téthys, 10:274.

Falconetti, C., 1980, Bionomie benthique des fonds situés à la limite du plateau continental du banc du Magaud (Iles d'Hyères) et de la région de Calvi (Corse), Université de Nice, Thèse Doctorat d'Etat.

Feldmann, J., 1937, Recherches sur la végétation marine de la Méditerranée. La côte des Albères, Rev. Algol., 10:1.

Gamulin-Brida, H., 1967, The benthic fauna of the Adriatic Sea. Oceanogr. mar. Biol., Ann. Rev., 5:535.

Giaccone, G., 1977, Significato bionomico delle comunità fotofile e sciafile nel sistema fitale del Mediterraneo, Atti IX Congr. Soc. ital. Biol. mar., 277.

Giaccone, G., 1983, Rapport bibliographique sur le phytobenthos de la Méditerranée et des provinces biogéographiques raccordées, Rapp. Comm. int. Mer Médit., 28:73.

Hong, J.S., 1980, Etude faunistique d'un fond de concrétionnement de type coralligène soumis à un gradient de pollution en Méditerranée nord-occidentale, Univ. Aix-Marseille II, Thèse 3ème Cycle.

Kisseleva, M.I., 1961, Katchestvennoïe i Kolitchestvennoïe raspredelenie bentosa b pridardanellskom raïone Egeiskovo Moria, Trudy Sevastopol biol. Sta., 14:135.

Kisseleva, M.I., 1964, Katchestvennoïe i Kolitchestvennoïe raspredelenie bentosa v Egeiskom More. Trudy Sevastopol biol. Sta., 16:192.

Kudinova-Pasternak, R.K., 1982, Deep sea Tanaidacea (Crustacea, Malacostraca) from Mediterranean Sea (in Russ.), Trudy Inst. Okeanol. P.P. Shirshov, Moskva, 117:151.

Laborel, J., 1981, Peuplements fossiles des niveaux marins surélevés holocènes dans l'arc égéen, Journ. Etud. Syst. evol. CIESM, Cagliari, 151.

Laubier, L., 1966, Le coralligène des Albères. Monographie biocénotique, Annls. Inst. océanogr., Paris, 43:137.

Laubier, L., 1972, Découverte du genre abyssal Fauvelipsis (Annélide Polychète) en Méditerranée occidentale, C.R. Acad.Sci. Paris, 274:697.

Le Campion, T., 1978, Les Cyanophycées endolithes marines. Systématique, ultrastructure, écologie et biodestruction, Univ. Aix-Marseille II, Thèse Doctorat Etat.

Molinier, R., 1960, Etude des biocénoses marines du Cap Corse, Vegetatio Acta geobotan., 9:121.

Pasternak, F.A., 1982, Composition, origin and peculiarities of distribution of the Mediterranean deep-sea isopod fauna (in Russ.), Trudy Inst. Okeanol. P.P. Shirshov, Moskva, 117:163.

Pérès, J.M., 1967, The Mediterranean benthos, in: "Oceanogr. mar. Biol. Ann. Rev.," H. Barnes, ed.

Pérès, J.M., 1982, Ocean mangement, in: "Marine Ecology," 5 (Part 1), O. Kinne, ed., Wiley & Sons.

Pérès, J.M., et Picard, J., 1958, Manuel de bionomie benthique de la mer Méditerranée, Rec. Trav. Sta. mar. Endoume, 23:1.

Pérès, J.M., 1964, Nouveau manuel de bionomie benthique de la mer Méditerranée, Rec. Trav. Sta. mar. Endoume, 47:1.

Picard, J., 1965, Recherches qualitatives sur les biocénoses marines des substrats meubles dragables de la région marseillaise, Rec. Trav. Sta. mar. Endoume, 52:1.

Picard, J., 1983, Réflexions sur le benthos méditerranéen des substrats meubles de hauts-niveaux dans l'étages Infralittoral, Rapp. Comm. int. Mer Médit., 28:179.

Relini, G., 1981, Distribution of Chthamalus montagui (Crustacea, Cirripedia) in the Mediterranean Sea, Rapp. Comm. int. Mer Médit., 27:149.

Relini, G., 1983, Remarks on the ecology of chthamalids in the Ligurian sea, Rapp. Comm. int. Mer Médit., 28:273.

Reyss, D., 1971, Résultats scientifiques de la campagne Polymède II. Polychètes Aphroditidae de profondeur en Méditerranée, Vie Milieu, 22:243.

Reyss, D., 1972, Résultats scientifiques de la campagne du "Jean Charcot" en Méditerranée Occidentale, mai, juin, juillet 1970. Cumacés, Crustaceana, 3:362.

Reyss, D., 1973a, Résultats scientifiques de la campagne Polymède II en mer Ionienne et en mer Egée. Cumacés, Crustaceana, 27:216.

Reyss, D., 1973b, Distribution of Cumacea in the deep Mediterranean, Deep Sea Res., 20:1119.

Reyss, D., 1974, Les canyons sous marins de la mer Catalane. Le rech du Cap et le rech Lacaze-Duthiers. 4. Etude synécologique des peuplements de macrofaune benthique, Vie Milieu, 23:101.

Romero-Martinengo, J., 1981, Empleo de metodos numericos en la typificacion de comunidades algales, Oecologia aquatica, Barcelona, 5:21.

Salen-Picard, C., 1982, Contribution à l'étude dynamique de peuplements marins de substrats meubles: les peuplements macrobenthiques circalittoraux soumis à l'envasement dans la région provencale, Univ. Aix-Marseille II, Thèse Doctorat Etat.

Vacelet, J., 1981, Les affinités du peuplement de Spongiaires de la Méditerranée. Journ. Etud. Syst. evol., CIESM, Cagliari: 29.

Vinogradova, N.G., Zezina, O.N., Levenstein, R. Ya., Pasternak, F.A., and Sokolova, M.L., 1982, Studies of deep water benthos of Mediterranean sea, Trudy Inst. Okeanol. P.P. Shirshov, Moskva, (in Russ.), 117:135.

ON THE BIOGEOGRAPHY OF THE BENTHIC ALGAE OF THE MEDITERRANEAN

Francesco Cinelli

Istituto di Zoologia e Anatomia Comparata
Università di Pisa, Via A. Volta 4, Pisa, Italy

Examination of the current literature shows that floristic knowledge of the benthic algal vegetation of the Mediterranean is still fragmentary. Nevertheless, since the second world war, researches of Mediterranean phycologists have substantially increased our knowledge in this field. Unfortunately, some parts of the Mediterranean are still unknown floristically (e.g. the Calabrian and Apulian coasts of Italy) or have not been sufficiently studied, such as the Aegean Sea. One may note that floristic comparisons assume that species given the same name by different authors are actually the same taxon. Another important consideration is that the intensity of collecting and the availability of suitable substrate differs throughout the region and that, in most groups, our understanding of the ecology, genetics, life-histories and distribution of the component species is quite inadequate.

It is likely, therefore, that suggestions made here and elsewhere in an attempt to summarise the algal phytogeography of large regions will require substantial modification once detailed studies reveal more problems with taxonomy and ecology than originally suspected. The above problems, taken together, compound the difficulties of reaching valid general conclusions about the phytogeography of even quite common algae. In the Atlantic, recent studies of the genetics of Laminaria (Chapman, 1973,1974,1975; Lüning et al., 1978) have gone some way toward the solution of confusing taxonomic problems, but have raised new difficulties. Where a great deal of information is available, for example, on life histories, this is often for a restricted region only, and there are comparatively few studies of the breadth of those of Muller (1979) on Ectocarpus, or Rietema (1975) on Bryopsis. A historical synopsis of Mediterranean phytogeography requires a knowledge of its geological record. During the Tertiary period, the Mediterranean and Indo-Pacific Oceans were linked, allowing the immigration of numerous Indo-Pacific species into the Mediterranean. There was also communication between the Mediterranean and the Atlantic.

Changes that occurred in the succeeding geological periods have caused frequent and large variations of factors, such as salinity and temperature, limiting the distribution of species. These changes have permitted, at different times, the introduction of floristic elements from boreal and tropical zones. These fluctuations in ecological factors were also of fundamental importance in the dispersal and distribution of indigenous species and in their potential for speciation. The two factors (geological events and genetic variability) influence the origin of species which are tropical or boreal or endemic in a sea now considered, in a floristic sense, warm-temperate. Some very important genera of algae, such as the genus Cystoseira, are characterised by a very strong genetic plasticity both in morphology and in ecology. Because the Straits of Gibraltar do not represent an important geographical boundary, the Mediterranean Sea must be considered not a distinct unit but part of a large system which includes the adjacent parts of the Atlantic: in the north the Lusitanian region and, in the south, the Mauritanian region including the Cape Verde Islands, Canaries, Madeira and the Azores. Several biogeographical elements are present in the Mediterranean flora: cosmopolitan, circumtropical, Atlantic (boreal and tropical), endemic or "Mediterranean" (neo- and paleoendemic), Indo-Pacific. In general, most of the species which comprise the present day marine flora of the Mediterranean originate in the Atlantic. Elements of Indo-Pacific origin are very few and have probably increased since the opening of the Suez Canal. The endemic elements may be divided into two parts: paleoendemic (of Indo-Pacific origin) and neoendemic (of Atlantic origin).

COSMOPOLITAN ELEMENTS

The cosmopolitan species in the Mediterranean are fiew. Most of them are considered, from an ecological point of view, eurythermal, euryhaline and ubiquitous. Some of them may be divided into two sharply distinct main groups: warm-water and bipolar cosmopolitan. The former includes a very few species but a relatively large number of genera.

The bipolar cosmopolitan species are cold or temperate algae which have a widespread distribution but which avoid the warm surface water of the tropic and subtropic by submergence.

One typical Mediterranean example is the brown alga Scytosiphon lomentaria widely distributed in the Arctic ocean, North Atlantic, Mediterranean, North Pacific and, in the southern hemisphere, along the South American coasts, Kerguelen Islands, Tasmania and New Zealand.

CIRCUMTROPICAL ELEMENTS

Few species of the Mediterranean flora may be called circumtropical. Most of them have a wide distribution in tropical and subtropical regions. Because of subtropical surface temperatures during summer, these species enjoy suitable conditions for reproduction, which in general, occurs during summer. Several Mediterranean tropical algae have been regarded as relicts from the Tertiary Tethys Sea. Indeed, the whole tropical and subtropical coastal flora in the East Atlantic, West Indies, Mediterranean, Indian and Pacific Ocean, is derived from it. Before the glacial period, most circumtropical species in the Mediterranean area, were destroyed by the low temperatures and now their distribution is more restricted than in the past. Our data on the present day flora of the Mediterranean seem to confirm the presence of tethyan relicts in contradiction to the idea of total dryness of this basin in the Mesozoic age.

Neomeris annulata, tropical Dasycladacea, now present in the West Indies, is known from fossiliferous lutetian deposits of the basin of Paris and in Bretagne. Some other species are actually confined to the south-eastern part of the Mediterranean (as *Caulerpa racemosa* or *C. scalpelliformis*, circumtropical Chlorophyta) where sea-water temperatures remain high throughout the year.

In a recent work, Furnari (in press) has shown that the circumtropical species of ten floristic zones of the Mediterranean are about 3% of the total floras and reach 3.82% in Greece.

In the genus *Halimeda* the species *H. tuna* is a typical representative of the circumtropical element as well as *Hypnea musciformis*, *Centroceras clavulatum*, *Digenea simplex*, *Valonia macrophysa* etc.

On the other hand there are many pantropical genera: *Acetabularia*, *Udotea*, *Caulerpa*, *Zonaria*, *Pockochiella*, *Sargassum*, *Cystoseira*, *Liagora*, *Galaxaura*, *Wrangelia*, *Amphiroa* etc.

INDO-PACIFIC ELEMENTS

Notwithstanding the many similarities between the eastern Mediterranean basin and the Red Sea and the Indian Ocean as regards physicochemical properties, very few algal species are common to the two regions. The majority of them have a restricted distribution area and entered the Mediterranean after the opening of the Suez Canal. Furnari (in press) reports that the flora of eastern Sicily has the highest number of Indo-Pacific species (completely lacking in the Golfe du Léon, France, and in Costa Brava, Spain). It is true that there is a small number of circumtropical species common to the Red Sea and the Mediterranean, but the majority of the species of Indo-Pacific origin reflect an earlier connection between the Mediterranean and the Indian Ocean. In the Tertiary period the main connection between these two seas was to be found to the north, through south-west Asia. It was many times broader and it lasted longer. Thus, we find that the similarity between the Mediterranean and the Indo-West Pacific flora does not depend upon the neighbouring Red Sea as much as regions further to the east.

In the Sicily Channel, the large area between Sicily and Tunisia, this group represents only 4.2% (7 species) of the total flora of this region (Cinelli, 1981). In this case the species of Indo-Pacific origin are reduced due to the strong influence of the north African surface current and particularly so for the species entering the Mediterranean within the Lessepsian migration. Examples of these species are: Caulerpa racemosa, Lophocladia lallemandii, Acetabularia parvula. This last species, common in all tropical Indo-Pacific waters, was found in the Mediterranean by Schussnig (1930) who described it as a new species: A. wettsteinii. The subsequent research of Feldmann and Feldmann (1947) reported the discovery of specimens in Algeria and showed the synonymy of Schussnig's alga with A. moebii described by Solms-Laubach in 1895. Valet (1968-1969) proposed to make A. moebii and A. parvula synonymous, giving priority to A. parvula. Feldmann thinks that this species of Indo-Pacific derivation entered the Mediterranean in the form of resistant organs in fairly recent times by way of the Red Sea. More recently, Rayss (1955), proposed that this could have also been a species already present in the Mediterranean during the Tethys age which remained localized there in the warmer zones. The present distribution is consistent with both hypotheses: either that A. parvula entered from the Red Sea and spread toward the central Tyrrhenian, or that the sites of discovery are those where it has continued to develop since the closing of the Mediterranean basin.

This is true of other species and the number of common genera in naturally larger: the general Ethelia, Mesospora, Rhyzophyllis, Codium, Vidalia are common to the two seas.

ATLANTIC ELEMENTS

Most of the species which constitute the present day marine flora of the Mediterranean originated in the Atlantic. The boreal and temperate Atlantic group is the richest one, in general, and the Mediterranean represents a sort of refuge area and the southern limit of their distribution. As for tropical-Atlantic species, there are species common to the Mediterranean in the vicinity of the Straits of Gibraltar and nearby areas (Morocco, Mauritania, southern Spain and Portugal) and to the Atlantic in the Caribbean region. This is the case for Caulerpa prolifera, Dasycladus vermicularis, Halicystis parvula, Galaxaura adriatica, Crouania attenuata etc. The atlantico-boreal species can be divided into two groups: the first is restricted to the Mediterranean and along the Atlantic coasts of Europe. The second is restricted to the Atlantic coasts of Europe, the western Mediterranean and the North American coast. Sphaerococcus coronopifolius, Taonia atomaria and Halopytis incurvus are good examples of the first group. In the second we find Stilophora rhizodes, Cutleria multifida, Gymnogrongus norvegicus, Chondria dasyphylla etc. It is very interesting to consider some Mediterranean areas where ecological conditions are now the same as those in the boreal Atlantic. The Straits of Messina and the northern Adriatic Sea have large number of species which now occur in the northern part of the Atlantic: the species Fucus virsoides and Catenella repens for the Adriatic and Laminaria ochroleuca, Saccorhiza polyschides, Phyllaria reniformis, P. purpurascens for the Straits of Messina are a few examples of this situation.

In two works of 1972 and 1976 Giaccone gives the results of his researches on laminarians communities of the Straits of Messina and of the Alboran Sea. In these exhaustive studies, Giaccone infers that the Straits of Messina and the Sea of Alboran are a centre of refuge and formation and, successively, of diffusion of Atlantico-boreal elements in the Mediterranean. This is confirmed also by Cinelli (1981) in his study of the Sicily Channel where 53.5% of the total flora of this region is represented by Atlantic elements.

ENDEMIC OR "MEDITERRANEAN" ELEMENTS

In all the Mediterranean region for which we have sufficient data, the number of endemic species is very high. Here some of the more important species only are mentioned. Some of them are unknown outside the Mediterranean, while others are found also in the northern and southern part outside the Straits of Gibraltar. According to the available information (Feldmann and Feldmann, 1947; Giaccone, 1972; Furnari, 1983; Lipkin and Sefriel, 1971; Gerloff and Geissler, 1971; Haritonidis and Tsekos, 1974,1975,1976; Nizamuddin and Lehnberg, 1970; Tsekos and Haritonidis, 1974,1977; Cinelli, 1981; etc.) it is possible to separate the endemic elements in two categories: paleoendemic and neoendemic. There are very few species in the first category: Rissoella verruculosa, Beckerella mediterranea, Rodriguezella strafforelli among the red algae and Laminaria rodriguezii and Mesospora mediterranea among the brown. These species of Indo-Pacific origin, originated probably in the Mesozoic age in the western part of the Tethys Sea. On the other hand neoendemic species are of Atlantic origin and, in general, from the Lusitanian region. According to Giaccone (1972) 29% of the endemic species are in Phaeophycaeae and, in particular, in the genus Cystoseira which, in the Mediterranean, has 29 species, 80% of which are considered neoendemic. The central Tyrrhenian Sea, the Balearic Islands, the eastern Mediterranean and the Adriatic Sea show the highest number of endemic species. Some of them are also considered as biogeographic indicators because of their geographical restricted distributions. For example, Rissoella verruculosa occurs only in the western, and Beckerella mediterranea in the eastern Mediterranean. The same situation respectively for Lithophyllum tortuosum, L. trochanter and Tenerea undulosa.

CONCLUSIONS

From the phytogeographic point of view the algal flora of the Mediterranean contains several distinct elements of different origin. On the basis of the available data we can also recognize the division of the Mediterranean into north-western, central-southern and eastern sub-regions.

According to Furnari (in press) it is possible to distinguish two major groups: on the basis of the comparison of the chorological spectra: the first with "Mediterranean" endemic character and the second with a more marked Atlantic character.

The northern Adriatic Sea, the Straits of Messina, the Sicily Channel and the Sea of Alboran form distinct "atlantico-boreal" islands in the Mediterranean.

It is true that in all studied areas the physiognomic and structural aspects of the ecosystems are diverse and it is due in particular to the vicariance of the guide-species of the communities.

SUMMARY

Our present knowledge of the chorology and ecology of marine algae in the Mediterranean is still fragmentary. The intensity of collecting and the availability of suitable substrate differs thoughout the region and the understanding of the ecology, genetics, life-histories and distribution, in most groups, is quite inadequate. The biogeography of the Mediterranean is strictly linked to its geological record, and the changes which occurred in the succeeding periods have caused frequent and large variations in salinity and temperature. These fluctuations in ecological factors were of fundamental importance also in the dispersal and distribution of the indigenous species and their relative potential for speciation. The two factors (geological events and genetic variability) influence the origin of species which are tropical or boreal or endemic in a sea now considered, in a floristic sense, warm-temperate. Several phytogeographical elements are present in the Mediterranean: cosmopolitan, circumtropical, Atlantic (boreal and tropical), Indo-Pacific, endemic or "Mediterranean" (paleo- and neoendemic). On the basis of the available literature it is possible to divide the Mediterranean into three regions: north-western, central-southern and eastern.

The flora is divided into two major groups: "Mediterranean" endemic and Atlantic.

The northern Adriatic Sea, the Straits of Messina, the Sicily Channel and the Sea of Alboran form distinct 'atlantico-boreal' islands in the Mediterranean.

REFERENCES

Chapman, A.R.O., 1973, Phenetic variability of stipe morphology in relation to season, exposure and depth in the non-digitate complex of Laminaria Lamour. (Phaeophyta, Laminariales) in Nova Scotia, Phycologia, 12:53.

Chapman, A.R.O., 1974, Genetic basis of morphological differentiation in some Laminaria population, Mar. Biol., 24:85.

Chapman, A.R.O., 1975, Inheritance of mucilage canals in Laminaria (section Simplices) in eastern Canada, Br. Phycol.J., 10:63.

Cinelli, F., 1981, Biogeography and ecology of the Sicily Channel. I. The algae of the Banks, in: 'XTH Intern. Seaweed Symp., Tore Levring Ed.

Feldman, J., and Feldmann, G., 1947, Addition à la Flore des algues marines de l'Algérie, Bull. Soc. Hist. nat. Afr. Nord, 38:80.

Furnari, G., (in press), The benthic algae of southern Italy. Floristic and geobotanic considerations, Atti Congresso O.P.T.I.M.A., Palermo.

Gerloff, J., and Geissler, U., 1971, Eine revidierte liste der Meeresalgen Griechenlands, Nova Hedwigia, 22:721.

Giaccone, G., 1972, Struttura, ecologia e corologia dei popolamenti a Laminarie dello Stretto di Messina e del Mare di Alboran, Mem. Biol. Mar. Ocean., 2:37.

Giaccone, G., and Rizzi-Longo, L., 1976, Revisione della flora dello Stretto di Messina (note storiche, bionomiche e corologiche), Mem.Biol.Mar.Ocean., 6:69.

Haritonidis, S., and Tsekos, I., 1974, A survey of the marine algae of Thassos and Mytilene Islands, Greece, Bot.mar., 17:30.

Haritonidis, S., and Tsekos, I., 1975, Marine algae of the northern Greece, Bot.mar., 18:203.

Haritonidis, S., and Tsekos, I., 1976, Marine algae of the Greek West coast, Bot.mar., 19:273.

Lipkin, J., and Safriel, U., 1971, Intertidal zonation on rocky shores at mikhmoret (Mediterranean, Israel), J.Ecol., 59:1.

Lüning, K., Chapman, A.R.O., and Mann, K.H., 1978, Crossing experiments in the non-digitate complex of Laminaria, from both sides of the Atlantic, Phycologia, 17:293.

Müller, D.G., 1979, Genetic affinity of Ectocarpus siliculosus (Dillw.) Lyngb. from the Mediterranean, North Atlantic and Australia, Phycologia, 18:312.

Nizamuddin, M., and Lehnberg, W., 1970, Studies on the marine algae of Paros and Sikinos islands, Greece, Bot.mar., 13:116.

Rayss, T., 1955, Les algues marines des côtes Palestiniennes. I. Chlorophyceae, Bull. Sea Fish.Res.Stn. Israel, 9:1.

Rietema, H., 1975, Comparative investigations on the life histories and reproduction of some species in the siphonous green algal general Bryopsis and Derbesia. Thesis. Rijksuniversiteit Groningen.

Schussnig, B., 1930, Phycologische Beiträge. III. Acetabularia wettsteinii n.sp. in Mittelmeer, Osterreichisch. Bot. Zeitschr., 79:333.

Tsekos, I., and Haritonidas, S., 1974, The marine algae of Rhodos island, Greece, Br. phycol. J., 9:399.

Tsekos, I. and Haritonidis, S., 1977, A survey of the marine algae of the Ionian islands, Greece, Bot.mar., 20:47.

Valet, G., 1968-1969, Contribution à l'étude des Dasycladales. I. Morphogenèse. 2. Cytologie et réproduction. 3. Révision systématique, distribution geographique et relation phylogénétique, Nova Hedw. 16:21; 17:551.

DISTRIBUTION AND ECOLOGY OF ENDEMIC ELEMENTS IN THE MEDITERRANEAN FAUNA (FISHES AND ECHINODERMS)

Enrico Tortonese

c/o Istituto Zooprofilattico, Lungo Bisagno Dalmazia 45 A
16141 Genova, Italy

INTRODUCTION

The endemic part of a fauna deserves particular attention not only for discussions of biogeographical problems but also in connection with the ecosystems occurring in the area. As far as the Mediterranean is concerned, endemism has often been considered by authors dealing with different groups of animals. The ratio of endemic species, of course, is variable according to them but as a whole it is high, having been estimated at 15-16% (Menzies, 1973). An excellent report has been presented by Fredj (1974) who discussed several problems after having examined the situations observed in some groups of benthic invertebrates.

Recent advances in the knowledge of the Mediterranean fauna are now stimulating new reviews and I have attempted to present one based on two groups - fishes and echinoderms - that are important elements in the populations of well-delineated biotopes. A few notes are added on some other organisms interesting for their distribution and ecology.

A species considered to be endemic in an area may lose such qualification with improvement of the knowledge of its distribution and taxonomy: it may be recognized to occur elsewhere, or to be identical to another, not endemic species. A second important remark is needed in this case. For a long time it has been known that the Mediterranean fauna is closely related to that of the Atlantic ocean, being practically a part of it. Many species are often considered to be "Mediterranean" but exist also in the nearest oceanic zone and are therefore endemic in a wider sense. So, we may distinguish endemics "sensu stricto" (found only in the Mediterranean) and "sensu lato" (found also in the Atlantic and/or in the Black Sea). Here, separate lists are given of fishes and echinoderms belonging to both categories. Surely, the numbers of species and their ratios are expected to change for the reasons listed above. Anyway, the review of endemics is interesting because they express the present, quantitative situation and help our understanding of the faunistic relations with the adjacent waters. The Red Sea is omitted as no species (at least of the groups now considered) are known to be restricted to it and to the Mediterranean.

The general aspects of endemism (distribution of the species, participation in ecosystems, possible origin) are connected with the history of the Mediterranean, that is with fascinating problems which in recent years have attracted biologists, paleontologists, geologists: a mention of them also will be necessary in these pages. Questions of classification and nomenclature are left aside.

MEDITERRANEAN ENDEMICS

Fishes

Some years ago (1976) I presumed that about 500 species of fishes exist in the Mediterranean and suggested that 77 (15.4%) of them are endemic. A total of 550 is now admitted and 60 (10.9%) appear to be endemic in the restricted sense, having not yet been reported outside this sea. Six of them (8.5%) are Chondrichthyes (out of 70) and 54 (11.2%) are Osteichthyes (out of 480). A list is submitted here and a review of the endemics "sensu lato" will follow in the next pages.

W = western Mediterranean basin, to the Sicily channel
A = Adriatic
E = eastern basin

Rajidae	*Raja asterias* (W A E)
	Raja rondeleti (W)
	Raja melitensis (W E)
	Raja polystigma (W A E)
	Raja radula (W A E)
Dasyatidae	*Dasyatis tortonesei* (W)
Acipenseridae	*Acipenser naccarii* (A)
Bathypteroidae	*Bathypterois mediterraneus* (W)
Paralepididae	*Paralepis speciosa* (W)
Heterenchelydae	*Panturichthys fowleri* (E)
Ophichthidae	*Ophichthus rufus* (W E)
Cyprinodontidae	*Aphanius fasciatus* (W A E)
	Aphanius iberus (W)

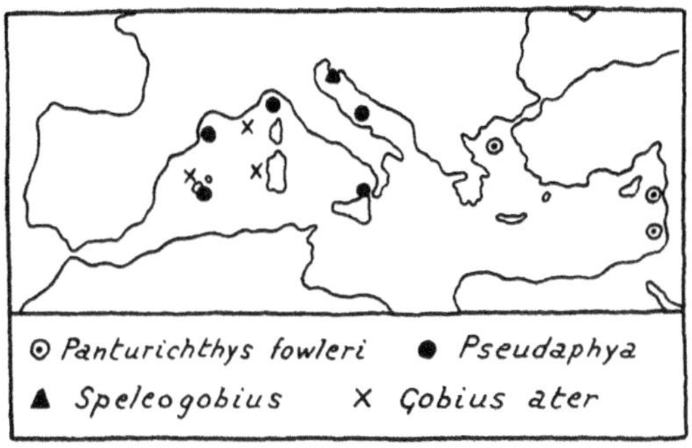

Fig. 1. Known localities of fishes now considered endemic in the Mediterranean.

Gadidae	Antonogadus megalokynodon (W A E)
Moridae	Lepidion lepidion (W) Eretmophorus kleinenbergi (W) Rhynchogadus hepaticus (W)
Syngnathidae	Syngnathus phlegon (W A E) Syngnathus taenionotus (A)
Apogonidae	Microichthys sanzoi (W)
Serranidae	Epinephelus haifensis (W E)
Labridae	Symphodus doderleini (W A E) Symphodus melanocercus (W A E)
Istiophoridae	Tetrapturus belone (W A E)
Gobiidae	Gobius arenae (W) Gobius ater (W) Gobius fallax (W A) Gobius geniporus (W A) Gobius luteus (W) Gobius strictus (W A) Gobius vittatus (W A E) Buenia affinis (W) Chromogobius zebratus (W A) Corcyrogobius lichtensteini (A) Didogobius bentuviai (E) Gammogobius steinitzi (W) Knipowitschia panizzai (A) Lesueurigobius suerii (W A) Millerigobius macrocephalus (A) Pomatoschistus canestrinii (A) Pomatoschistus kneri (W A E) Pomatoschistus quagga (W A) Pomatoschistus tortonesei (W E) Pseudaphia ferreri (W)

	Speleogobius trigloides (A) Thorogobius macrolepis (W A) Zebrus zebrus (W A)
Blenniidae	Lipophrys adriaticus (W A E) Lipophrys dalmatinus (W A E) Lipophrys nigriceps (W A E) Lipophrys basiliscus (W A E) Lipophrys sabry (E)
Tripterygiidae	Tripterygion melanurus (W A E)
Ophidiidae	Parophidion vassali (W)
Bythytidae	Oligopus ater (W A)
Carapidae	Carapus acus (W A E)
Scorpaenidae	Scorpaenodes arenai (W)
Liparidae	Paraliparis leptochirus (W)
Lepadogasteridae	Gouania wildenowi (W A E) Opeatogenys gracilis (W A)

No families are endemic to the Mediterranean, but nine genera are: Eretmophorus, Rhynchogadus, Millerigobius, Gammogobius, Speleogobius, Corcyrogobius, Didogobius, Gouania, Opeatogenys. The small apogonid Microichthys cannot be added because a fish found in the North Atlantic is quite probably M. coccoi, reported only from the strait of Messina (Sicily).

As far as the species are concerned, most belong to: Rajidae: Medit. 16, endemic 6 (37.5%), Gobiidae: Medit. 50, endemic 23 (46%), Blenniidae: Medit. 22, endemic 5 (23.8%). If infraspecific units are considered, quantitative evaluations become more difficult as a consequence of their often-discussed systematic value. However, they also are interesting for a more complete survey of endemism and related problems. A first group of them includes Mediterranean forms of species present in the Atlantic:

Alosa fallax nilotica (remarks will follow)
Paralepis coregonoides coregonoides (W A E)
Lestidiops jayakari pseudosphyraenoides (W E)
Notoscopelus elongatus elongatus (W A E)
Diplodus sargus sargus (W A E)
Bothus podas podas (W A E)

Of two other species the Mediterranean has the nominate form and a subspecies:

Solea vulgaris aegyptiaca (A E, partim): The exact distribution is not well known and so neither are the morphological characters,

Lepadogaster lepadogaster purpureus (W A E): in the westernmost part of the Mediterranean and in the near ocean it is replaced by the nominate form.

Different authors observed the geographic variations of *Syngnathus* so that polytypic conditions are perhaps to be recognized. Mediterranean subspecies would be the following:

Syngnathus acus rubescens (W A E)
Syngnathus typhle rondeleti (W A south, E)
Syngnathus typhle rotundatus (A north)

Lack of agreement also concern the shad (*Alosa*) and the hake (*Merluccius*). Three forms of *A. fallax* have been described:

A.f. nilotica (A E?)
A.f. rhodanensis (W: Italy, France, Spain?)
A.f. algeriensis (W: Africa, Sardinia?, Sicily?)

The two latter were usually included in the former but differ for meristic features and for distribution (see above).

Merluccius merluccius mediterraneus has been separated from the Atlantic, nominate form. However, it is still uncertain if the variable populations really support this view.

A second group includes subspecies, recently recognized, which in the Mediterranean accompany (allopatric) the nominate forms of a few endemic species:

Chromogobius zebratus zebratus (A)
" " *levanticus* (E: Israel)
Lipophrys nigriceps nigriceps (W A E)
" " *cypriacus* (E: Cyprus)
" " *portmahonis* (W: Balearic islands)
Tripterygion melanurus melanurus (part of W and E)
" " *minor* (A, part of W and E)

The distribution of *T. melanurus* is clearly shown in the map of Zander and Heymer (1976). The Mediterranean fishes that are now considered to be so restricted, have a very different horizontal distribution: some are present almost everywhere, others seem limited to one or a few zones. 17 species have hitherto been reported only from the western basin, the whole fauna of which is richer. Seven species are known only from the Adriatic, three only from the eastern basin (*Panturichthys fowleri*, *Lipophrys sabry*, *Didogobius bentuviai*). Our information on the marine fauna along some north African coasts (Libya) is still very inadequate.

Many other fishes are endemic "sensu lato" because the

Mediterranean is their essential residence, but the habitat is also extended to the near parts of the Atlantic and/or to the Black Sea. The total number of these species is 45, the lists of which are given below according to the existing, reliable data on the distribution.

Atlantic. The oceanic area of interest here (and better defined in the last chapter) lies chiefly between Portugal and Maureitania. It is often populated by many fishes that are usually called Mediterranean in the widest sense. They are partially sympatric with a greater number of species which in the ocean don't remain so restricted, but are widespread towards northern (Scandinavia, etc.) or southern latitudes (Angola, South Africa). Some fishes (e.g. Serranus atricauda, Pagellus coupei) may be found in the SW Mediterranean but live essentially in the near Atlantic: therefore they are excluded from the following list.

The latter includes 36 species, 12 of which (marked with asterisk) cross the whole Mediterranean and reach the Black Sea. It is necessary to recall once more how provisional may be the indications given here: new investigations are going on, surely involving changes, but leaving the general biogeographic picture as now outlined.

Mustelus mediterraneus
Somniosus rostratus
Dalophis imberbis
Apterichthys caecus
Belone svetovidovi
Nerophis maculatus
Hippocampus hippocampus
Oedalechilus labeo
*Sciaena umbra
*Spicara maena
*Spicara flexuosa
*Spicara smaris
*Labrus viridis
Lappanella fasciata
*Symphodus cinereus
*Symphodus roissali
*Symphodus tinca
Symphodus mediterraneus
*Uranoscopus scaber
*Callionymus pusillus
Parablennius rouxi
Lippphrys canevai
Tripterygion xanthosoma
*Gobius bucchichi
Gobius auratus
Deltentosteus colonianus
Lesueurigobius sanzoi
Odondebuenia balearica
Odondebuenia pruvoti
Tetrapturus georgei
Bellottia apoda
Echiodon dentatus
Lepidotrigla cavillone
Lepidotrigla dieuzeidei
*Solea impar
Arnoglossus rueppelli

Three subspecies are to be added: Belone belone gracilis, Trisopterus minutus capelanus, Gadiculus argenteus argenteus. All these fishes variably occur west of Gibraltar (Portugal, coasts of NW Africa from Morocco to Mauritania, Madeira, Canary). 23 species are distributed chiefly southwards, only or partly. At least three of the others reach the bay of Biscay (Somniosus rostratus, Hippocampus hippocampus, Symphodus cinereus). Callionymus pusillus has been reported only from the gulf of Cadiz.

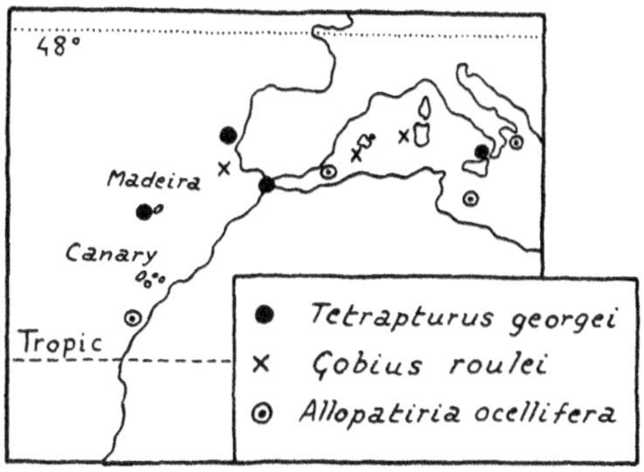

Fig. 2. Known localities of fishes and echinoderms fround in the Mediterranean and the nearest Atlantic zones.

Black Sea. A much smaller group (9 species) is composed of Mediterranean endemic fishes that extend also to the Black Sea, at least to its western part:

Symphodus ocellatus
Symphodus rostratus
Chromogobius quadrivittatus
Zosterisessor ophiocephalus
Parablennius sphynx
Parablennius zvonimiri
Parablennius incognitus
Ophidion rochei
Arnoglossus kessleri

A Mediterranean subspecies of clupeid (Alosa fallax nilotica) is found (rare) in the Black Sea. In the latter, two subspecies replace the corresponding, Mediterranean nominate forms: P. zvonimiri ponticus and Solea impar nasuta.

An opposite situation to that of the list above is offered by a few Pontic fishes present also in limited areas of the Mediterranean (North Aegean, Adriatic): Huso huso, Syngnathus tenuirostris, Merlangius merlangius euxinus, Knipowitschia caucasica, Platyichthys flesus luscus. Only in the north Aegean has the Pontic goby Proterorhinus marmoratus been collected.

The submitted quantitative data are assembled in the following table:

Classes	N.sp.	End.s.s.	End.s.l.	% s.s.	% s.l.
Chondrichthyes	70	6	8	8.5	11.4
Osteichthyes	480	54	97	11.2	20.2
Totals	550	60	105	10.9	19

A remarkable variety appears if the ecology within the Mediterranean is considered. The anadromous fishes are represented by two species: a sturgeon (Acipenser naccarii: Adriatic) and a shad (Alosa fallax, of which at least one subspecies is Mediterranean).

The pelagic fishes belong to two sections:

a) Epipelagic - These are only two: Syngnathus phlegon (for this genus is an exception) and Tetrapturus belone. The latter is the largest endemic fish in this sea and its geographic restriction is surprising, as will be explained later on.

b) Meso- to bathypelagic - The species are six:

Bathypterois mediterraneus: 300-1000 m
Paralepis speciosa: no bathymetric data
Rhynchogadus hepaticus: id.
Eretmophorus kleinenbergi: 30-500 m
Lepidion lepidion: 150-2230 m
Paraliparis leptochirus: 500-700 m

Some others were formerly known only from the Mediterranean, but then have been found even very far in the Atlantic (Bathophilus nigerrimus, Chalinura mediterranea, etc.).

The other endemics are littoral and occur in different ecosystems. The benthic ones form a high percentage, as it can be inferred by the number of them belonging to the three families already quoted (Rajidae, Gobiidae, Blenniidae). Therefore they require a particular attention with reference to the biotopes. These are situated at infra- and circalittoral levels: the lower limit of the infralittoral is that reached by the phanerogams, the lower limit of the circalittoral corresponds to that of the non photophilous algae. The endemic fishes occupy various ecological niches, which in many cases are still poorly known. Zander's investigations on blenniids (1973) had very important results on this subject.

Rocky bottoms along the shore, about 0-10 m, are partly covered with algae or interrupted by stones, gravel and small sandy spaces. Common fishes are gobiids and blenniids. The endemic Gobius bucchichi is interesting for its association with a sea anemone (Anemonia sulcata). Lipophrys adriaticus and the related species have microhabitats in such environments are often seen resting on rocks or hiding in cavities. According to the species, there are also preferences in food (harpacticoid copepods, algae, etc.). Tripterygion tripteronotus and other species of this genus are more or less common. One of them (T. xanthosoma) is more photophilous.

Other infralittoral environments are the beds of phanerogams (Posidonia, etc.) distributed to 40 m at least. The ground is sandy and partially muddy, populated by animals among the rhizomes and the algae that often are abundant (e.g. Peysonnelia). The leaves offer good shelter to the fishes, which can obtain food from them or from syntopic organisms. This is a proper biotope for the species of Syngnathus and Symphodus. Among the representatives of gobiids are Gobius geniporus and Zebrus zebrus, together with Zosterisessor ophiocephalus.

The lagoons are particularly extended in the north Adriatic and along the French coasts. On their bottom, sandy or muddy, the vegetation is variably-developed. The lagoons are populated by euryhaline and eurythermous fishes, some of which are small, endemic species. Knipowitschia is a Ponto-Caspian genus represented in the north Adriatic; it dwells in brackish estuaries and lagoons, but also in fresh water. Pomatoschistus canestrinii is another species of the same area and P. tortonesei has been discovered in a wide lagoon of western Sicily. Aphanius may occur in great numbers and is very euryhaline, being found even in fresh water.

Several endemic fishes live at circalittoral levels on hard and soft bottoms. Rocks may be more or less covered with calcareous algae and support the so called "coralligenous", a biotope that may show different features and always has a great interest. It is present with interruptions at about 10-50 depth, offshore or on deeper parts of rocky littorals, chiefly under edges, in caves, on steep slopes where light is scarce. The benthic community differs from that of the soft bottoms because it is usually formed by a greater number of species and a lesser number of individuals. The animal population is rich, largely prevailing on the vegetation, and includes many sponges, gorgonians, madrepores, bryozoans; the red coral may be present or not.

An endemic fish living here is Gobius vittatus, very similar to Parablennius rouxi in colour pattern (a black band along the sides). They form a pair of "ecological equivalents" or organisms very alike having a similar way of life in identical biotopes. However, according to Zander and Heymer (1977) these species are not exactly syntopic because their ecological niches are not wholly similar. The fishes of the genera Odondebuenia and Tripterygion also occur on coralligenous bottoms.

Such bottoms often exist in caves, where several Mediterranean endemics may be observed. Oligopus ater is remarkable for its eurybathy: it reaches more than 700 m but in some caves of Dalmatia has been observed at 1-2 m only. Lipophrys nigriceps shows very evident adaptive characters as a cave dweller (reduced melanophores, larger eyes, etc.) and is very similar to Tripterygion melanurus, a species that is also speleophilous (syntopic). The Mediterranean gobids have been recently described and till now are known only from caves near Marseille (Gammogobius steinitzi) and Rovinj (Speleogobius trigloides).

Sandy, muddy, detritic, gravelly bottoms extend to various levels and are populated by organisms adapted to their particular features. On sands, the granulometry is important for many invertebrates and also for benthic fishes. As far the endemics are concerned, such bottoms are of course the residence of the rays and of Dasyatis tortonesei. On sands near shores are found Callionymus pusillus, Pseudaphia ferreri, Gobius arenae and Gouania wildenowi (also on gravel). Didogobius bentuviai is a dweller of muddy sands offshore and probably a burrower. Lesueurigobius suerii, living on mud to 100 m, is one of the Mediterranean gobies reaching greater depths. Panturichthys fowleri is a benthic anguilliform found on muddy bottoms at 30-60 m.

Echinoderms

The number of Echinoderms known from the Mediterranean has increased remarkably during the recent years. After 1960, more and more species have been collected in different regions and several have been described as new. As a consequence, the ratio of endemics also changed:

Pérès, 1967	: 107 species, 26.1% endemic
Fredj, 1974	: 134 species, 23.9% endemic
Tortonese, 1980	: 144 species, 22.2% endemic

The total number of species is still 144, because here Amphiura stepanovi (Black Sea) is left aside, but it is replaced by an Atlantic holothuroid (Aslia lefebvrei) recently collected off eastern Spain. 35 Mediterranean species (about 1/4) can be said to be endemic s.s. and are distributed in all the classes, as is shown by the following list. For some of these echinoderms (Phyllophorus, Amphiura, etc.) the distribution is still poorly known and cannot be exactly indicated. Abbreviations: W = western basin, A = Adriatic, E = eastern basin.

Crinoidea	Antedon mediterranea (W A E)
	Leptometra phalangium (W A E)
Holothuroidea	Holothuria stellati (W A E)
	Pseudocnus syracusanus (W A E)
	Pseudocnus grubei (W A)
	Neocnus incubans (W)
	Thyone cherbonnieri (W E)
	Pseudothyone sculponea (W A)
	Phyllophorus urna (W E)
	Phyllophorus granulatus (W E)
	Phyllophorus drachi (W)
	Neocucumis marioni (W A)
	Irpa ludwigi (E)

	Hedingia mediterranea (W) Labidoplax thomsoni (W A E) Trochodota furcipraedita (W) Trochodota venusta (W) Myriotrochus geminiradiatus (A) Prototrochus meridionalis (W) Prototrochus mediterraneus (W)
Asteroidea	Astropecten jonstoni (W A E) Astropecten spinulosus (W A E) Asterina pancerii (W A E) Anseropoda lobiancoi (W)
Ophiuroidea	Amphiura apicula (W) Amphiura cherbonnieri (W) Amphiura delamarei (W) Amphiura lacazei (W) Amphiura mediterranea (W A E) Pectinura vestita (E)
Echinoidea	Psammechinus microtuberculatus (W A E) Spatangus inermis (W) Echinocardium fenauxi (W) Echinocardium mortenseni (W) Schizaster canaliferus (W A E)

The two crinoids are common and widespread. The holothuroids (18 species) form half of the endemic echinoderms. Recently the total number of species belonging to the order Apodida has increased: 14, including three new members (all endemic) of the family Myriotrochidae, formerly unknown in this sea. Neocnus is now the only endemic genus of Mediterranean echinoderms. The North Atlantic Thyone fusus is represented by a subspecies: Th. fusus mediterranea (W A E).

Among the asteroids, the two Astropecten are 1/3 of the total Mediterranean species of this genus. The ophiuroid genus Amphiura has a peculiar interest, being the richest in the Mediterranean: 10 species. Those not yet found outside this sea have nearly all been described since 1950. Some have been collected only in the NW zone (French and Spanish coasts, Ligurian Sea) and very few data are presently available on their distribution and ecology. Nearly all the endemic echinoids are spatangoids and Schizaster is probably the more abundant.

Some of the named genera are very frequent as fossils in the circumediterranean lands. Astropecten, so rich of species distributed in all temperate and warm seas, has many Tertiary representatives: at least 15 have been recognized in Italy alone and are partly related to the living A. jonstoni and A. bispinosus. Fossil Psammechinus are found in south Italy (Calabria, Sicily: Pliocene

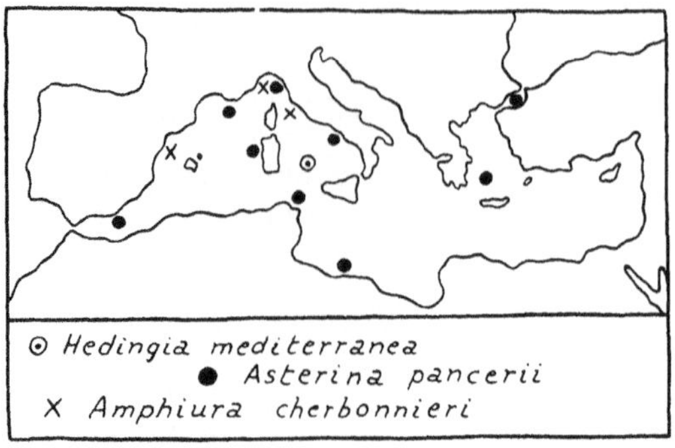

Fig. 3. Known localities of echinoderms now considered endemic in the Mediterranean

and Pleistocene). Spatangus and Echinocardium exist in the Eocene and Oligocene rocks around the Mediterranean. Even more abundant as Tertiary and Pleistocenic fossils are Schizaster of many species, including S. canaliferus, still living.

If we consider that the western Mediterranean has a richer fauna, it is not surprising that 15 endemic echinoderms (42.8%) have been reported only from it. However, several species formerly supposed to be "western" (Astropecten jonstoni, Psammechinus microtuberculatus, etc.) really exist also in the eastern basin (Aegean, Cyprus, etc.). From the latter were reported (both only once) Irpa ludwigi (bathyal) and the "mytical" Pectinura vestita: no endemic littoral species are known.

The participation of the endemic echinoderms in the ecosystems, that is their populations in a series of biotopes, are very different. Antedon and Psammechinus are present on a variety of bottoms hard and soft, respectively, to 200 and 100 m. From these eurytopic species differ those that are restricted to soft bottoms (sand, mud, gravel) where a great density of individuals is possible, constituting important elements of biocoenosis.

Almost all the endemics are littoral and distributed in the phytal system (0-200 m). Some of them dwell at upper levels in three infralittoral biotopes:

a) Rocks with algae - Neocnus incubans is a peculiar, very small holothuroid discovered in Tunisia where it abounds on Cystoseira sedoides, one meter depth. It has not yet been collected elsewhere or in other conditions. On the same kind of bottom live the two species of Pseudocnus, present also on sands.

b) Beds of phanerogams (Posidonia, etc.) - About 15 endemic species occur, most of them also extended to other biotopes. Two asteroids are characteristic members of this community but differ ecologically. Astropecten spinulosus resides on the ground among rhizomes and basal leaves (lower, not photophilous biocoenosis) and also among algae (Caulerpa, Vidalia). Asterina pancerii dwells on the leaves of Posidonia, belonging to the upper, photophilous biocoenosis.

c) Sandy bottoms - Astropecten jonstoni is strictly limited to fine sand at about 2-12 m, where it is often abundant, associated with Echinocardium mediterraneum and small bivalves as Donax trunculus. Schizaster canaliferus abounds in wide areas of north Adriatic, where depth does not exceed 50 m. It is a member of two communities: Schizaster-Turritella and Schizaster-Amphiura. The latter is A. chiajei, forming dense populations. It remains unknown if and how the endemic ophiuroids of this genus are also important part of other biocoenoses. On coarser sand (1-5 m) have been discovered the two species of Trochostoma, a genus of very small holothuroids.

The following echinoderms live at lower, circalittoral levels:

Pseudothyone sculponea (sand, mud, detritic: 80-150 m)

Neocucumis marioni (id.: 25-100 m)

Myriotrochus geminiradiatus (mud: 49-225 m))

Spatangus inermis (sand, coralligenous: 40-50 m)

Echinocardium mortenseni (sand and mud, sand and Peyssonnelia, 10-140 m)

A second, much smaller group belongs to the aphytal system, being part of the bathyal benthos living on mud below 200 m. According to the more reliable data, the vertical distribution of the crinoid Leptometra phalangium is about 200-500 m and dense populations exist also on detritic, circalittoral bottoms (Pérès & Picard, 1958). Of Prototrochus meridionalis there is a single report (gulf of Lion, 540 m). Two other holothuroids have been found only once, more than 60 years ago: Irpa ludwigi (Aegean Sea, 755-1292 m) and Hedingia mediterranea (Tyrrhenian Sea, 800-1005 m).

As was done for the fishes, it is necessary now to consider the species of echinoderms present also in the adjacent waters and therefore endemics "sensu lato". 13 live in the nearest part of the Atlantic and may be arranged in three groups:

a) Species known from north of Gibraltar:

Holothuria tubulosa (to the Bay of Biscay)
Holothuria helleri (Northern Spain)
Trachythyone tergestina (Gulf of Cadix, Portugal)
Astropecten platyacanthus (Algarve, South Portugal)
Echinocardium mediterraneum (to southern Bay of Biscay)

b) Species known from south of Gibraltar:

Holothuria mammata (Canary Islands)
Astropecten bispinosus (Morocco, Azores?)
Hacelia attenuata (Portugal, Azores)
Allopatiria ocellifera (Mauritania, Morocco?)
Cryptopelta brevispina (Morocco)

c) Species known from north and south of Gibraltar:

Holothuria polii (Bay of Biscay to Canary Islands)
Astrospartus mediterraneus (Portugal, Morocco)
Ophioconis forbesi (Bay of Biscay to Morocco, Azores)

It is uncertain if Allopatiria really is a Mediterranean endemic s.l. or is a chiefly Atlantic asteroid.

Several pairs are formed by an Atlantic and a Mediterranean species which may be the only components of the genera (Leptometra, Irpa, Hedingia, Psammechinus). The members of each pair reported below are closely related and their separation is not always sure. Some of the Atlantic species extend into the Mediterranean (one asterisk) at least in the SW part (two asterisks):

Atlantic	Mediterranean
**Leptometra celtica	L. phalangium
Irpa abyssicola	I. ludwigi
*Pseudothyone raphanus	P. sculponea
*Labidoplax digitata	L. thomsoni
Astropecten spiniphorus	A. jonstoni
**Amphiura incana	A. mediterranea
Psammechinus miliaris	P. microtuberculatus

These invertebrates are usually said to be absent from the Black Sea. However, an endemic ophiuroid (Amphiura stepanovi) and two small holothuroids: Stereoderma kirschbergi and Paracucumis hyndmani, both widespread in the Mediterranean and the Atlantic have been collected at least in its SW part.

In the following table is reported the general situation of the endemism in the Mediterranean echinoderms:

Classes	N.sp.	End.s.s.	End.s.l.	% s.s.	% s.l.
Crinoidea	5	2	2	36.40	40
Holothuroidea	50	18	23	36	46
Asteroidea	31	4	8	12.9	25.8
Ophiuroidea	32	6	10	18.7	31.2
Echinoidea	26	5	6	19.2	23
Totals	144	35	48	24.3	3.3

Other Groups. To the more detailed reports on fishes and echinoderms, a few data are added on different groups of animals containing endemics.

No cetaceans are endemic in the Mediterranean, but a subspecies (relicta) of *Phocoena phocoaena* lives in the Black Sea. The Monk Seal (*Monachus monachus*) is a typical endemic "sensu lato" because it occurs (now very scarce) in all the basins and also along the Atlantic shores from Portugal to Madeira and Spanish Sahara. Here are evaluations of endemic species ("sensu stricto") in some groups of invertebrates:

Hydroida:	27% (Pérès and Picard, 1958)
Ascidiacea:	50% id.
Polychaeta:	23.7% (Fredj, 1974)
Decapoda:	18.2% id.
Hemichordata:	40% id.
Porifera:	42.2% (Vacelet, 1981)

Even if some are not very exact, these numbers prove a high ratio of endemism. A few Porifera, Cnidaria and Mollusca, presently known only from the Mediterranean, are listed below.

Porifera. The Demospongiae, the largest class, are represented by about 453 species, of which 202 are endemic: 44.6% (Vacelet, 1981). The following are examples, living in different biotopes: *Cliona rhodensis* (near the surface), *Acanthella acuta*, *Axinella cannabina*, *Agelas oroides*, *Crambe crambe*, *Haliclona mediterranea*, *Calyx nicaensis*, *Hippospongia communis* (all infralittoral), *Anchinoe tenacior*, *Reniera sarai*, *Ircinia oros* (infra- and circa-littoral), *Rhizaxinella gracilis* (muddy bottoms, 45-200 m).

Gorgonaria. According to Carpine and Grasshoff (1975) there are 21 Mediterranean species, 8 of which (38%) are restricted to this sea. Paramuricea clavata and P. macrospina, Eunicella singularis (=stricta) and E. cavolinii are often plentiful as members of coralligenous communities. They bear a series of epibionts (Parerythropodium, Pteria, Neosimnia, bryozoans, ascidians, etc.) having different kinds of relations with the gorgonias (Neosimnia eats the polyps, ophiurans standing on the branches easily capture prey). Other endemics are: Lophogorgia sarmentosa (eurytopic, on sand etc., 15-300 m) and Isidella elongata (epibathyal, on mud, 500-1000 m) and two rarer species (Echinomuricea klavereni, Placogorgia massiliensis). Corallium rubrum is a well-known gorgonarian present also in the Atlantic from Portugal to the Cape Verde islands.

Madreporaria. As Zibrowius reported (in press and pers. comm.) it is probable that no endemic species exist in the Mediterranean. A single madreporarian (Cladopsammia rolandi) has not yet been reported from the Atlantic. In the latter have recently been collected three species that formerly were considered as strictly Mediterranean:

Cladocora caespitosa (Portugal, Morocco)

Astroides calycularis (Gulf of Cadix, Morocco)

Balanophyllia europaea (Gulf of Cadix)

Nudibranchia. 35 species appear to be endemic in the Mediterranean and almost all are found only in the western basin. Many (e.g. Peltodoris atromaculata, Phyllidia pulitzeri, Marionia blainvillea, Cratera peregrina) live in the coralligenous biotope (circalittoral) and have trophic relations with hydroids and sponges. Others are spread also in the Atlantic: one of them is Platydoris argo (Canary Islands).

In order to add a few data concerning the animals interpreted as "Mediterranean sensu lato", it is useful to quote some other species which belong to different groups of invertebrates and are variably distributed. Some of them are present north and south of Gibraltar: Portunus hastatus, Maja verrucosa, Pilumnus spinifer, Chlamys glabra, Pecten jacobaeus, Brachydontes minimus are examples. Others seem to be present only north of the Strait (Lysmata seticaudata, Sepiola rondeleti, etc.) and a much smaller number only south: the zoanthid Gerardia savaglia, observed in several Mediterranean localities, has been reported from Madeira and the Azores. As examples of species restricted to the Mediterranean and the Black Sea may be mentioned the crab Macropodia aegyptia and the cirriped Chtamalus depressus. Other crabs have a much wider distribution, from the Black Sea to the Atlantic (Eriphia verrucosa, Carcinus mediterraneus, Pachygrapsus marmoratus).

HISTORICAL AND EVOLUTIONARY ASPECTS

Of the changes that occurred in the Mediterranean and its fauna, only the main lines are recalled here, omitting the details that are explained in a rich literature (Ruggieri, 1967; Hsu et al., 1973, etc.). This sea, formerly included in the Tethys, lost communication with the Indian ocean during the Miocene and then, near the end of the same period, was separated also from the Atlantic and became a closed basin. During the Messinian (Upper Miocene), extensive regressions occurred and the waters became largely brackish: wide areas dried up completely in the western and eastern regions, as it is proved by the evaporitic layers recently discovered.

At the beginning of the Pliocene, communication with the Atlantic was established again, with the opening of the strait of Gibraltar and a general transgression started. The Mediterranean was refilled and gradually assumed its present aspect, while the climate turned to temperate conditions. During the Quaternary, remarkable climatic fluctuations were accompanied by those in the salinity and thermal stratifications of the waters. According to Mars (1963) the direction of currents in the strait of Gibraltar was periodically reversed: during the glacial periods the surface current was directed towards the Atlantic and the deeper one towards the Mediterranean, during the interglacial periods the two movements were in the opposite directions just as it is today.

All these events in the sea had a basic influence on the history of the fauna. The effects of the Messinian crisis were more serious: the animal populations were greatly reduced and most of the paleo-mediterranean elements disappeared, as did many tropical previously-existing fishes (Eocene and Oligocene). However, the abundant paleontological evidence proves that the Miocene fauna included several genera and species of fishes still present in the Mediterranean. According to Arambourg (1965) the modern ichthyofauna of this sea has an essential, direct origin from that Oligo-Miocenic.

Even if such statements may appear too absolute, it is true that some actual species are "paleoendemic" because they originated in such Tertiary times or are related to ancestors that old. During the Miocene the lagoons were populated by a series of euryhaline fishes of genera still present: _Syngnathus_, _Gobius_, _Aphanius_ (=_Pachylebias_), etc. Sorbini and Tirapelle-Rancan (1979) consider the recent species of _Aphanius_ as relicts of the Messinian fauna which "because of its characteristic of euryhalinity, was able to subsist in the Mediterranean area, in spite of salinity variations, to differentiate and to give rise to endemic forms". The researches of the same authors support their hypothesis that "marine fishes repopulated the Mediterranean during the Pliocene and Quaternary after a crisis that had totally or almost totally eliminated pure marine forms, while the forms which were able to evolve continuously on the spot".

Among the paleoendemics that have been resident in the Mediterranean since the Miocene, can be quoted Callionymus pusillus. Evidently derived from miocenic ancestors are also the two modern species of Aphanius, a genus that has nothing to do with the Atlantic fauna.

When the Mediterranean was communicating again with the Atlantic, the penetration began of many animals from the temperate-boreal zones of this ocean or, at different times, from the southern ones. Northern and southern immigrants were alternately favoured in their displacements and came to form a good part of the modern population in which boreal and subtropical or clearly tropical species coexist. Rather than an "arrival" to the east of Gibraltar accomplished by some animals coming from the west African region, it is better stated as a "return" because that part of the ocean had been a refuge during the deterioration of climate and the other physical changes in the Mediterranean. Since the Pliocene the development of endemism was facilitated (Cappetta-Ledoux, 1970). The ancestors came from both "cold" and "warm" groups, as is proved by a number of fishes and echinoderms of different families. The qualification of "neoendemic" is appropriate for these new species or subspecies that enriched the Mediterranean fauna.

An interesting case is that of the epipelagic genus Tetrapturus, belonging to a family (Istiophoridae) whose members generally have a wide distribution, very often panoceanic. Three species are present in the Mediterranean: T. belone (often confused with other species), T. georgei and T. albidus. The former is endemic s.s., the second is endemic s.l. and the third is well spread across the Atlantic. The restricted habitat of the two first is suprising but - as far as I know - the researches on these large fishes have confirmed their endemism. The genus, recorded as fossil from the Eocene, might be one of the "pure marine forms" that disappeared when the Messinian crisis arrived. Presumably it populated the Mediterranean again during the Pliocene or Quaternary, coming from the Atlantic (rather low latitudes).

The two main Mediterranean basins (western and eastern separated by the channel between Sicily and Tunisia and by the strait of Messina) differ for physical and biological features that were summarized by various authors (Tortonese, 1969, etc.). The western one has been much more intensively explored and of course more species have been found in its regions: a good number of them are still unknown elsewhere.

The Adriatic is joined to the eastern basin (Ionian Sea). In old times it was connected with the Pontic zone, that is with the area now corresponding to the Black Sea. That zone, which was formerly part of the wide Paratethys, was populated by the so called Sarmatic or Ponto-Caspian fauna. Relations now evident between the latter and the Adriatic are expressed by several fishes already mentioned: let us recall the sturgeons, which constitute a genus (Acipenser) dating from the Upper Cretaceous. A. naccarii, endemic in the Adriatic, surely derived from Sarmatic ancestors.

About 10% of the benthic animals living in the Adriatic are endemic (Gamulin Brida, 1967): Gibbula adriatica, Distoma adriaticum, Syngnathus taenionotus are a few examples. Species with a so restricted habitat are chiefly found in the northern part of this sea, that is shallower and has lower temperature and salinity.

The Ionian, the Aegean and the Levant Seas form the eastern Mediterranean basin and it is important to note that temperature, salinity and depth (5020 m) reach their maximum in this complex. The northern Aegean communicates with the Black Sea, that is so different in its physical and biological characters. The present extension was reached by the end of the Tertiary era (Caspers, 1953).

In this sea the animal population changed its primitive aspect because initially eurytherm and euryhaline species penetrated from the Mediterranean, followed later on (after Würm glaciation) by others facilitated by the increased salinity. Some animals of Sarmatic origin are present today in the Black Sea, together with Mediterranean ones partly quoted in the preceeding pages. Such a mixture is well seen in the gobiid fishes which are interpreted as significant elements in the ichthyofauna of both seas (Miller, 1965). Didogobius bentuviai of the eastern Mediterranean (Israel) is considered related to genera of the Black Sea. In the SW part of the latter is found an ophiuroid (Amphiura stepanovi), the Mediterranean origin of which cannot be doubted: its affinities are with A. chiajei.

The Mediterranean has been a center of evolution that was essentially realized in littoral animals. In a benthic group such as the echinoderms are, the endemic species "sensu stricto" are about 1/4 of the total, but among much more mobile animals, like fishes, the ratio is only about 1/11 with a large prevalence of benthic species. The relations and origins of the Mediterranean benthos have been explained with details by Pérès (1967). An important cause of endemism was certainly the geographic isolation of forms coming from the near ocean.

According to Zander (1973) the speciation in blenniid fishes was accomplished very late: the ancestors came to populate different biotopes and here their descendents adapted to new conditions (light, substrata, water movements, etc.) and diverged in some morphological characters. However, adaptive aspects consequent on speciation cannot be easily detected in pelagic (Tetrapturus) or littoral fishes (Epinephelus, Symphodus, etc.) or in echinoderms like Astropecten, Amphiura and others: their differences have a genetical basis and are not the result of environmental factors.

Further problems concern the endemic species living at deep levels, where their number shows a rapid and high reduction (Menzies, 1973; Fredj, 1974). It is difficult to submit hypotheses on their origin. If really "the Mediterranean basins were already deep before the salinity crisis," (Hsü et al., 1973), the life of bathyal animals has been possible for a long time. It has been stated (Cappetta and Ledoux, 1970) that the bathyal fauna in the Mediterranean during the Miocene was more abundant than the present one. The close affinity between congeneric species of this sea and the Atlantic (e.g. the holothuroids quoted in the preceeding pages) testifies to their phylogenetic relations and suggests that also the bathyal endemics now living in the Mediterranean had ancestors of similar ecology, arrived from the ocean when the conditions were favourable. At least for the benthic species this seems to be a simple hypothesis more acceptable than the one supposing their origin to be from species resident in shallower waters. Perhaps a much longer time was necessary for such species to undergo morphological and biological changes.

Meso- and bathypelagic fishes were present in the Mediterranean during the Pliocene and Pleistocene. A good number of fossils discovered in south Italy have been reported (Landini and Menesini, 1978). They belong to many Atlanto-Mediterranean species, still existing (Maurolicus, Argyropelecus, Chauliodus, etc.). One of these fishes is the small apogonid Microichthys coccoi, type of a genus that generally has been considered as a mediterranean endemic but is not so (see above): of the two species, only M. sanzoi is unknown out of this sea.

Finally, the possibilities of future evolution require our attention: speciation and subspeciation are going on and surely will originate new endemics. It is presumed that these will derive from ancestors having a high degree of variability often related to the distribution. New taxa may arise after the evolution of benthic fishes. The speciation is expected to go on, for instance, in the genus Tripterygion (Zander and Heymer, 1976) and probably also in other blenniids and in Syngnathus. At least some members of the latter genus are rather variable geographically.

Among the echinoderms, Astropecten irregularis is a very common asteroid worthy of notice. The Mediterranean population is usually referred to the subspecies pentacanthus, extended also to the Atlantic where it is mixed with other forms (as it happens, at a much lesser degree, in the Mediterranean). It is evidently a fluctuating condition, suggesting that pentacanthus is not a subspecies but simply the expression of a polymorphism that varies in different regions: a phase of it prevails largely in the Mediterranean and corresponds to pentacanthus. Another case is observed in Holothuria stellati: it is not yet clear if it really differs from H. tubulosa. Maybe there is an evolving situation that will eventually separate stellati as a good species.

According to Rampal (1981) speciation occurs in the thermophilous elements of the plankton existing in the Levant Sea. For its biological condition, the latter is a particular, possible source of new Mediterranean endemics: these can be expected to appear as a consequence of the Lessepsian immigration. The echinoderms are to be considered again. Asterina burtoni and Ophiactis savignyi came from the Red Sea and have been well-established for several years along the coasts of Egypt, Israel, Lebanon. It is possible that these populations will eventually diverge from those present in the previous areas where the species live: the environmental conditions are not the same. On this subject we shall return in the following chapter because it concerns also the biogeography.

ENDEMISM AND BIOGEOGRAPHY

The study of the endemism is generally recognized to be important in relation to biogeographical problems. The Mediterranean fauna and that of the near Atlantic occupy what is said to be "the widest, warm temperate area in the world" (Briggs, 1974). In the ocean its northern limit is at the entrance of the Channel (48°N) and the southern one is on the west coast of Africa, where it is variably fixed: Cap Boiador (26°N), Cap Blanco (21°N) or Cap Verde (15°N). In the first two cases, a "Senegalian" province would follow southwards. Anyway Madeira, Canary and Azores islands are included. A major faunistic change occurs at Cap Verde, where the truly tropical region begins. The subtropical and warm-temperate zone is from there to the Portugal (surface mean annual temperature 15-25°C). From the Bay of Biscay begins the transition towards the boreal fauna.

It must be emphasized that the strait of Gibraltar cannot be accepted as a boundary separating different faunae: a large number of Mediterranean fishes and invertebrates are distributed east and west of it. Some species present in the ocean are found chiefly or only northwards (Portugal, Bay of Biscay), others are spread only in southern or even western directions. The latitudinal or longitudinal distribution may be more or less extended and sometimes the Azores are reached: this well isolated archipelago is faunistically related to the eastern zones.

The fauna present along the shores of Portugal and Morocco

includes a great number of Mediterranean species. Several of them were formerly held to be endemic in the Mediterranean but later on - often quite recently - have been observed also in the Atlantic, at least in the gulf of Cadix. The investigations that presently are going on will certainly change some patterns of distribution that now are indicated: we may expect a reduction in the number of Mediterranean endemics "sensu stricto" and a corresponding increase of those qualified "sensu lato".

The extent of oceanic distribution appears to differ if whole groups of animals are considered and not just the species. According to Fredj (1974) a much higher number of bivalves and polychaetes are spread only north of Gibraltar.

From the biogeographical studies on the areas concerned here, different provinces have been recognized. Forbes (1856) was the first to fix the entrance of the Channel as an important boundary and established two provinces: Lusitanian (from the Channel to Canary and Azores) and Mediterranean (including the Black Sea). In the same year, the malacologist Woodward joined them in a single, wider Lusitanian province. Strangely, Ekman (1953) and Hyman (1955) fixed Gibraltar as a limit between a Mediterranean province and two Atlantic ones: Lusitanian (Gibraltar-Channel) and Mauritanian, (Gibraltar-Cap Verde). After geological and malacological investigations, Mars (1963) proposed a Mediterranean-Atlantic province, divided in three regions: Franco-Iberian (Channel-South Portugal: Cap S. Vicente), Mediterranean (including the gulf of Cadix and the Black Sea) and Moroccan (Gibraltar-Cap Boiador). According to Klausewitz (1968) there are an East-Atlantic subtropical region (Bay of Biscay-Cap Blanco) and a Mediterranean subregion (Black Sea excluded). Briggs (1974) admits a Mediterranean-Atlantic region divided into two provinces: Lusitanian and Black Sea.

This short review shows that the Atlanto-Mediterranean region is well-recognized, even if the southern limit has been variably fixed. But the results of the modern investigations in all its parts have increased the difficulty of dividing it into smaller zoogeographical units (provinces, intended to be of lower rank than the regions). Clear geographic limits separating these may be accepted for explaining and representing situations occurring in one or a few groups of animals. If the whole fauna is considered, we notice such a mixing of different patterns of distribution, that those limits can hardly be kept. Great variations - geographic and ecological - are observed within a single group: the fishes offer a good example.

What is the meaning of the Mediterranean endemics? It is to be recalled that the endemics, generally, are to be appreciated and evaluated from two points of view: qualitative and quantitative. Their quality consists in being peculiar elements of the fauna, representing quite characteristic genera and families. This situation occurs more often in terrestrial environments (e.g. Madagascar, Galapagos) where the endemism largely contributes to define the fauna and flora. The same, however, may be repeated for the seas when wide areas are considered.

The quantitative aspect, of course, is important and has received great attention. However, the evaluations are subject to easy changes for the reasons expressed in the preceeding pages. As far as the Mediterranean fauna is concerned, the endemics are not prominent qualitatively (Monachus may be an exception) and evidently are very "provisional" quantitatively. At least among the fishes and echinoderms there are no endemic families and the species are largely sympatric with many others often not too different. Many of those described from the Mediterranean are now reported also from the Atlantic and their sure increase will confirm a faunistic unity: it is well known that the strait of Gibraltar is not a boundary between different faunae.

The present investigation on the two zoological groups just mentioned confirms such a statement and gives a support to the division of the Atlanto-Mediterranean region only in two provinces: Lusitanian and Pontic. The first includes the whole Atlantic section and the Mediterranean, the second is much smaller being restricted to the Black Sea. This is the division adopted by Briggs (1974) and some previous authors.

However, this may appear an excessive simplification for two reasons. Firstly, the Black Sea is so-defined for its history, conditions and living world, that its position in this region may be objected to and its place in another (Ponto-Caspian) might be preferred. Secondly, the lack of uniformity within the Mediterranean cannot be overlooked. Gamulin Brida and Span (1981) suggested that the Adriatic is recognizable as a distinct "sub-province" in which proper physical conditions occur and a good series of endemics is present. However, in both cases the northern zone is essentially concerned; in the other areas the fauna and the environments do not differ appreciably from those commonly observed in the Mediterranean.

The more oriental part of the latter (Levant Sea) deserves a particular attention, for a series of facts and problems considered in many papers (Tortonese, 1973, etc.) and chiefly dealt with by Por (1978). The subtropical character is going to become more evident because the fauna continues to receive many immigrants from the Red Sea, mainly sublittoral. These "Lessepsians" are already 10-15% of the present species and their number is increased by new arrivals. Por (1978) listed 13 species of fishes supposed to be "the most successful migrants".

The same author (1981) suggested that a separate biogeographical unit (Lessepsian province) will require our recognition. However, at the present moment, it would be the only one acceptable in the Mediterranean, where faunistic, local differences scarcely support biogeographical divisions. Probably there will be some difficulty in fixing the western limits of the proposed province, quoted above. As it was said in the previous chapter of this paper, new endemics can be expected there as a consequence of future speciation or sub-speciation, connected with gradual insertions in the ecosystems.

SUMMARY

The Mediterranean fauna includes a rather high number of endemic species "sensu stricto" (s.s.) and many others endemic "sensu lato" (s.l.) that are found also in the nearest Atlantic zones and/or in the Black Sea. The results of modern explorations suggest a rapid reduction of the first category. However, the present situation deserves to be investigated to establish the number of endemic fishes and invertebrates, their distribution and the relations to the ecosystems. Fishes and echinoderms are considered here, with a few notes on other animals.

Out of 550 Mediterranean fishes, 60 (10.9%) are endemic s.s., included in 105 (19%) endemic s.l. There are no endemic families, but 9 genera have this qualification. Of the species, 6 belong to Chondrichthyes (out of 70: 8.5%), 54 to Osteichthyes (out of 480: 11.2%). The Gobiidae have the highest number (23). Some species are geographically differentiated and may have 2-3 subspecies. The number of endemics hitherto found only in each basin is very different: 17 in the western, 3 in the eastern, 7 in the Adriatic.

A sturgeon and a subspecies of shad are anadromous. Two species are epipelagic and 6 meso- to bathypelagic. Two of the latter belong to monotypic, endemic genera of Gadiformes. Littoral benthic forms largely prevail and are distributed in different biotopes: these are infralittoral (rocks, beds of phanerogams, lagoons, sands) and circa-littoral (coralligenous, caves, soft bottoms of various kinds).

36 endemics s.l. extend from the Mediterranean to the near Atlantic zones, but only between Portugal (sometimes the Bay of Biscay) and Mauritania, where the distribution is more or less broad and often includes Madeira and Canary islands, sometimes the Azores. East of Gibraltar, 12 of these species reach the Black Sea. Nine other species are restricted to the latter and the Mediterranean.

Out of 144 echinoderms, 35 (24.3%) are endemic s.s., being a part of 48 s.l. (33.3%). They belong to all the classes but chiefly to the Holothuroidea (18 s.s., 23 s.l.), which has the single endemic genus (Neocnus). The few Echinoidea are nearly all spatangoids. Some genera abound as fossils in the circummediterranean regions.

Several species are important in the ecosystems, forming rich populations chiefly at infra-littoral levels. 4 species are bathyal: one crinoid (200-500 m) and 3 holothuroids (500-1300 m). 13 species have different distributions in the Atlantic area mentioned above. There are pairs formed by closely-related species, one Mediterranean and one Atlantic. An ophiuroid is endemic in the Black Sea.

Historical, evolutionary and biogeographical aspects are then briefly considered. Paleoendemic and neoendemic species exist in the Mediterranean. Both categories can be recognized in fishes. The representatives of the former date from the Miocene, when large zones were desiccated (Messinian crisis): a part of the previous fauna disappeared (numerous tropical fishes) but genera and species mostly euryhaline survived along the littorals (Callionymus pusillus) and in the lagoons (Aphanius) becoming elements of the modern fauna. The neoendemics derive from ancestors that, during the Pliocene and Quaternary, penetrated from the Atlantic, coming from zones north or south of Gibraltar.

The Mediterranean has been an important evolutionary center, as proved by fishes (rays, gobies, blennies, etc.) and echinoderms (holothuroids, Astropecten, Amphiura, etc.). New endemics may appear as a consequence of future speciation occurring not only in indigenous and very variable forms, but also in the organisms arrived from the Red Sea (Lessepsian immigrants) and well established in the eastern basin.

The Mediterranean belongs to the Atlanto-Mediterranean region, which includes also the oriental part of the ocean from the entrance of the Channel to Africa (Mauritania or Cap Verde), the near islands, the Azores and, according to some authors, the Black Sea. The character is warm-temperate and partly subtropical. This wide region has been divided into provinces that cannot be well-defined if the whole fauna is considered: very different types of distribution of species are mixed in variable proportions according to the groups of animals. It must be confirmed that Gibraltar is not at all a faunistic boundary. After the study of the endemics reported here it would be preferable to accept only two provinces: Lusitanian (Atlantic and Mediterranean) and Pontic (Black Sea). However it is necessary to remark that the Black Sea surely has very peculiar characters and that faunistic differences are often evident within the Mediterranean. It has been presumed that a separate biogeographical unit (Lessepsian province) will have to be recognized for the more oriental part of this sea, where the Indopacific elements coming from the Red Sea will accentuate its subtropical character.

ACKNOWLEDGEMENTS

For bibliographical materials and interesting comments I am glad to thank Dr. G. Barletta (Aquarium, Milan), F. Bentivegna (Zool. Stat. Naples), A.M. Clark (Brit. Mus. Nat. Hist., London), B. Hansen (Zool. Museum, Copenhagen), W. Landini (Lab. Paleontol.,Univ. of Pisa), V. Monteiro (Mar. Biol. Stat. Cascais), G. Pulitzer (Zool. Inst. Univ. Genova), L. Sorbini (Mus.Nat.Hist. Verona), H. Zibrowius (Stat. Mar. Endoume, Marseille). I feel particularly obliged to Prof. G. Fredj (Lab. Ocean. Biol. Univ. Nice) for a welcome meeting and discussions.

REFERENCES

Arambourg, C., 1965, Considérations nouvelles au sujet de la faune ichthyologique paléoméditerranéenne, Senk. leth.,46:13.

Briggs, J.C., 1974, "Marine Zoogeography," McGraw-Hill Co., New York.

Cappetta, H. et Ledoux, J.C., 1970, Comparaison de la faune ichthyologique miocène avec la faune actuelle de la Méditerrannée, Journ. ichthyol. CIESM, Rome:21.

Carpine, Chr. et Grasshoff, M., 1975, Les Gorgonaires de la Méditerranée, Bull. Inst. Océan., 71, 1430:140.

Caspers, H., 1953, Black Sea and Sea of Azof, Marine Ecology, Paleoecology, Geol. Soc. America, Mem. 67, 7:801.

Ekman, S., 1953, "Zoogeography of the Sea," Sidgwick-Jackson, London.

Forbes, E., 1856, Map of the distribution of marine life, in: "The physical atlas of natural phenomena," A.K. Johnston, Edinburgh-London.

Fredj, G., 1974, Stockage et exploitation des données en écologie marine. C. Considérations biogéographiques sur le peuplement benthique en Méditerranée, Mém. Inst. Océan., 7:88.

Gamulin Brida, H., 1967, The Benthic fauna of the Adriatic Sea, Ocean. Mar. Biol. Ann. Rev., 5:535.

Gamulin Brida, H. et Span, A., 1981, Situation biogéographique de la Méditerranée et de l'Adriatique en tant que sous-region distincte, Journ. études syst. évol. etc. CIESM, Cagliari:73.

Hsü, K.J., Ryan, W.B.F., and Cita, M.B., 1973, Late Miocene Dessiccation of the Mediterranean, Nature, 242:240.

Hyman, L.H., 1955, "The Invertebrates: Echinodermata," McGraw-Hill, New York.

Klausewitz, W., 1968, Remarks on the zoogeographical situation of the Mediterranean and the Red Sea, Ann.Mus.St.Nat.Genova, 77:323.

Landini, W. e Menesini, E., 1978, L'ittiofauna plio-pleistocenica della sezione della Vrica (Crotone, Calabria), Boll.Soc. Paleont. It., 17, 2:143.

Mars, P., 1963, Les faunes et la stratigraphie du Quaternaire méditerranéen, Rapp. Proc. Verb. CIESM, 17, 3:1029.

Menzies, R.J., 1973, Biological history of the Mediterranean Sea with reference to the abyssal benthos, Rapp.Proc.Verb.CIESM, 21, 9:717.

Miller, P.J., 1965, Relictogobius kryzhanowskii and the penetration of Mediterranean Gobies into the Black Sea, Nature, 208, 559: 474.

Pérès, J.M., 1967, The Mediterranean Benthos, Oceanogr. Mar. Biol. Ann. Rev., 5:449.

Pérès, J.M. et Picard, J., 1958, Manuel de bionomie benthique de la mer Méditerranée, Rec. Trav. St. Mar. Endoume, 24:122.

Por, F.D., 1978, "Lessepsian Migration," Springer, Berlin-New York.

Por, F.D., 1981, The Lessepsian biogeographic province of the eastern Mediterranean, Journ. études syst. évol. etc. CIESM, Cagliari, 81.

Rampal, J., 1981, Biogéographie méditerranéenne d'aprés l'étude des Thécosomes (Mollusques pélagiques), Journ. études syst. évol. etc. CIESM, Cagliari, 45.

Ruggieri, G., 1967, The Miocene and later evolution of the Mediterranean Sea, Syst. Assoc. Publ., 7:283.

Sorbini, L. e Tirapelle Rancan, R., 1979, Messinian fossil fish of the Mediterranean, Paleogeogr. Paleoclimat. Paleoecol., 29:143.

Tortonese, E., 1969, La fauna del Mediterraneo e i suoi rapporti con quelle dei mari vicini, Pubbl. St. Zool. Napoli, 37:369.

Tortonese, E., 1973, Facts and perspectives related to the spreading of Red Sea organisms into the eastern Mediterranean, Ann.Mus. St.Nat.Genova, 79:322.

Tortonese, E., 1976, Endemic elements in the Mediterranean fish fauna, Rév.Trav.Inst.Pêches Mar., 40:771.

Tortonese, E., 1980, Review of present status of the Mediterranean Echinoderms, in: "Echinoderms present and past," Jangoux, M., ed., Balkema, Rotterdam.

Vecelet, J., 1981, Les affinités du peuplement des spongiaires de la Méditerranée, Journ.études syst.évol.etc.CIESM, Cagliari:29.

Woodward, S.P., 1851-56, "A Manual of the Mollusca," J. Wheale, London.

Zander, C.D., 1973, Evolution of Blennioidei in the Mediterranean Sea, Rév. Trav. Inst. Pêches Mar., 37, 2:215.

Zander, C.D., 1980, Morphological and ecological investigations on sympatric Lipophrys species (Blenniidae, Pisces), Helgol. Meeres., 34:91.

Zander, C.D. and Heymer, A., 1976, Morphologische und ökologische Untersuchungen an den speleophilen Schleimfischartigen Tripterygion melanurus Guichenot, 1850 und T. minor Kolombatovic, 1892 (Perciformes, Blenniioidei, Tripterygiidae), Zeit. Zool. Syst. Evol., 14:41.

Zander, C.D. und Heymer, A., 1977, Analysis of ecological equivalents among littoral fish, in: "Biol. benthic organisms",11 Europ. Symp. Mar. Biol. Galway, 1976, Pergamon Press, Oxford.

Zibrowius, H., (in press), Nouvelles données sur la distribution de quelques Scleractiniaires "méditerranéens" à l'est et à ouest du détroit de Gibraltar, Rapp. Comm. int. Mer Medit., Cannes.

MEDITERRANEAN SEA MEIOBENTHOS

Jacques Soyer

Université Pierre et Marie Curie, Laboratoire Arago -
URA 117, 66650 Banyuls-sur-Mer, France

The term "meiobenthos" was introduced by Mare (1942) to describe fauna of intermediate size-animals, smaller than those traditionally classed as "macrobenthos", but larger than "microbenthos" including bacteria and most protozoa.

Despite a certain vagueness in the definition, it is traditionally accepted that in operational terms, meiobenthos refers to those organisms which pass through a 0.5-1 mm sieve and are retained on a sieve with mesh width smaller than 0.1 mm (0.06 or 0.04 mm).

In this essay, we shall attempt to review our knowledge on Mediterranean meiobenthos, and compare its quantitative distribution, seasonal fluctuations, and factors of productivity, to those of other areas. From these data, the importance of meiobenthos within the benthic Mediterranean communities will be re-evaluated.

ORIGIN OF MEDITERRANEAN MEIOBENTHOS: A UTOPIAN TASK

Taxonomic studies on marine meiobenthic organisms were going on in the middle of the 19th century. In 1845, Dujardin recorded nematodes in marine algae near Toulon, Sète and in the Thau Lagoon; in 1866, Claus described some harpacticoid copepods from the shore of Nice. From 1850 to now, many zoological and taxonomic studies have been published; but Filipjev (1921) on nematodes, and Monard (1928) on harpacticoids were the first to point out meiobenthic assemblages. In 1960, Delamare-Deboutteville synthesized the known generalities on the mesopsammon discovered by Remane, (1933) with many data from the Mediterranean shores.

We shall not present an exhaustive review of the literature concerning Mediterranean meiobenthos. Papers are numerous and of unequal value. Many biotopes are quite completely unknown, e.g. phytal fauna. Most of the studies were developed in the Northwestern Basin of the Mediterranean. So, it is quite impossible to give a good biogeographical analysis of Mediterranean meiobenthos, and it seems that it will remain so for some time.

In 1964, Por studied the marine copepod assemblages from Israel and identified 140 species - 26 had never been described before, and 27 were new for the Mediterranean. Soyer (1970) recorded 254 species from the soft substrates of the French Catalonian shelf; 56 had never been described before, and 50 were new for the Mediterranean. Altogether 340 species were recorded from 1928 to 1970 in the area of Banyuls alone.

The problem is even more intricate for nematodes. In the area of Marseille, Vitiello (1972) identified 348 species of which 117 had never been described before.

As pointed out by Wells (1965) and Coull (1968), our present knowledge on meiobenthos is too poor to carry out a valuable biogeographical analysis.

MEDITERRANEAN MEIOFAUNA POPULATIONS

General Distribution

Although Moore (1931) was the first to produce some quantitative data on meiofauna, Mare (1942) first stressed the importance of marine meiofauna within benthic communities.

Early meiofauna quantitative investigations took place in Europe and were rapidly extended to other areas, in particular to the North American shore.

Deep-sea meiobenthos was also studied: in 1966, Thiel recorded high meiobenthic abundances below 1,000 m in the Indian Ocean.

A number of reviews dealing with the subject have been published from 1969 to 1979 (McIntyre, 1969; Gerlach, 1971, 1978; Coull, 1973; Thiel, 1975; Coull and Bell, 1979).

In the Mediterranean, the first quantitative data were collected in 1946 off Banyuls by Bougis. Since then, around twenty quantitative papers have been published. Table 1 summarizes those papers that enumerate "all" meiobenthic organisms in the sediment, most of them concerning the Northwestern Basin of the Mediterranean.

Table 1. Total number of meiofauna per m^2 in the Mediterranean Sea ($\bar{m}$: mean)

Depth	N/10 m^2	Localization	Procedure	Authors
Lagoons				
1.5-4 m	6.17×10^5 11.34×10^6	Mouth of the Pô	Box-coring	Ceccherelli and Ceridelli, 1978
0.80 m	5.7×10^6	Lagoon of Tunis	Box-coring(?)	Vitiello and Aissa, 1979
Beaches				
0	4.9×10^4 2.8×10^5	Tunisia	Hand-coring (10 cm)	Hulings, 1971a
0.40	5×10^3 2.5×10^4			
0	1.9×10^5	Tunisia	Hand-coring (10 cm)	Hulings, 1971b
0	1.5×10^5	Marocco ($\bar{m}$ 15 beaches)		
0	2.7×10^5	Lebanon ($\bar{m}$ 5 months)		
0 40 cm	$4.5 \times 10^5 - 2.9 \times 10^6$ $4.6 \times 10^4 - 9.2 \times 10^5$	Lebanon	Hand-coring (annual cycle)	Hulings, 1974
0	1.9×10^5 3.3×10^6	Gulf of Tunis	Hand-coring (10 cm)	Vitiello and Triki, 1978
Continental Shelf				
30 m	8.2 10^5	Banyuls (mud)	Kullemberg	Bougis, 1946
30 m	6.7 10^5	Banyuls (mud)	Kullemberg	Bougis, 1950

(continued)

Table 1 (Cont.)

Depth	N/10 m	Localization	Procedure	Authors
15–91 m 14–87 m	3.5×10^5–3.8×10^4 6.8×10^5–9.3×10^4	Banyuls	Kullemberg	Guille and Soyer, 1968
10 m	2.5×10^6	Marseille (sand)	Hand-coring	Vitiello, 1968
10 m	5.4×10^5	Banyuls	Hand-coring	Soyer, 1971
3–10 m	2.5×10^5	Banyuls	Hand-coring	
30–70 m 70–130 m	6.1–2.7×10^5 1.3×10^5–5.6×10^4	Banyuls	Kullemberg	
32.5 m	2.9–8.8×10^6	Banyuls (mud)	Hand-coring (annual cycle)	de Bovée and Soyer,1974
13–17 m	2.1–3.6×10^5 0.3–5.2×10^5	Adriatic (polluted area)	Kullemberg (summer-winter)	Marcotte and Coull,1974
32.5	1.7–8.8×10^6	Banyuls (mud)	Hand-coring (annual cycle)	de Bovée, 1981
Continental Slope and Abyssal Plain				
150–550 m	4.9–3.8×10^4	Banyuls	Kullemberg	Soyer, 1971
168–810 m	1.6×10^4–8.0×10^5	Marseille: Cassidaigne	Reineck	Vivier, 1976 (1978)
260–550 m	2.3–6.4×10^5	Planier		
130–1209 m	8.7×10^4–4.2×10^5	Aegean Sea	Reineck	Dinet, 1976
2100–2855 m	3.3–7.8×10^4	W. Mediterranean		Dinet et al., 1973
35 m	3.09×10^6	Banyuls	Hand-coring	Boucher, 1971

The Russian papers (Kiseleva, 1961, 1964; Kiseleva and Tchukhtchin, 1965; Tchuckhtchin, 1964) dealing with the Eastern Mediterranean were not included because the methods used were too different.

It should be pointed out that the data recorded in the different parts of the Mediterranean fit very well, although sampling or sorting methods differ slightly from one author to another.

In the littoral parts of the continental shelf, densities of meiofaunal organisms range from 10^5 to 10^6 m^{-2}. Obviously, these numbers vary according to season, locality, water depth, as well as sampling devices used: gravity cores (Bougis, 1946, 1950; Guille and Soyer, 1968; Soyer, 1971; Marcotte and Coull, 1974), or hand coring (Vitiello, 1968; Boucher, 1972; de Bovée and Soyer, 1974; de Bovée, 1981). The above densities are found on beaches (Hulings, 1971a,b, 1974; Vitiello and Triki, 1978), or in lagoons (Ceccherelli and Cevidalli, 1978; Vitiello and Aissa, 1979).

Numbers and biomass decrease by at least one order of magnitude (10^4-10^5 ind.m^{-2}) with increasing depth and distance from shore.

In the Gulf of Lion, Guille and Soyer (1968) and Soyer (1979), using gravity coring, found a striking decrease of meiofaunal organisms in the deepest parts of the continental shelf, at 50-70 m depth. Vivier (1976) off Marseille, and Dinet (1976) in the Aegean Sea, recorded relatively high standing stocks (10^5 ind.m^{-2}) on the continental slope. The difference could be due to the sampling device (gravity cores, box-cores), or to the width of the French Catalonian shelf.

From the deep plain of the Mediterranean, below 2,000 m, Dinet et al. (1973) reported very low densities, ranging from 82×10^3 to 195×10^3 ind.m^{-2}.

In other areas, the average meiofaunal densities on the continental shelf are of 10^6 ind.m^{-2} (Coull and Bell, 1979). Some higher values are known from estuaries or intertidal mudflats (2.6×10^7 m^{-2}, Warwick, or. com. in Coull and Bell, 1979; 4.7×10^7 m^{-2}, Delille et al., 1983b). Generally, numbers decrease with increasing depth or distance from shore, reaching an average of 10^5 ind.m^{-2}. In the deepest parts of the ocean, Thiel (1975) and Dinet (1980) recorded such densities.

From the above analysis, it appears that on the Mediterranean continental shelf, meiofaunal densities are comparable to those recorded in other oceans. However, they decrease strongly with increasing depth and distance from shore, with extremely low values in the deepest parts. Generally, this decrease is explained by the low levels of food availability on the deep-sea floor (Sanders and Hessler, 1969; Rowe and Staresinic, 1979).

Mediterranean data compare to those of Thiel (1979) from the deep parts of the Red Sea, a typical oceanic blue water.

In the Mediterranean, the lower densities could also be related to the oligotrophic nature of this sea. Furthermore, the rates of recycling under relatively high temperature (13-14°C) could be rather important. Dinet (1973, 1979), Thiel (1978), Dinet and Khripounoff (1982), showed that the densities of meiobenthos in the deep-sea were related to the pelagic productivity of the euphotic layer. In the Eastern Mediterranean, at comparable depths, Dinet (1976) found higher densities in the Northeastern Basin of the Aegean Sea than in the Northwestern Basin. This could be due to the fertilizing influence of the waters originating from the Black Sea and flowing through the Marmara Sea.

Paradoxically, the relations of sediment organic content (in terms of carbon and nitrogen) to meiobenthos densities appear difficult to establish, in both coastal and deeper systems (Coull, 1970; Tietjen, 1977; Dinet and Vivier, 1977; de Bovée et al., 1979).

Nevertheless, Khripounoff et al. (1980) pointed out that the quality of sediment organic matter could not be neglected, and showed a significant correlation between the sediment labile particulate protein content and macrobenthos densities in the abyssal system. Some relations have also been established for meiobenthos (Dinet and Khripounoff, 1982).

On the shallow parts of the shelf, some dynamic relations were reported. Delille et al. (1983a) observed temporal sequences in peaks of carbon, nitrogen, microbial densities and meiofauna. In the sublittoral muds of the Catalonian coast, de Bovée (1981) showed a close relation between the peaks of nematodes, particularly during the warmer months, and the decrease of carbon and nitrogen. Similar observations were made on the copepod assemblage of the sublittoral fine sands in the same area (Bodiou and Delille, or. comm.).

Most of the studies suggest that meiofauna populations are controlled by the habitat's physical properties (McIntyre, 1971). However, biological interactions, and particularly the quantity and quality of available food, seem to play a significant role in the regulation of the meiofauna; but at the present time, we still don't know exactly what and how much it consumes.

Seasonal Changes

Temporal evolution. The few data published on quantitative variations of meiofauna in the shallow Mediterranean waters, show seasonal cycles of abundance of total meiofauna (Soyer, 1971;

de Bovée and Soyer, 1974; Hulings, 1974; de Bovée, 1981), nematodes (de Bovée, 1981) and harpacticoids (Soyer, 1971,1980; Dinet, 1972a; Fava and Volkmann, 1975; Nodot, 1977,1978; Dinet et al., 1982). During the cold months, densities are low, sometimes with irregular peaks (Soyer, 1980; de Bovée, 1981; Bodiou, pers. comm.). From late spring to early autumn, densities are high with two major peaks in spring and autumn.

From the results of only four annual samplings, Soyer (1971) suggested that on the Catalonian continental shelf, seasonal cycles decreased with increasing depth down to 70 m, below which densities stayed practically constant throughout the year due to reduced temperature variations.

The reduced influence of temperature with increasing depth was indirectly shown by Bodiou and Chardy (1973). Analysing the qualitative evolution of the sublittoral fine sand harpacticoid assemblage off Banyuls, they reported a spatio-temporal continuum in the community from 10 to 20 m. First, the summer structure shifted to the autumn structure as depth increased. Second, the qualitative structure of the assemblage was more stable at 20 m than in the shallower parts of the community.

Other studies, generally based on irregular or few samples, reported little, if any variations in the annual abundance cycle of Mediterranean meiofauna. Sampling on a polluted Lebanese beach Gowing and Hulings (1976) suggested that this stability could be due to a constant food availability throughout the year. Boucher (1972) observed no significant numerical difference between two winter and summer samples from the 32 m sublittoral muds off Banyuls.

In a coarse sand harpacticoid assemblage off Marseille, Dinet (1972a) observed a reversed quantitative cycle - the maximum appearing in the autumn and winter, and related it to the upward summer migration of the RPD layer in response to rising temperatures and low hydrodynamism. Dye and Furstemberg (1978) reported comparable quantitative cycles on the South African shore, and Dye (1979) found a direct relation between decreasing redox potentials and decreasing meiofaunal abundances.

In a general manner, as shown by Coull and Flegger (1977) in South Carolina, these quantitative annual cycles are concurrent with important changes in the species composition and population structure. In the Mediterranean, de Bovée (1981) on nematodes, and Dinet (1972a), Bodiou and Chardy (1973), Fava and Volkmann (1975), Nodot (1977, 1978), Soyer (1980), Dinet et al. (1982) on harpacticoids, observed that the general quantitative evolution of the assemblages was the net result of the biological cycles of the main species. These cycles were either synchronous or more frequently displaced in time.

Considering the strong diversity of meiofauna in sublittoral shallow waters, these phenomena become somewhat difficult to analyse. Great differences in the biological behavior can be found among the various taxa or species: important variability in reproductive patterns has been observed within the same trophic group, the same family, the same genus, and even between sympatric species.

Even when a certain degree of synchronism exists, e.g. triggering or activation of spring reproductive activity, differences in reproduction processes (number of broods, of eggs), in biological cycles (development rate, longevity), induce a displacement in the abundance peaks of the different species.

Regulation mechanisms. Past emphasis has been on the overriding influence of temperature as the primary control of quantitative meiofaunal seasonal fluctuations. It is a fact that temperature and meiofaunal abundances are directly correlated, and that the extent of density fluctuations increases with the variability of climatic conditions. In the intertidal zone, estuaries or lagoons, Smidt (1951), Wiesen and Kanwisher (1961), Schmidt (1968, 1969, 1972a,b), Tietjen (1969), Skoolmun and Gerlach (1971), Schmidt and Westheide (1971), Harris (1972), Nyhlom and Olsson (1973), Lasserre et al. (1976), Clasing (1976), Castel and Lasserre (1977), observed very distinct quantitative seasonal variations in meiofauna.

In the subtidal, these seasonal fluctuations are not so obvious (Boucher, 1980; McIntyre, 1964; Warwick and Buchanan, 1971; McIntyre and Murrisson, 1973; Lorenzen, 1974). Nevertheless, Coull (1970) in the shallow Bermuda waters, did find distinct seasonal fluctuations similar to those from the Mediterranean.

A number of authors invoke strong temperature dependence for the regulation of reproduction of copepods and nematodes in the field. The data reported by Dinet (1972a), Nodot (1978) and Soyer (1980) on Mediterranean copepods, compare to those of Muus (1967), Barnett (1968 and 1970), Coull (1970), Harris (1972), Heip (1974), Heip and Smol (1976), and Mielke (1976). Hicks and Coull (1983), in their exhaustive analysis reported that for a large majority of copepods, abundance maxima occur in the warmer months of the year, and particularly late spring and early summer.

Nematode populations seem to be less temperature-dependent, and even in the shallow waters, nematode density/temperature relations are not obvious (Smidt, 1951; Skoolmun and Gerlach, 1971; Brenning, 1973). Nevertheless, Gerlach and Schrage (1971, 1972), Tietjen and Lee (1972,1977a); Hopper et al. (1973); Heip et al. (1978) and Laybourn (1979), did show a direct correlation between reproductive performance and temperature.

Temperature stability would also be important to consider; however, our present knowledge of meiofaunal reproductive activity in relatively constant conditions is quite poor. Mathews (1964) observed that the reproductive activity of copepods in the deeper parts of a Norwegian fjord took place in winter. In the intertidal zone of Kerguelen Islands, Delille et al. (1983a) noted that the maximal density of a population in which nematodes were largely dominant (99%) did not occur in the summer, but in the winter, when air and water temperatures were almost equal: they suggested that the reproductive capacity of meiofauna was limited in the summer by strong diurnal temperature fluctuations on the beach.

The second factor usually invoked in meiofauna regulation is food availability. This is however difficult to assess in the field: since food quality and quantity vary according to season, the influence of food is superimposed on that of temperature fluctuations. Furthermore, our knowledge of in situ meiofaunal feeding is far from satisfactory. Observations on the components of the diet, made essentially from laboratory culturing, showed that the meiofauna was capable of using a broad spectrum of food sources (nematodes: Meyers et al., 1970; Lee et al., 1977; Tietjen and Lee, 1977b; Tietjen, 1980; Alongi and Tietjen, 1980; - copepods: see Hicks and Coull, 1983). But food selection in the natural habitat, as a function of nature, energy content or availability has not yet been explored.

Direct relations between quantity and quality of available food and in situ meiofaunal abundance are not easy to assess either. Westheide (1968), Hickel (1969), Rheinheimer and Kullmann (1972), Bianchi (1973), Pugh et al. (1974), as well as Cahet and Mouneimnê (1976) for the Mediterranean, showed that bacterial numbers and/or activity increased during the warmer months. Benthic primary production also increases greatly in spring and autumn (Hopper and Meyers, 1967; Steele and Baird, 1968; Boucher, 1975; Davies, 1975). In the Mediterranean, Le Bouteiller-Bougnol (1980) reported the same pattern.

Boucher (1972) and Tietjen (1969, 1980) noted important changes in the structure of nematode assemblages in relation to food sources. In the Mediterranean, de Bovée (1981), using data from Cahet and Mouneimnê and Le Bouteiller-Bougnol observed complex modifications within a nematode assemblage from sublittoral muds. Just after the winter, as the water temperature rises, microphytobenthic biomass, chemosynthetic productivity and densities of the four trophic groups of nematodes increased. In the same manner, at the end of autumn, all these factors decreased. However, during the periods of temperature stability, i.e. summer and winter, the two most abundant groups of nematodes (epistrate-feeders and non-selective deposit-feeders) exhibited inverse trends; even in the same trophic group, de Bovée noted that the quantitative fluctuations were not homogeneous.

Studying the effects of a thermal effluent on a harpacticoid assemblage, Dinet et al. (1982) found that despite an early production of larvae in the heated biotope, the mean standing crop of the population did not differ significantly from that of a non-exposed area used as reference. This fact was explained by a high larval mortality probably related to the lack of suitable food in the late winter.

Many authors have suggested that predation by higher trophic levels was the factor controlling meiofaunal populations (McIntyre, 1969; Coull and Bell, 1979), and an increasing literature demonstrates the importance of harpacticoids for larval, juvenile or small fishes (see Hicks and Coull, 1983). Although more numerous, nematodes appear to be less consumed. Coull and Bell (1979) invoked their distribution within the sediment to explain this lower utilization. The direct effect of predation has, however, usually been observed only in experimental enclosures (Lasserre et al., 1976; Bell and Coull, 1978; Bell, 1980; Castel and Lasserre, 1982). Some exceptions apart (see Gerlach and Schrage, 1969; Bell and Coull, 1978; Caine, 1978; Schaffer, 1979), the meiofauna, other than harpacticoids for young fishes, does not appear to be a significant food source for higher trophic levels. Even in the case of harpacticoids, predation seems to be very selective: Sibert (1979) suggested that the drastic decrease of Harpacticus uniremis recorded in June in Nanaimo River estuary, was due to the recruitment of young fishes, and particularly chum salmon, all of them predators on this one species, otherwise uncommon in this area.

In the Mediterranean, Villiers (1979, 1980) and Tito de Morais (1983) who studied the diets of juveniles of the Gobiidae Deltentosteus quadrimaculatus and the pleuronectiforms Arnoglossus thori and Buglossidium luteum, gave some interesting data on the selective character of predation, and on its intensity. During the first months of their benthic life, the young feed on meiofauna and particularly harparcticoids. However, only four species of copepods make up to 70-90% of the ingested prey (Bodiou and Villiers, 1978-79). The species preyed upon change with growth, and the diets progressively switch to macrobenthos. After 2-3 months of benthic life, only B. luteum still derives some of its energy from meiofauna, and copepods.

Predation would therefore be very selective, as well as time-limited, since recruitment and diet-changes do not occur simultaneously in the three species. It cannot, in their present case, be the factor responsible for the important fluctuations in total meiofaunal densities.

Productivity considerations

Assessment of production in benthos, and especially meiobenthos in the field is impeded by logistic problems (Warwick, 1980). Sampling at frequent intervals, sorting, identification, counting, measuring and weighing meiofauna individuals, involve intensive work. Reproductive patterns are very diverse in meiofauna, and following the history of recognizable cohorts in a time-series of samples, is made difficult by excessive overlap of generations.

Thus many studies on meiofaunal production rely on experimental work, or field data on intertidal or opportunistic organisms. Most of the authors use short-cut approximations such as the annual $P/\bar{B}$ ratio, and it is still customary to use the value $P/\bar{B} = 9-10$, proposed by Gerlach (1971-1978) on the basis of the *in vitro* life cycle characteristics of two nematode species.

In situ meiofauna production estimates are still in their initial stages, and such studies are an important avenue of research to evaluate the energetic role of meiobenthos.

Our purpose in the following section, is to check from the few known biological cycles, if the production in the Mediterranean is comparable to that reported in other areas.

Data concern only the two main taxa, i.e. copepods and nematodes.

Copepods. The studies of Dinet (1972a,b), Nodot (1978,1982), Soyer (1970), and Dinet et al. (1982) show that the great variety in the patterns of reproductive activity in harpacticoids noted by Hicks (1979), Hicks and Coull (1983) on an oceanic scale exist also in the Mediterranean.

Dinet (1972b), studying an organogenic sand community off Marseille, found that out of the two dominant species, one - *Bulbamphiascus imus*, had an univoltine pattern with restricted breeding; the other - *Halectinosoma herdmani*, had a multivoltine yet non-continuous pattern with protracted breeding. In a fine sand community in the same geographic area, Nodot (1978) reported that three species had a restricted breeding pattern, seven a protracted pattern; in this case, the reproductive activity of *H. herdmani* was continuous. Soyer (1980) observed that the five dominant species of a sublittoral mud community all had restricted or protracted breeding patterns, except for *Cletodes pusillus* that had a continuous reproductive activity.

From this variability within the same assemblages, it appears difficult to consider a population as a set of homogeneous forces. This does not facilitate the assessment of overall production. The reproductive potential of a species varies according to intrinsic factors, fecundity, longevity, rate of development, as well as external factors, temperature, nutrition, salinity, etc. (see Hicks and Coull, 1983).

Higher temperature in the Mediterranean should favour higher production. Soyer (1980) showed that the generation time of Cletodes pusillus in the field decreased when temperature increased, ranging from 84 to 163 days throughout the year. In theory, from the simplified form of Belehradeck's formula (Heip, 1974), C. pusillus that has three annual generations in natural conditions, should only have one at 8.7°C. Dinet et al. (1982) showed that an artificial increase of the water temperature could induce a significant precocity in reproductive phenomena.

Unfortunately, only one datum is available on the in situ productivity of a Mediterranean harpacticoid copepod. Soyer (1979), using the empirical method of standing crop comparison, estimated a $P/\bar{B}$ ratio of 8 for C. pusillus, in spite of a very low intrinsic rate of natural increase ($r = 0.0032$). The annual production of this species (not very common in the assemblage) would then approximate to 0.130 $gC \cdot m^{-2}$.

These few Mediterranean results do not differ noticeably from those of other geographic areas. In their review, Hicks and Coull (1983) emphasized the great diversity of production activity and strategy within the same community, but also of variation both within and between sampling localities for the same species. Thus it appears difficult to compare the results on a species or assemblage level. To our knowledge, only three in situ productivity estimations, including juveniles, exist in the literature. Feller (1980,1982) calculated an annual production of 0.7-1.7 $gC \cdot m^{-2}$ for Huntemania jadensis with a $P/\bar{B}$ ratio of 3.8. For Microarthridione littorale, Fleeger and Palmer (1982) reported a production of 0.87 $gC \cdot m^{-2} \cdot y^{-1}$, with a very high $P/\bar{B}$ ratio of 18. Preliminary results from Heip et al. (1982) gave a $P/\bar{B}$ ratio of 9.34 for Trachidius discipes. Although probably underestimated, the $P/\bar{B}$ ratio calculated for C. pusillus in the Mediterranean is comparable to the above values. The lower annual production in this species could be explained by its lower mean densities, and/or its biological patterns.

Nematodes. For this taxon, it is customary to use the annual $P/\bar{B}$ ratio of 9-10 established by Gerlach (1971, 1978), on the basis of the life cycle characteristics of two laboratory cultured species.

De Bovée (1981) was the first to estimate the in situ production of a marine nematode. Analysing the population dynamics of three species, he showed that an important part of the population was made up of adults - 61% for Sabatiera proabyssalis, 75.6% for Dorylaimopsis mediterranea, and 83% for Axonolaimus arcuatus, and that reproductive patterns differed greatly from one species to the other. De Bovée estimated the $P/\bar{B}$ ratios of the two species, S. proabyssalis and D. mediterranea using two different methods: - the one used by Gerlach (1971) for Chromadorita tenuis, and based on the demographic structure and mortality of larval stages, and - the classical method of standing crop comparison (in Yeates, 1972). The first gave $P/\bar{B}$ ratios of 10.4 and 9.4 respectively. With Yeate's method, the values were somewhat lower: 6.4 and 7.7 respectively. From these field results, which compare well to those of Gerlach, de Bovée calculated a production of 7.3 $gC{\cdot}m^{-2}y^{-1}$ for all the nematodes of the sublittoral muds off Banyuls.

In a general manner, it is clear that our knowledge on the life history and the reproductive patterns of meiofauna is too scarce to yield definite conclusions. It is certain that the use of overall $P/\bar{B}$ ratios for meiobenthos is spurious. Field and experimental observations show that many species have flexible life cycles adapted to local regimes of food availability, predation pressure, etc. Nevertheless, from the few data concerning meiobenthic organisms, we cannot infer that meiofauna biology is very different in the Mediterranean from that of other areas.

DISCUSSION AND CONCLUSION

The assessment of meiofauna's role in the benthic ecosystem is obviously based on little evidence. From the previous results it would be expected to be more important in the shallow than in the deep Mediterranean. In 1978, Gerlach, refering to our own data, noted the high relative biomass of meiobenthos on the continental shelf. In other areas, macrofauna/meiofauna biomass ratios range from 10/1 to 5/1. In the Mediterranean, these ratios would be lower due to low macrofauna density and biomass. For example in the sublittoral muds off Banyuls, biomass ranged from 2.92 to 6.70 $g{\cdot}m^{-2}$ (mean 3.7 $g{\cdot}m^{-2}$) for macrofauna (Guille, 1971) and from 0.99 to 5.12 $g{\cdot}m^{-2}$ (mean 2.6 $g{\cdot}m^{-2}$) for meiofauna (de Bovée, 1981). The use of more sophisticated macrofauna sampling gear, e.g. orange-peels or Smith-McIntyre grabs (Bhaud and Duchêne, 1977,1979) did not significantly modify the results in terms of biomass (Reyss, 1968; Desbruyères et al., 1973).

McIntyre (1969), then Gerlach (1971,1972,1978) by coupling standing crop biomass with turn-over rates, suggested that meiofauna, including Foraminifera (often left aside) could have a production as high as that of macrofauna, in systems quantitatively dominated by the latter; in systems such as the deep-sea shallow waters or upwellings, meiofaunal production could exceed by far macrofaunal production.

The low macrofauna numbers and biomass, and the high meiofauna abundances on the Mediterranean continental shelf, could be related to limited food availability, as in oligotrophic parts of the ocean. However, one must remain cautious. If meiofauna production seems to be quite similar to that of other areas, macrofauna production could be higher.

In his review on secondary benthic production, Warwick (1980), using the assessment of Allen (1971), Hibbert (1976) and Waters (1977) showed that macrobenthos $P/\bar{B}$ ratios were highly dependent on growth and mortality; as longevity increases, $P/\bar{B}$ ratios become very low, with sometimes negative values (Wolff and de Wolf, 1977; Warwick et al., 1978). From the data gathered by Warwick (1980), it is clear that the current $P/\bar{B}$ ratio of 2, proposed by Gerlach (1971) is somewhat speculative. Warwick (1980) demonstrated that a $P/\bar{B}$ ratio of 2.5 for an annual species would decrease to about 1 for a species of 4 years longevity - these values varying according to the life cycle pattern of the species.

In the shallow waters of the Mediterranean, Massé (1972) showed that the main species (echinides, pelecypods) had short lives, fast growth and early sexual maturity. In fine sands, most of the species live only one or two years, and grow twice as much as in boreal seas during the first year. Duchêne (1982) studying the biology of the Polychaete Terebellides stroemi in the sublittoral muds off Banyuls, reported that the populations comprised three age classes at most; growth was important during the first year and $P/\bar{B}$ ratio reached about 1.44. For the same species in Greenland, Curtis (1977) found populations of five year classes and a $P/\bar{B}$ ratio of 1.04.

We should point out that data on macro- and meiobenthos turnover rates are scarce in the Mediterranean, and that an overall generalization of these rates would be tenuous.

In the Mediterranean benthic ecosystem, meiofauna appears to be a significant part of total biomass, possibly used as food by higher trophic levels. However, this role seems limited. A large proportion of meiofauna individuals seem to be free from predation, and there are few known predators. Most of them, juvenile and small fishes, prey upon meiofauna for short periods of their benthic life and meiofauna provided only a small part of their energy requirement (Villiers, 1979; Tito de Morais, 1983). Playing an important part in recycling nutrients, good competitors with large animals for food, meiofauna thus seems to be at the top of a food chain.

SUMMARY

In the past years meiofaunal research has emphasized the ecological role of meiofauna from all parts of the world. In this paper, we attempt to give an overview of our present knowledge on Mediterranean meiobenthos, and compare its quantitative distribution, seasonal fluctuations, the factors affecting its productivity..., to those of other areas. From these data, the importance of meiobenthos within the Mediterranean communities should be re-evaluated.

REFERENCES

Allen, K.R., 1971, Relation between production and biomass, J.Fish. Res. Bd Can., 28:1573.

Alongi, D.M., and Tietjen, J.H., 1980, Population growth and trophic interactions among free-living marine Nematodes, in: "Marine Benthic Dynamics," K.R. Tenore and B.C. Coull, eds., Univ. of S. Carolina, Columbia, S.C.

Barnett, P.R.O., 1968, Distribution and ecology of Harpacticoid copepods of an intertidal mudflat, Int. Rev. ges. Hydrobiol. Hydrogr., 53:177.

Barnett, P.R.O., 1970, The life cycles of two species of Platychelipus Brady (Harpacticoida) on an intertidal mudflat, Int. mudflat, Int. Rev. ges. Hydrobiol., 55:169.

Bell, S.S., 1980, Meiofauna-macrofauna interaction in a high salt marsh habitat, Ecol. Monogr., 50:487.

Bell, S.S., and Coull, B.C., 1978, Field evidence that shrimp predation regulates meiofauna, Oecologia, 35:141.

Bhaud, M., and Duchêne, J.-C., 1977, Observations sur l'efficacité comparée de deux bennes, Vie Milieu, 27:35.

Bhaud, M., and Duchêne, J.-C., 1979, Données quantitatives sur les fonds meubles de 90 m au large de Banyuls-sur-Mer, Vie Milieu, 28-29:21.

Bianchi, A., 1973, Variations de la concentration bactérienne dans les eaux et les sédiments littoraux, Mar. Biol., 22:23.

Bodiou,J.-Y., and Chardy, P., 1973, Analyse en composantes principales du cycle annuel d'un peuplement de copépodes harpacticoides des sables infralittoraux de Banyuls-sur-Mer, Mar. Biol., 20:27.

Bodiou,J.-Y., and Villiers, L., 1978-79, La prédation de la méiofaune par les formes juvéniles de Deltentosteus quadrimaculatus (Teleostei, Gobiidae), Vie Milieu, 28-29:143.

Boucher, G., 1972, Distribution quantitative et qualitative des Nématodes d'une station de vase terrigène côtiere de Banyuls-sur-Mer, Cah. Biol. Mar., 13:457.

Boucher, D., 1975, Production primaire saisonnière du microphytobenthos des sables envasés en baie de Concarneau, Thèse Doct. 3e cycle, Univ. Bretagne occidentale.

Boucher, G., 1980, Facteurs d'équilibre d'une peuplement de Nématodes des sables sublittoraux, Mem. Mus. natn Hist. nat. Nvelle serie, A, 144:1.

Bougis, P., 1946, Analyse quantitative de la microflore d'une vase marine à Banyuls-sur-Mer, C. r. hebd. Séanc. Acad. Sci. Paris, 222:1122.

Bougis, P., 1950, Méthode pour l'étude quantitative de la microfaune des fonds marins (Méiobenthos), Vie Milieu, 1:23.

Bovée, F. de, 1981, Ecologie et dynamique des Nématodes d'une vase sublittorale (Banyuls-sur-Mer), Thèse Doct. Etat, Univ. P. et M. Curie, Paris.
Bovée, F. de, and Soyer, J., 1974, Cycle annuel quantitatif du méiobenthos des vases terrigènes côtières. Distribution verticale, Vie Milieu, 24:141.
Bovée, F. de, Delille, D., and Soyer, J., 1979, Essai d'analyse en composantes principales des relations entre densités en méiofaune et facteurs du milieu dans l'Archipel de Kerguelen, Mem. Mus. Natn. Hist. nat., nlle série C (43):197.
Brenning, U., 1973, The distribution of littoral nematodes in the Wismarbucht, Oikos, supp. 15:98
Cahet, G., and Mouneimné, 1976, Production carbonée microbentique en sédiments margino-littoraux (Golfe du Lion), Ann. Inst. Oceanogr., 52:189.
Caine, E.A., 1978, Habitat adaptations of North American carpellid amphipode (Crustacea), Biol. Bul. Mar. Biol. Lab. Woods Hole, 155:288.
Castel, J., and Lasserre, P., 1977, Colonisation et distribution spatiale des copépodes dans les lagunes semi-artificielles, in: "Biology of Benthic organisms," B.F. Keegan, P.O. Cedig and P.J.S. Boaden, eds., Pergamon Press, Oxford, New York.
Castel, J., and Lasserre, P., 1982, Régulations biologiques du méiobenthos d'un écosystème lagunaire par an alevinage expérimental en soles (Solea vulgaris), Oceanologica acta, No sp.:243.
Ceccherelli, V.V., and Cevidalli, F., 1978, Osservazioni preliminari sulla bionomia dei popolamenti meiobentonici della socca di Scardovari (Delta del Po), con particulare riferimento ai nematodi ed ai copepodi, Atti X Congr. S.I.B.M., Ancona, Mazzio 1978:1.
Clasing, E., 1976, Annual fluctuations of the meiofauna in the Chinguihue marsh (Puerto Montt Chile), Stud. Neotrop. Fauna Environ., 11:179.
Claus, C., 1866, Die copepoden-Fauna von Nizza. Ein Beitrag zur Charakteristik fer Formen und deren Abänderungen "im Sinn Darwins," Schr. ges. Naturn. Marsburg, suppl. 1:1.
Coull, B.C., 1968, Shallow water meiobenthos at the Bermuda platform, Ph.D. Thesis, Lehigh University.
Coull, B.C., 1970, Shallow water meiobenthos at the Bermuda platform, Oecologia, 4:325.
Coull, B.C., 1973, Estuarine meiofauna: review,trophic relationships and microbial interactions, in: "Estuarine Microbial Ecology," L.H. Stevenson and R.R. Colwells eds., Univ. S. Carolina, Columbia, S.C.
Coull, B.C. and Fleeger, J.W., 1977, Long term temporal variation and community dynamics of meiobenthos copepods, Ecology, 58: 1136.
Coull, B.C., and Bell, S.S., 1979, Perspectives of marine meiofaunal ecology, in: "Ecological Processes in coastal and marine systems," R.J. Livingston, ed., Plenum Press, New York.

Curtis, M.A., 1977, Life cycles and population dynamics of marine benthic polychaetes from the Disco Bay area of West Groenland, Ophelia, 16:9.

Davies, J.H., 1975, Energy flow through the benthos in a Scottish sea loch, Mar. Biol., 31:353.

Delamare-Deboutteville, C., 1960, "Biologie des eaux souterraines littorales et continentales," Herman Paris.

Delille, D., Bovée, F. de and Soyer, J., 1983a, Bacteria meiofauna relationship in a subantarctic sand beach (Kerguelen), Proceed. International Sandy Beaches as ecosystems, Junk, ed., Amsterdam, (in press).

Delille, D., Guidi, L., and Soyer, J., 1983b, Experimental approach of nutrition of Allotanais hirsutus in Kerguelen Islands, Fourth Symposium on Antarctic Biology, Wilderness, South Africa, September, 1983.

Desbruyères, D., Guille, A. and Ramos, F., 1973, Bionomie benthique du plateau continental de la côte catalane espagnole, Vie Milieu, 23:335.

Dinet, A., 1972a, Etude écologique des variations quantitatives annuelles d'un peuplement de copépodes harpacticoïdes psammiques, Tethys, 4:95.

Dinet, A., 1972b, Reproduction, développement et croissance de Bulbamphiascus inus (Brady) et Halectinosoma herdmani (T. et A. Scott), Copepoda, Harpacticoida, Tethys, 4:437.

Dinet, A., 1973, Distribution quantitative du méiobenthos profond dans la région de la dorsale de Walvis, Mar.Biol.,20:20.

Dinet, A., 1976, Etude quantitative du méiobenthos dans le secteur nord de la mer Egée, Acta Adriatica, 18:83.

Dinet, A., 1979, A quantitative survey of meiobenthos in the deep Norwegian sea, Ambio, Spec. Rep., 6:75.

Dinet, A., 1980, Répartition quantitative et écologie du méiobenthos de la plaine abyssale atlantique, Thèse Doctorat Etat, Univ. Aix-Marseille III.

Dinet, A., Laubier, L., Soyer, J., and Vitiello, P., 1973, Résultats biologiques de la campagne Polymède. II. Le méiobenthos abyssal, Rapp. Comm. int. Mer Médit., 21:701.

Dinet, A., and Vivier, M.H., 1977, Le méiobenthos abyssal du Golfe de Gascogne. 1. Considérations sur les données quantitatives, Cah. Biol. Mar., 28:85.

Dinet, A., and Khripounoff, A., 1982, Rapports quantitatifs entre le méiobenthos et la matière organique sédimentaire en zone abyssale, Colloques internationaux du CNRS No 293, Biogéochimie de la matière organique à l'interface eau-sédiment marin:319.

Dinet, A., Nodot, C., Vitiello, P. and Vivier, M.H., 1982, Impact d'un effluent thermique sur une communauté de copépodes harpacticoïdes benthiques, Tethys, 10:355.

Duchêne, J.-C., 1982, Etudes comparées de la biologie d'annélides polychètes considérées en différentes zones climatiques, Thèse Doctorat Etat, Univ. P. et M. Curie.

Dujardin, F., 1845, Histoire Naturelle de Helminthes ou vers intestinaux, Roret Ed., Paris.
Dye, A.H., 1979, Aspects of the ecology of meiofauna in Morgazana estuary Transkei, S. Afr. Tydskr. Dierk., 14:67.
Dye, A.H., and Furstenberg, J.-P., 1978, An ecophysiological study of the meiofauna of the swartkops estuary. 2. The meiofauna: composition, distribution, seasonal fluctuation and biomass, Zool. afr., 13:19.
Fava, G., and Volkmann, B., 1975, Tisbe (Copepoda: Harpacticoida) species from the lagoon of Venice. I. Seasonal fluctuations and ecology, Mar. Biol. 30:151.
Feller, R.J., 1980, Development of the sand-dwelling meiobenthic harpacticoid Copepod Huntemannia jadensis Poppe, in the Laboratory, J. Exp. mar. Biol. Ecol., 46:1.
Feller, R.J., 1982, Empirical estimates of carbon production for a meiobenthic harpacticoid copepod, Can. J. Fish. Aquat. Sci., 39:1435.
Filipjev, L., 1918-1921, Free-living marine nematodes of the Sebastopol Area, Israël program for scientific translations, M. Raveh, ed., Jerusalem, part 1, part 2.
Fleeger, J.W., and Palmer, M.A., 1982, Secondary Production of the estuarine meiobenthic copepod Microarthridion littorale, Mar. Ecol. Prog. Ser., 7:157.
Gerlach, S.A., 1971, On the importance of Marine Meiofauna for benthos communities, Oecologia, 6:176.
Gerlach, S.A., 1972, Die Produktiosleitung des Benthos in der Helgoländer Bucht, Verh. d. Dtsch Zool. Ges., 65:1.
Gerlach, S.A., 1978, Food chain relationship in subtidal silty sand, marine sediments and the role of meiofauna in stimulating bacterial productivity, Oecologia, 33:55.
Gerlach, S.A. and Schrage, M., 1969, Freilebende Nematoden als Nahrung der Sandgarnde Crangon crangon. Experimentelle Untersuchungen über die Bedeutung der Meiofauna als Narhung für das marine Makrobenthos, Oecologia (Berl.), 2:362.
Gerlach, S.A. and Shrage, M., 1971, Life cycles in marine meiobenthos experiments at various temperatures with Monhystera disjuncta and Theristus pertenius (Nematoda), Mar. Biol., 9:274.
Gerlach, S.A., and Schrage, M., 1972, Life cycles at low temperatures in some free-living nematodes, Veröff. Inst. Meeresforsch. Bremerh., 14:5.
Gowing, M.H., and Hulings, N.C., 1976, A spatial study of the meiofauna on a sewage polluted lebanese sand beach, Acta Adriatica 18:341.
Guille, A., 1971, Bionomie benthique du plateau continental de la côte catalane française. IV. Densités, biomasses et variations saisonnières de la macrofaune, Vie Milieu, 22 (1B):93.
Guille, A., and Soyer, J., 1968, Contribution à l'étude comparée des biomasses du macrobenthos et du méiobenthos des substrats meubles au large de Banyuls-sur-Mer, 3e Symposium Européen de

Biologie marine, Arcachon, Vie Milieu, Suppl. 22:15.
Harris, R.P., 1972, Seasonal changes in the meiofaunal population of an intertidal sand beach, J. Mar. Biol. Ass. U.K., 52:389.
Heip, C., 1974, A comparison between models describing the influence of temperature on the development rate of copepodes, Biol. Jaarb. 42:121.
Heip, C., and Smol, N., 1976, Influence of temperature on the reproductive potential of two brackish-water harpacticoids (Crustacea: Copepoda), Mar. Biol., 35:327.
Heip, C., Smol, N., and Absillis, V., 1978, Influence of temperature on the reproductive potential of Oncholaimus oxyuris (Nematoda: Oncholaimidae), Mar. Biol. 45:255.
Heip, C., Herman, P.M.J., and Coomans, A., 1982, The productivity of marine meiobenthos, Med. Konink Acad. Weteusch., Lett. en Schan Kunsten Belgü, 44:1.
Hibbert, C.J., 1976, Biomass and production of a bivalve community on an intertidal mud flat, J. exp. mar. Biol. Ecol., 25:249.
Hickel, W., 1969, Sedimentbeschaffenheit und Bakteriengehalt im Sediment eines zukünftigen Verklappungs Gebietes von Industrieabwässern nordwestlich Helgolands, Helgoländer wiss. Meeres., 19:1.
Hicks, G.R.F., 1979, Pattern and strategy in the reproductive cycles of benthic harpacticoid copepods, in: "Cyclic Phenomena in Marine plants and animals," E. Nayla and R.G. Hartnoll, eds., Pergamon Press, Oxford.
Hicks, G.R.F., and Coull, B.C., 1983, The ecology of marine meiobenthic harpacticoid copepods, Oceanogr. Mar. Biol. Ann. Rev., 21:67.
Hopper, B.E., Fell, J.W., and Cefalu, C., 1973, Effect of temperature on life cycles of nematodes associated with the mangrove (Rhizophora mangle) detrital system, Mar. Biol., 23:293.
Hopper, B.E., and Meyers, S.P., 1967, Population studies on benthic nematodes within a subtropical sea-grass community, Mar. Biol., 1:85.
Hulings, N.C., 1971a, A quantitative study of the sand beach meiofauna in Tunisia. Preliminary report, Bull. Inst. Oceanogr. Pêche, 2:237.
Hulings, N.C., 1971b, A comparative study of the sand beach meiofauna of Lebanon, Tunisia and Marocco, Thalassia Jugoslavica, 7:117.
Hulings, N.C., 1974, A temporal study of lebanese sand beach meiofauna, Cah. Biol. Mar., 15:319.
Kiseleva, M.I., 1961, Répartition qualitative et quantitative du benthos dans la région pré-dardanelle de la Mer Egée, Trudy Sevast. Biol. St., 14:135.
Kiseleva, M.I., 1964, Distribution prédative et quantitative du benthos de la Mer Egée, Trudy Sevast. Biol. St., 16:192.
Kiseleva, M.I. and Tchukhtchin, V.D., 1965, Quelques données quantitatives sur le macro- et le microbenthos de la Méditerrannée orientale, Izv. Akad. Nank. CCCP Okean, K.:192.

Khripounoff, A., Desbruyères, D., and Chardy, P., 1980. Les peuplements benthiques de la faille Vema: données quantitatives et bilan d'énergie en milieu abyssal, Oceanol. Acta, 3:187.
Lasserre, P., Renaud-Mornant, J., and Castel, J., 1976, Metabolic activities of meiofaunal communities in a semi-enclosed lagoon. Possibilities of trophic competition between meiofauna and mugilid fish, 10th European Symposium on Marine Biology, vol. 2, G. Persoone and E. Jaspers, eds., Universa Press, Wetteren.
Laybourn, J., 1979, The effects of temperature on the respiration and production of the freshwater nematode Anonchus sp., Oecologia, 41:329.
Le Bouteiller-Bougnol, C., 1980, Biologie et reproduction d'une population d'Amphipode Siphonocoetes dellavallei, Thèse Doct. 3e cycle, Univ. P. et M. Curie, Paris.
Lee, J.J., Tietjen, J.H., Mastropaolo, C. and Rubin, H., 1977, Food quality and the heterogeneous spatial distribution of meiofauna, Helgoländer wiss. Meeresunters 30:272.
Lorenzen, S., 1974, Die Nematodenfauna der sublitoralen region der Deutschen Bucht, insbesondere im Titan-Abwasser-gebiet bei Helgoland, Veröff. Inst. Meeresforsch. Bremerh., 14:305.
Marcotte, B.H., and Coull, B.C., 1974, Pollution, diversity and meiobenthic communities in the North Atlantic (Bay of Piran, Yugoslavia), Vie Milieu, 24:281.
Mare, M.F., 1942, A study of the marine benthic community with special reference to the micro-organisms, J. mar. biol. Ass. U.K., 25:517.
Massé, H., 1972, Contribution à l'étude de la macrofaune de peuplements des sables fins infralittoraux des côtes de Provence. VII. Discussion, comparaison et interprétation des données quantitatives, Tethys, 4:397.
Matthews, J.B.L., 1964, On the biology of some bottom living copepods (Aetideidae and Phaennidae) from Western Norway, Sarsia, 16:1.
McIntyre, A.D., 1964, Meiobenthos of sublittoral muds, J. mar. Biol. Ass. U.K., 44:665.
McIntyre, A.D., 1969, Ecology of marine meiobenthos, Biol. Rev., 44: 245.
McIntyre, A.D., 1971, Observations on the status of subtidal meiofauna research, Smithsonian Contrib. Zool., 76:149.
McIntyre, A.D., and Murisson, D.J., 1973, The meiofauna of a flatfish nursery ground, J. mar. Biol. Ass. U.K. 53:93.
Meyers, P.P., Hopper, B.E., and Cefalo, R., 1970, Ecological investigations of the marine nematode Metoncholaimus scissus, Mar. Biol., 6:43.
Mielke, W., 1976, Okologie der Copepoda eines Sandstrandes der Nordsee insel Sylt., Mikrofauna des Meeresbodens, 59:453.
Monard, A., 1928, Les harpacticoïdes marins de Banyuls, Arch. Zool. exp. gén., 67:259.

Moore, H.B., 1931, The muds of the Clyde sea area. III. Chemical and physical conditions, rate and nature of sedimentation and fauna, J. mar. Biol. Ass. U.K. 17:325.

Morais, de L. Tito, 1983, Etude de la nutrition et de ses aspects énérgetiques chez des formes juvéniles de Pleuronectiformes, Thèse 3e cycle, Univ. P. et M. Curie.

Muus, B.J., 1967, The fauna of Danish estuaries and lagoons: distribution and ecology of dominating species in the shallow reaches of the mesohaline zone, Medd. fra. Danmarks Fiskeri-og Havunders N.S., 5:3.

Nodot, C., 1977, Cycles biologiques du méiobenthos des sables fins infralittoraux. Analyse in situ et étude expérimentale des effets du facteur thermique, Thèse Doc. 3e cycle, Univ. Aix-Marseille.

Nodot, C., 1976;(1978), Cycles biologiques de quelques espèces de Copépodes harpacticoïdes psammiques, Tethys, 8:241.

Nodot, C., 1982, Reproduction et croissance des copépodes harpacticoïdes psammiques Halectinosoma herdmani (T. et A. Scott) et Tryphoema porca Monard, Tethys, 10:373.

Nyholm, K.G., and Olson, I., 1973, Seasonal fluctuations of the meiobenthos in an estuary on the Swedish West coast, Zoon, 1:69.

Por, F.D., 1964, A study of the Levantine and Pontic Harpacticoïda (Copepoda, Crustacea), Zool. Verh. Rijksmus. Natuurl. Hist. Leiden, 64:1.

Pugh, K.B., Andrews, A.R., Gibbs, C.F., Davis, J., and Floodgate, G.D., 1974, Some physical, chemical and microbiological characteristics of two beaches of Anglesey, J. exp. mar. Biol. Ecol., 15:305.

Remane, A., 1933, Verteilung und organisation der benthonischen Mikrofauna der Keiler Bucht, Wies. Meeresunters (Abt. Kiel), 21:161.

Reyss, J.-P., 1968, Quelques données quantitatives sur les biocénoses benthiques du Golfe de Marseille, Rapp. Comm. Int. Mer Med., 19:121.

Rheinheimer, G. von, and Kullmann, K.H., 1972, Untersuchungen über den Bakterienimd Hefegehalt von Wasser im sand an einem Badestrand der Ostseekuste, Kieler Meeresforsch. 28:204.

Rowe, G.T., and Staresinic, N., 1979, Sources of organic matter to the deep sea benthos, Ambio, spec. Rep., 6:19.

Sanders, H.L., and Hessler, R.R., 1969, Ecology of the deepsea benthos, Science, N.Y., 163:1419.

Schaffer, P.L., 1979, The feeding biology of Podarke pugettensis (Polychaete: Hesionidae), Biol. Bul. Mar. biol. Lab. Woods Hole, 156:343.

Schmidt, P., 1968, Die quantitative Verteilung und Population dynamik des Mesopsammons am gezeiten Sandstrand der Nordsee insel Sylt. I. Faktoringefüge und biologische gliederung des Lebensraumes, Int. Rev. ges. Hydrobiol., 53:723.

Schmidt, P., 1969, Die quantitative Verteilung und Population dynamik des Mesopsammons am gezeiten Sandstrand der Nordsee insel Sylt. II. Quantitative Verteilung und populations dynamik einzehner Arten, Int. Rev. ges. Hydrobiol. 54:95

Schmidt, P., 1972a, Zonierung und Jahreszeitliche Fluktuationen des Mesopsammons im Sandstrand von Schilksee (Kieler Bucht), Mikrofauna Meeresbodens, 10:353.

Schmidt, P., 1972b, Zonierun und jahreszeitliche Fluktuationen der interstitiellen Fauna im Sandstränden des gebietes von Tromsø (Norwegen), Mikrofauna Meeresboden, 12:81.

Schmidt, P., and Westheide, W., 1971, Etude sur la répartition de la microfaune et de la microflore dans une plage de l'île de Sylt (Mer du Nord), Vie Milieu, suppl. 22:449.

Sibert, J.R., 1979, Detritus and Juvenile Salmon production in the Nanaimo Estuary. II. Meiofauna available as food to juvenile Chum salmon (Onchorhynchus keta), Journ. Fish. Res. Board Canada, 36:497.

Skoolmun, P., and Gerlach, S.A., 1971, Jahreszeitliche Fluktuationen der Nematodenfauna in Bezeitenbereich des Weser Astuars (Deutsche Bucht), Veröff. Inst. Meeres. Bremerh., 13:119.

Smidt, E.L.B., 1951, Animal in the danish Waddensea, Meddr. Danm. Fisk.-og Havunders, 11:1.

Soyer, J., 1970, Bionomie benthique du plateau continental de la côte catalane française. III. Les peuplements de copépodes harpacticoïdes (Crustacea), Vie Milieu, 21:337.

Soyer, J., 1971, Bionomie benthique du plateau continental de la côte catalane française. V. Densités et biomasses du meiobenthos, Vie Milieu, 22:351.

Soyer, J., 1979, Cycle biologique et potentiel de reproduction d'un copépode harpacticide des vases terrigènes côtières de Banyuls-sur-Mer, Cletodes pusillus (Sars), Rapp. comm. int. Mer Médit. 25/26:225.

Soyer, J., 1980, Cycle biologique d'un copépode harpacticoïde des vases terrigènes côtières de Banyuls-sur-Mer: Cletodes pusillus (Sars), Vie Milieu, 30:35.

Steele, J., and Baird, I., 1968, Production ecology of a sandy beach, Limnol. Oceanogr., 13:13.

Tchukhtchin, V.D., 1964, Répartition quantitative du benthos dans la partie orientale de la Méditerranée, Trudy Sevast. Bid. st., 16:215.

Thiel, H., 1966, Quantitative Untersuchungen über die Meiofauna des Tiefseebodens. (Vorlanfiges Erge buis des "Meteor" - Expedition in dem Indischen Ozean), Veröff. Inst. Meeresforsch. Bremerh., 2:131.

Thiel, H., 1975, The size structure of the deep sea benthos, Int. Rev. ges. Hydrobiol., 60:575.

Thiel, H., 1978, Benthos in upwelling regions, in: "Upwelling ecosystems," R. Boje and M. Tomczak, eds., Springer Verlag, Berlin, Heidelberg.

Thiel, H., 1979, First quantitative data on Red Sea deep benthos, Mar. Ecol. Prog. Ser., 1:347.

Tietjen, J.H., 1969, The ecology of shallow water meiofauna in two new England estuaries, Oecologia, 2:251.

Tietjen, J.H., 1977, Population distribution and structure of the free-living Nematodes of Long Island, Mar. Biol., 43:123.

Tietjen, J.H., 1980, Population structure and species composition of the free-living Nematodes inhabiting sands of the New York Bight Apax, Estuarine coastal mar. Sci., 10:61.

Tietjen, J.H., and Lee, J.J., 1972, Life cycles of marine Nematodes, Influence of temperature and salinity on the development of Monhystera denticulata Timm, Oecologia, 10:167.

Tietjen, J.H., and Lee, J.J., 1977a, Life histories of marine Nematodes. Influence of temperature and salinity on the reproductive potential of Chromodorina germanica Bütschli, Mikrofauna Meeresboden, 61:263.

Tietjen, J.H., and Lee, J.J., 1977b, Feeding behaviour of marine nematodes, in: "Ecology of marine benthos," B.C. Coull, ed., Columbia Univ., South Carolina Press.

Villiers, L., 1979, Contribution à l'étude de la nutrition et de sea aspects énergétiques chez les formes juvéniles de Deltentosteus quandrimaculatus (Valenciennes) (Pisces: Gobiidae), Thèse 3e cycle, Univ. P. et M. Curie.

Villiers, L., 1980, Changes in predation by the juvenile goby Deltentosteus quadrimaculatus (Teleostei, Gobiidae), Netherlands Journ. Sea Res., 14:362.

Vitiello, P., 1968, Variation de la densité du microbenthos sur une aire restreinte, Recl. Trav. Stn. mar. Endoume, 59 (bull. 43):261.

Vitiello, P., 1972, Peuplements de Nématodes marins des fonds envasés de Provence occidentale, Thèse Doc. Etat, Univ. Aix-Marseille.

Vitiello, P., and Triki, S., 1978, Données écologiques sur la méiofaune interstitielle du bas des plages du Golfe de Tunis, Arch. Inst. Pasteur Tunis, 55:383.

Vitiello, P., and Aissa, P., 1979, Observations sur la méiofaune d'une lagune sud-méditerranéenne, Rapp. Comm. int. Mer Méditer. 25/26:137.

Vivier, M.H., 1976, Conséquences d'un dévasement de boue rouge d'alumine sur le méiobenthos profond (canyon de Cassidaigne, Méditerranée), Tethys, 8:249.

Warwick, R.M., 1980, Population dynamics and secondary production of benthos, in: "Marine Benthic depaning," K.-R. Tenore and B.C. Coull, eds., The Belle W. Baruch library in marine Sciences, 11, University of South Carolina Press, Columbia.

Warwick, R.M., and Buchanan, J.B., 1971, The meiofauna off the coast of Northumberland. II. Seasonal stability of the nematodes populations. J. mar. biol. Ass. U.K. 51:355.

Warwick, R.M., George, C.L. and Davies, J.R., 1978, Annual macrofauna production in a Venus community, Est. coastl. mar. Sci., 7:215.

Waters, T.F., 1977, Secondary production in inland waters, Adv. ecol. Res., 10:91.

Wells, J.C.J., 1965, Copépodes (Crustacea) from the meiobenthos at some Scottish marine sub-littoral muds, Proc. Roy. Soc. Edinburgh B., 69:1.

Westheide, W., 1968, Zur quantitativen Verteilung von Bakterien und Hefen in einem gezeiteustraud der Nordseekuste, Mar. Biol., 1:336.

Wieser, W., and Kanwisher, J., 1961, Ecological and physiological studies on marine Nematodes from a small salt marsh near Woods-Hole, Massachussets, Limnol. Oceanogr., 6:262.

Wolff, W.J., and de Wolf, L., 1977, Biomass and production of zoobenthos in the Grevelingen estuary, the Netherlands, Est. coastl. mar. Sci., 5:1.

Yeates, G.W., 1972, Nematoda of a Danish Beech Forest. Methods and general analysis, Oikos, 23:178.

THE DEEP MEDITERRANEAN BENTHOS

Gaston Fredj[1] and Lucien Laubier[2]

[1]Laboratoire d'Océanographie Biologique, Nice, France
[2]and Centre Océanologique de Bretagne, Brest, France

The Mediterranean Sea, where the religions, sciences and arts of the Western Europe were born, was also the site of the first explorations of the deep sea. Several historical syntheses have been published (Perrier, 1886; Rouch, 1946; Pérès, 1961; Merriman, 1968), but it seems useful to underline some major scientific results and to pay a tribute to the work of a forerunner.

Three main periods can be defined, from the early times to the present through the great expeditions and establishment of the biological marine stations.

In the early times of our knowledge on the deep mediterranean benthos, two important dates must be recalled. The opinion of Forbes (1843) that life disappears completely below 550 meters has been held as a pure and simple truth. This result came from the decreasing abundance of life with depth observed by Forbes during numerous dredgings in the Aegean Sea to a maximum depth of 420 meters. In the 1850's, the scientific community admitted these results and the so-called theory of the "azoic zone" in the deep ocean.

This is especially curious as 30 years before the Forbes expedition in the Aegean Sea, the apothecary and lecturer A. Risso, who was the first naturalist to describe deep mediterranean species, made references to deep fishes and crustaceans living in the Gulf of Genova and in the vicinity of Nice at depths between 600 to 1000 meters, in a series of papers published beginning in 1810 (1810, 1816, 1826, 1827). "How did Risso succeed in obtaining species, if not abyssal, of course, at least clearly from depth?" (translated from Monod and Hureau, 1977). In a paper dealing with the crustacean species described by Risso, Holthuis (1977) also writes

"Especially striking... is the fact that so many species from very deep water were described by him, like those of the genera Aristeus, Aristeomorpha, Funchalia, Acanthephyra, Ligur, Paromola and Geryon. He must have obtained these from professional fishermen. "In a few cases, Risso (1827) states how the deep species were caught. For Paromola, he writes "à la profondeur de mille mètres, où on la pêche au palangre", or also in a manuscript preserved in the library of the Paris Natural History Museum "on en prend... pourvu qu'on jette les hameçons dans les grandes profondeurs".

Nevertheless, the ecological notes of Risso have been widely ignored, and the error of Forbes was only corrected in 1861 by A. Milne Edwards when the submarine cable joining Cagliari (Sardinia) to Bône (=Annaba, Algeria), laid in 1857, was lifted two years later for repairs: several tens of specimens, mainly molluscs and corals probably living at 1 800 meters depth, were identified attached to the broken piece of cable (Milne Edwards mentions greater depths, from 2 000 to 2 800 meters - Zibrowius, 1980, reckons a lower actual depth).

Theories are hard to kill, and for some more years, a strong opposition developed against the indisputable evidence of the existence of life in the deep sea. The animals could have been collected during operation of lifting the cable, and so on... Forbe's theory was completely forgotten with the increase of deep sea catches, in particular those of Wyville Thomson in 1868 and 1869, and biological oceanographers moved in the exploration of the world ocean deep sea with the organisation of the great oceanographic expeditions.

In the Mediterranean, the period of the great expeditions and establishment of biological marine laboratories started in the 1880's. Rouch (1952) has recorded the oceanographic activities conducted in the Mediterranean before 1914. De Buen (1934) published a chronological list of the oceanographic cruises (which comprises several approximations). A more comprehensive report of all oceanographic cruises since the early times has been established by Ovchinnikov and Plakhin (1965): it includes 199 cruises, from which only a part collected deep sea organisms.

Among the major cruises, the first cruises after the North East Atlantic explorations were those of the Porcupine (1870) and Shearwater (1871), in the western and eastern basins. These expeditions led directly to the famous Challenger expedition (1872-1876) which did not involve the Mediterranean.

In 1881, the expedition of the Travailleur conducted in the style of Milne Edwards was an important contribution to the study of the western mediterranean basin. The ship worked during 24 days on the Corsica and Provence coasts. Dredgings below 1 000 meters were unsuccessful and the cruise, considered as a failure, was not renewed in the Mediterranean. During the same year, the italian cruise, using the ship Washington was a complete success, and up to very recent years, the biological samples obtained in the Tyrrhenian Sea at more than 3 000 meters were the deepest records for the whole sea.

The eastern basin was later investigated by the russian ship Vitiaz (1889), and especially by the austrian ship Pola (1890, 1891, 1892 and 1893), which collected several deep sea biological samples studied by Marenzeller and Adensamer.

Since 1885, the expeditions of Prince Albert 1st of Monaco on board l'Hirondelle and Princess Alice I and II were mainly devoted to the Atlantic. The danish expeditions lead by J. Schmidt with the Thor in 1908-1910 entering the Black Sea through the Mediterranean and completed in 1921 and 1922 with the Dana cruises, are also important.

This period was also the time of the establishment of the large biological marine stations which, even devoid of large ships, gave an important contribution to our knowledge of mid-depths.

Among the more active ones, the Zoological Station of Naples was the first, created by A. Dohrn in 1874. The Laboratoire Arago in Banyuls-sur-Mer was created by the Lacaze Duthiers in 1881, and enabled the noticeable investigations of the benthic fauna by Pruvot. The Endoume marine station at Marseille decided in 1879 was completed by Marion in 1888. The Oceanographic Institute created by Prince Albert of Monaco, with the Monaco Oceanographic Museum in 1906, was followed in 1910 by the foundation of the International Commission for Scientific Exploration of the Mediterranean Sea, which is still active and brings together the majority of the bordering countries.

The contemporary period starts at the beginning of the 1960's, except for a short incursion of the swedish ship Albatros (1948) into the Mediterranean. This period is characterized by a clear increase of the research effort in terms of man-power, funding and logistics, the advent in oceanographic research of newly independent countries, and the introduction of new technologies mainly due to Commander J.Y. Cousteau. As far as deep sea benthos is concerned, the oceanographers were able to use new sampling gear and direct observation tools. No doubt the first years of this period are marked by the cruises of the ship Calypso in the Mediterranean, the development of the Endoume marine Station with Professor J.M. Pérès and J. Picard (1958, 1964, 1967), and the launching of the bathyscaphs FNRS III and Archimède, the diving saucers SP 300 and Cyana, the submarine troika, etc. Later on several important expeditions took place: the russian expeditions of the Akademik Vavilov and Akademik S. Kowalevsky, the american ships Atlantis (1958, 1961), Chain (1961, 1965), the french cruises of Pdt. Theodore Tissier,

Thalassa, and more recently the deep sea expeditions of *Jean Charcot* (1970, 1972, cruises Polymede).

At the same time, the first modern monographs on the deep sea mediterranean benthos were published: papers by Carpine (1970) and Reyss (1970), dealing respectively with the bathyal fauna of three regions of the western basin (western and eastern coasts of Corsica, Monaco area), and the benthic bionomy of two submarine canyons of the Catalonia Sea. In the eastern basin, Vamvakas (1970) described the benthic fauna of bathyal muds in the South of the Aegean Sea. More recently, faunal assemblages living on the deeper part of the continental shelf in the north-western basin have been studied by Falconetti (1980).

In the same period, two cruises of the drilling ship *Glomar Challenger* in the Mediterranean (1971 and 1975) produced very important results: they returned the first direct evidence of the occurrence in the western and eastern basins of the layer of evaporite varying from 500 to 2 000 meters thickness beneath plio-quaternary sediments (non consolidated muds, semi-consolidated marls). This layer of evaporites (rock-salt, gypsum, anhydrite) was deposited during the Upper Miocene (Messinian). The results of the *Glomar Challenger* are essential for the understanding of the geological story of the Mediterranean; they also illuminate in a new way our understanding of the present mediterranean fauna, and particularly the deep sea fauna.

This historical insight could give the impression that the deep mediterranean benthos is at present rather well known. In fact, the actual data, numerous and recent in the north-western basin for mid-depths, are completely inadequate below 2 000 meters depth. Moreover, the present knowledge is limited to the qualitative and descriptive aspects of the deep sea fauna: how the deep mediterranean ecosystems act, compared with other parts of the world deep ocean, is largely unknown.

CHARACTERISTICS AND SPECIFIC FEATURES OF THE DEEP MEDITERRANEAN ENVIRONMENT

The deep mediterranean environment exhibits some general characteristics and a few specific features which must be recalled before any discussion of the benthic fauna. The following summary concentrates on the evaluation of the importance of the environmental parameters on the organisms. Till the first dives of the bathyscaph *FNRS III* in 1954, our knowledge of benthic life was based on dead organisms. *In situ* experimentation is in its infancy, and the present evaluation of environmental factors will probably be modified in the near future. Also, special consideration will be given to the specific features of the Mediterranean, by comparison with the world ocean. Conversely, the role of light or hydrostatic pressure has been neglected.

Physiographical Remarks

The Mediterranean Sea is at the same time an extended and restricted body of water, and a unique one.

It is a large area, covering nearly 2,500,000 km², with an average depth of 1 650 meters, leaving out the Black Sea and its particular deep waters, according to the standard figures of Kossinna (1921). It is a restricted body of water, since the Mediterranean *sensu stricto* covers less than 0.7 per cent of the world ocean surface, nearly 1/35th of the Atlantic Ocean. The average mediterranean depth, 1 650 meters, is also small compared with the world average estimated at 3 800 meters. The deepest part of the Mediterranean Sea, 5 093 meters according to Hersey (1965), is a little less than one half of the deepest pacific trenches.

Elongated from East to West, the Mediterranean is a narrow sea: its width does not exceed 800 km between Genova and Tunis, reaches some 400 km on the Matapan meridian, and is reduced to 138 km between Sicily and Tunisia. In fact, there is no point distant from more than 370 km from the coasts, and generally speaking distances from coasts are much smaller. To reach the great depths of the Mediterranean never takes more than one to a few hours of navigation (Stanley, 1972).

It is also interesting to compare the percentages of total surface at different depths of the Mediterranean with similar data from the world ocean. According to Stocks (1938), whose calculations exclude the Adriatic, the zone 0-200 meters represents 17 percent of the total surface in the Mediterranean, versus 7.6 percent for the world ocean. The continental shelf appears relatively larger in the Mediterranean, but these figures need to be reevaluated using modern bathymetric charts and including the Adriatic. While 3/4 of the oceans surface (76%) have depths between 3 000 and 6 000 meters (Bruun, 1957), representing more than one half of the Earth surface, the same area takes up slightly more than 13% for the Mediterranean. Truly abyssal depths are limited in the Mediterranean.

These peculiarities apart, the Mediterranean Sea is a unique sea due to its diverse topography and the origin of its deep basins. The Mediterranean Sea comprises two distinct units:

- The western Mediterranean on one hand, divided in two different basins: the algerian - provence basin, or balearic basin extending from 2 500 meters to a maximum of 3 420 meters, and the tyrrhenian basin, with a complex volcanic structure, and maximum depths of more than 3 730 meters between Sicily and Sardinia. On the french coast of the northwestern basin, many canyons or submarine valleys occur from the edge of the continental shelf to the deep flattened area at 2 000 meters. They were discovered nearly one century ago (Pruvot, 1894) and are relatively well known.

- On the other hand, the eastern Mediterranean, is twice as big in surface area and deeper, with an average depth of 1 656 meters compared to 1 615 meters in the western basin.

The eastern basin is composed of four different seas, the Adriatic Sea, the Ionian Sea, the Levant Sea and the Aegean Sea. The Adriatic reaches a maximum depth of 1 230 meters in its southern part, and is divided from the Ionian Sea by a sill of 810 meters depth in the Otranto strait. The Ionian Sea is composed of several deep basins with a northern boundary on the 3 000 meters line and a maximum depth of 3 847 meters in the Syrtis plain. The hellenic trench extends 330 km, from the 3 000 meters line, and comprises a series of depressions with the deepest part of the Mediterranean at 5 093 meters. The Aegean Sea has an average depth of 583 meters, while several deep basins occur with a maximum depth of 2 509 at Karpathos. The Levant Sea offers depths often over 3 000 meters, and several trenches. The maximum depth has been recorded in the Crete trench at 4 384 meters, while depths of 4 338 meters are known from trenches of Rhodes. This part of the Mediterranean is mainly characterized by the presence of a major topographic and geological structure, the so-called mediterranean ridge from its similarity with mid-oceanic ridges. In fact, from a structural point of view, the Mediterranean ridge is not similar to the typical mid-oceanic ridges (Ryan et al., 1970): the upper part of the ridge is only 700 meters higher than the surrounding plains (Emery et al., 1966).

Surprisingly enough, recent expeditions still produce new data on the topography of the Mediterranean Sea. In 1957, Alinat and Cousteau (1962) have discovered in the Ligurian Sea several irregularities of the ground, also found by Bourcart (1960) and later charted in detail by the Oceanographic Museum of Monaco (Alinat et al., 1969). The morphology of these features led Hersey (1965) to suppose that they were created by salt domes. More recent work in the Tyrrhenian Sea and Aegean Sea led to the discovery of numerous small structures and deeps, as shown by the U.S. Defense Mapping Agency Hydrographic Center (1972).

The Sea Bottom

While rocky substrate are not infrequent in the deep Mediterranean, soft sediments are by far more prevalent. The deep sedimentation process presents peculiar features which reflect the geological, climatic and hydrological specificities of the Mediterranean.

Two recent publications synthesize most of our knowledge on the Mediterranean, with contributions from contemporary specialists: Stanley (1972) and Nairn et al. (1978). The different types of deep sediments have been described in detail by Emelyanov (1972) as well as the characteristics of the clay sedimentation by Chamley (1972). The major points are the following:

- The deep sedimentation in the Mediterranean is different from other Inland Seas, and on the whole recalls that of the tropical Atlantic, due to reduced terrigenous deposits and slight concentration of suspension matter linked to a relatively low planktonic productivity.
- While the northern areas are located in a climatic zone with seasonal precipitation, the southern parts are subject to an arid climate, the basin of the algerian provence occupying an intermediate position.
- During the recent Quaternary, sedimentation rates varied from 10 to 40 cm/1 000 years (Herman, 1972). In the Baleares basin, the sedimentation rate is estimated at 30 cm/1 000 years, i.e. ten times higher than the neighbouring East Atlantic: the terrigenous deposits are mainly due to the river Rhone, of which the deep sea fan has recently been studied in detail (Bellaiche et al., 1983). In the Tyrrhenian Sea, the sedimentation rate is slower partly due to volcanic barriers impeding the terrigenous material. The sediment covering is not very thick, and discontinuous with local intercalations of volcanic ash.
- In the eastern Mediterranean, the Nile deposits were very important and appreciable up to Cyprus and the Syrian coast. The Asswan dam has totally modified the sedimentary regime of this area, where the wind deposits play also a role in the dryest part of the basin.
- The biogenic sediments are by far the more frequent sediments in the Mediterranean. Calcium carbonate content varies from 30 to 55% according to the basin: 30 to 50% in the Baleares basin, 30% in the foraminiferan oozes of south Adriatic, increasing to 45-55% in the Tyrrhenian Sea and from 50 to 69% between Crete, Cyrenaica and Alexandria. These deposits are mainly due to pelagic Foraminifera, with some benthic Foraminifera together with coccolithophorids and pteropods (Emelyanov, 1972).

The composition of the ooze depends upon the depth, due to the different solubilities of aragonite and calcite: the former dissolving the faster, the latter the slower. The carbonate compensation depth, estimated to be 3,500 meters in the Pacific and 5,500 meters in the Atlantic, is also related to the temperature, which is unusually high in the Mediterranean depths. As a result, the pteropods, whose skeleton contains a high percentage of aragonite, are abundant for depths less than 1,000 meters, but gradually disappear with depth and are totally absent below 2,500 meters. This phenomenon of CCD level probably has consequences for the calcification process in benthic organisms.

The deep mediterranean sediments are characterized by a middle to high carbonate content and a remarkable deficiency in amorphous silica. Nevertheless, variations in sediment composition and texture are not sufficient to play a role comparable to the situation in shallow water sediment.

The rocky substrates are still very poorly known, mainly for technological reasons. Not until the 1960's, with the development of submarine cameras fitted on the Commander Cousteau troika and the use of deep submersibles (FNRS III, SP 3 000), did significant advance in our knowledge occur (Petit and Laubier, 1962; Laban et al., 1963; Vaissière and Fredj, 1964).

Hydrological characteristics and circulation of deep waters

It is generally agreed that the Mediterranean Sea can be defined as a concentration basin, mainly fueled by surface atlantic waters coming through Gibraltar strait, deeply inserted between large continental land masses, some of which receive very little precipitation in any season.

The works of Lacombe and Tchernia (1972), Wust (1961), Allan et al. (1966) have demonstrated that the renewal time of the Mediterranean Sea is in the order of one hundred years, the average deficit without the atlantic inflow being 0.76 cm/year. This permanent deficit, estimated by different methods, is such that, in the present climate conditions, a period of one thousand years would be sufficient to drain completely the whole basin.

A complex hydrological mechanism led to the formation of deep dense water starting from the inflow of atlantic water. The deep water fills more than 70% of the whole basin and finally pours into the Atlantic through Gibraltar. According to Lacombe and Tchernia (1972), dense deep water formation occurs in three main areas: southeast of the Aegean Sea and the Rhodes-Cyprus area: the Adriatic Sea; and northern part of the western basin.

These phenomena are probably the origin of the most striking peculiarities of the deep Mediterranean environment.

Deep Waters Temperature

As early as 1706, Marsilli, the forefather of mediterranean oceanography, had yet suggested that at depths of 180-200 meters off the coast of Provence, the temperature of the Mediterranean Sea remains fairly constant. Later observations showed that important differences occur in the different parts of the Mediterranean: during the Porcupine and Shearwater cruises, Carpenter established that the depth from which the water temperature remains nearly constant varies from 200 meters in the western basin to 400 meters in the eastern basin (Picotti, 1968).

Nielsen (1912), working on board the Thor, was the first to establish that even below those depths, very small vertical differences occur in a given area, the temperature of the deep water being always slightly higher in the eastern basin. Starting from 12.8°C in the western basin, the homogeneous layer increases by 0.2°C in the Tyrrhenian and reaches 13.5°C in the Ionian Sea and 13.7°C in the Levant basin (Rouch, 1946).

Nevertheless, these fluctuations of a few tenths of a degree probably have no direct consequences on the deep sea organisms, for which the major and most specific features are the homothermy and the high temperature and their environment. The 10°C and 4°C isotherms of which Bruun (1957) has shown the important role for the distribution of bathyal and abyssal species of the world ocean, do not occur in the Mediterranean. In such conditions, the absence in the deep mediterranean benthos of cold stenotherm species, so common in other oceans, is not surprising.

Salinity

The water deficit, and its principal compensation by the atlantic surface waters (S=36.15 o/oo) is responsible for the high salinity of the Mediterranean. Surface water salinity increases eastward, reaching 39 o/oo in the eastern basin with maximum of 39.5 o/oo in the vicinity of Cyprus and Rhodes. The deep waters are topped by an intermediate layer with a salinity maximum.

In the deep homogeneous layer, the salinity ranges from 38.50 to 38.60 ‰ in the center of the western basin, and varies between 38 and 38.50 ‰ off the coast of Provence (Gostan, 1961; Carpine, 1970).

The deep eastern water is made of a mixture of 87.5% adriatic water and 12.5% levantine water; its salinity is around 38.67 ‰.

For the whole Mediterranean, throughout the year, the range of salinity fluctuations is no more than 0.7 ‰ (from 38 to 38.70 ‰).

This is to be compared with world deep ocean average figures which vary by 0.2 per thousand around a mean of 34.80 per thousand (Bruun, 1957). In the present state of knowledge, it does not appear than within the deep Mediterranean, salinity fluctuations can have any biological importance. The high difference in salinity between the Mediterranean and the Atlantic seems to be particularly discriminant.

Dissolved Oxygen

The first measurements by Carpenter (1871-72) led to the erroneous conclusion that the deep Mediterranean was very poor in dissolved oxygen and consequently very unfavourable for life. The deep samples of the Washington in the Tyrrhenian Sea (Giglioli, 1884) and new observations by Natterer (1892,1895) showed that these first results were wrong, the quantity of dissolved oxygen varying from 66 to 75% of saturation, i.e. about 4 ml/l for a saturation value of 6 ml/l at 13.7°C. The Dana cruise in 1930 demonstrated that the concentration was relatively high at all depths in the western basin, and above 4 ml/l with only a single exception. Even east of Sicily, no records fall below 3.5 ml/l (Richards, 1957). More recent works confirmed these observations (Gostan, 1961; de Maio et al., 1968; Carpine, 1970).

Brouardel and Fage (1954) made several measurements at various levels within 1 meter of the sediment, at 460 meters off the coast of Provence: they found lower oxygen content near the bottom, with a sharp decrease in the first 15 cm above the bottom.

In any case, the presence of high dissolved oxygen content at all depths in the Mediterranean gives evidence for strong deep currents, and is similar to the situation encountered in the deep world ocean.

Deep Sea Currents

The circulation in the homogenous deep water and near the bottom is very poorly known: a few observations using the bathyscaphe, and scattered measurements are not sufficient to give a clear idea of the water circulation. However, there is various indirect evidence that deep circulation occurs in the Mediterranean Sea:

- Deep circulation on sills and within the basins must be influenced by seasonal fluctuations, due to the deep water formation mechanism.
- From the location of the major areas of deep water formation and the importance of the outflow at Gibraltar (according to different estimates, the outflow ranges from 36,000 to 45,000 km³/year), the newly created deep water must travel 1 300 to 1 500 km.
- During the search for the atomic bomb lost off the coast of

Palomarès (Spain), the submersible Alvin recorded bottom currents of 0.5 knot.

- During the formation of the deep eastern water resulting from the mixing of water masses from Adriatic and Levant Sea, the flow of water through the Otranto Strait and between Rhodes and Crete can be rapid.
- Water from trenches below 5,000 meters is not anoxic, and renewal phenomena must occur there too (Miller et al., 1970).

The upwelling of deep water in the Messina strait long known for their fauna, add further indirect evidence of deep circulation (Picotti, 1968).

In the present state of our knowledge, the deep mediterranean circulation must be linked with seasonal sporadic events.

Organic Matter in the Water Column and Organic Carbon Sediment Content

The 800 samples collected by filtration on board the *Akademik S. Vavilov* between 1959 and 1968 in the whole Mediterranean are probably the most comprehensive set of data available. Only a few samples were made below 500 meters (Emelyanov and Shimkus, 1972).

Particulate organic matter is composed of phytoplankton, zooplankton and organic matter, the latter being responsible for most of the organic flux in the water column to the bottom.

At the surface, 88 to 92% of the organic matter is made of detritus and plankton that can avoid the nets. The quantity of organic matter decreases with depth, from 0.22 to 0.63 mg/l at 100 meters to 0.10 to 0.48 mg/l at 500 meters. The proximity of the coasts plays an important role, for the lower values (0.1 to 0.2 mg/l) occur in the central parts of the Levant and Baleares basins.

To these data can be added some preliminary measurements of the organic matter fluxes performed by the International Agency for Atomic Energy off Monaco, using sediment traps.

The organic carbon content of the sediments varies in the Mediterranean from 0.2 to 1.6% dry weight, with lower values in surface sediment from the central Levant basin. Carpine (1970) recorded average figures of 0.88% in the Corsica channel, 1.14% west of Corsica and 0.90% nearly Monaco. If these figures show that except in some parts of the Ionian and Levant Seas the organic matter is not very abundant, but never lacking, they also show that depth has less importance than proximity of the coast and surface productivity. Up to now, there are no data available on the quantity of organic compounds utilisable by benthic organisms.

The figures on organic carbon content can also be compared with those found in sapropelic layers discovered in the eastern basin: they contain from 2 to 8% of organic carbon, 10 to 40 times higher than the oxidized sediments (Ryan et al., 1966; Sevastyanov, 1968; Herman, 1972). The sapropel levels indicate that anoxic conditions due to a deficient circulation of the deep waters occurred at various times during the quaternary in the Eastern basin.

Recent geological history of the Mediterranean

It is not possible to discuss the deep fauna of the Mediterranean Sea without considering its geological history. The present Mediterranean is a very small remnant of a much larger sea, the Tethys of Suess, coming from the Mesogean of Wegener. The Tethys Sea existed till the Miocene. At the end of the Miocene, a closing of the communication with the Atlantic, in several stages, had destroyed the conditions of sedimentation, creating circumstances favourable for deposition of evaporites. These depositions occurred in shallow waters between 7 and 5.5 Millions Years, and were progressively settled down by the action of subsidence. This hypothesis had been suggested by Ruggieri (1967) some years before the drillings of the Glomar Challenger through the evaporite layers gave direct evidence of this crisis. Whatever the cause, this crisis involved the complete disappearance of a former deep benthic fauna of which there are no paleontological traces. Hence the present deep mediterranean fauna must originate from successful atlantic migrations since 5.5 M Y. More recently, the quaternary climatic fluctuations make up the second major event: transgressions succeed to regressions, with changes in temperature, salinity and circulation conditions.

Thus, the deep Mediterranean Sea appears as a much younger sea than any other part of the world ocean. The Tyrrhenian basin, with a subsidence rate of 1.1 mm/year, could very well be the youngest deep sea of the world (Selli and Fabbri, 1971). In this manner, the Mediterranean constitutes a remarkable natural laboratory for study of the process of colonisation at depth since a well known and recent date, with specific histories for each of the great deep basins.

In conclusion, the deep Mediterranean offers a series of major original features:

- a deep homothermy, at a relatively high temperature;
- a deep circulation controlled by more or less sporadic seasonal movements;
- a rapid sedimentation rate with a biogenic dominance;
- a deep fauna of recent origin, coming from the Atlantic.

THE DEEP SEA BENTHOS

What can we consider as deep sea benthos in the case of the Mediterranean Sea ? It is not possible to utilize the depth limits now well-admitted in the deep world where faunal changes generally correspond to clearcut variations or boundary values of different factors of the environment. As a matter of fact, those factors which are discriminant for the communities of the mediterranean continental shelf, do not vary significantly below 200 meters in the western basin and 400 m in the eastern basin.

Thus, it seems appropriate to first consider a quantitative approach, i.e. number of species existing below different increasing depths. If there are objective bathymetric boundaries, they will be recognized through corresponding changes in the composition of the fauna. Due to the very scattered nature of the data, bacteria and benthic protozoans as well as most of the meiofaunal organisms, are not considered in this paper, which concentrates on the macrofauna, using the mediterranean fauna data bank developed since several years at the University of Nice (Fredj, 1972) (1). These data have been used to establish the peculiarities of the vertical distribution of different groups of mediterranean Invertebrates (Fredj, 1974).

Since the beginning, the data bank has been brought up to date by use of a computer system (additions, withdrawals, complements). Several other zoological groups have recently been added to the data bank (Cnidaria, sedentary polychaetes.Entoprocta, Ecroprocta Ctenostomata, chordates). Others such as gastropods and fishes, will be introduced in the near future.

Finally, since this year (Fredj, 1983), the system enables instataneous selection of species using one or more given criteria using remote interrogation.

Table 1 gives the total number of species in the mediterranean benthos, without consideration of their vertical distribution.

Comparisons with earlier data from Antipa (1941) and Fredj (1974) show that the benthic mediterranean macrofauna comprises more than 6 000 species, and one can assert without fear that it is far from being completely known.

Although the mediterranean fauna is probably one of the best-

(1) Recent achievements of the data bank were made possible through the financial support of the Direction des Bibliothèques, Musées et de l'Information Scientifique et Technique du Ministère français de l'Education Nationale (DBMIST, grant 15/82).

Table 1. Number of Species of the Major Group of the Mediterranean Benthos

	Total number Antipa (1941)	Number of benthic species Fredj (1974)	Present Paper
PORIFERA	110	-	502
CNIDARIA	156	-	328
Hydrozoa	109	-	183(B)
Anthozoa	47	-	145
ECHINODERMATA	53	134	144
Crinoidea		4	5
Holothurioidea		43	50
Asteroidea		28	31
Ophiuroidea		33	32
Echinoidea		26	26
PRIAPULIDA	0	1	3
ANNELIDA	516	668	791(B)
Errantia		371	384
Sedentaria		297	407
ECHIURIDA		6	6
SIPUNCULA		20	20
BRACHIOPODA	23	15	11
ENTOPROCTA	11	-	19
ECTOPROCTA	306	-	657
Cyclostomata		-	100
Ctenostomata		-	57
Anasca		-	300
Ascophora		-	200
MOLLUSCA			
Solenogastres		26	26
Caudofoveata	22	6	6
Polyplacophora		15	15
Scaphopoda	14	21	21
Pelecypoda	358	302	302
Cephalopoda	63	34	34(B)
Gastropoda	965	-	965(B)
ARTHROPODA			
Decapoda	251	286	306
Others	960	-	1635(E)
POGONOPHORA	0	1	1
PHORONIDEA	1	4	4
HEMICHORDATA	3	5	5
CHORDATA	200	-	200
PISCES	549	-	346(B)
TOTAL	4 561	2 760	6 346

(B) Benthic species only; (E) Estimation

studied marine fauna, the fast and recent increase of our knowledge, especially for the deep species, is striking. The total number of species newly-recorded for the mediterranean, or new to science, during the last 25 years amount to more than 15% of the present total figure. For the fauna living strictly below 200 meters, the ratio is much higher, nearly 30%.

Reasons for this fast increase of recorded deep species are multiple:

- The use of manned submersibles and generally speaking of visual technics enabled observation of large organisms and collection of new species: for instance, the utilisation of SP 300 led to the discovery of the polychaetes Acanthicolepis cousteani (Laubier, 1961) and Lagisca drachi (Reyss, 1961) and the rediscovery of the unique mediterranean stylasterid Errina aspera (Fredj and Giermann, 1982). Using deep sea TV cameras from the laboratory buoy enabled study of the behavior of the crab Geryon tridena at 2 200 meters in the northwestern basin.

- The development of new sampling gears such as the spatang dredge (Carpine, 1970) and the epibenthic sled used for the first time on board Jean Charcot during cruise Polymed I (Reyss, 1971). The large trawl used to recover pieces of the jet plane Caravelle lost off Nice in 2 200 to 2 400 meters depth (Carpine, 1970) caught large numbers of biological samples, leading to the description of the serpulids Filogranula stellata and Hyalopomatus sp. (Zibrowius, 1977) and the hydroid Egmundella valdiviae (Marinopoulos, 1981).

- The development of sorting techniques adapted to small organisms greatly contributed to the increase in our knowledge with the adoption of sieves of 0.5 mm mesh size. The utilisation of the Reineck corer also led to the discovery of the mediterranean deep meiofauna.

- Finally, recent monographs (Logan, 1979; Zibrowius, 1980; Ruffo et al., 1982) and large biological collections also facilitated study of the mediterranean fauna.

Quantitative Analysis of the vertical distribution

Table 2 recalls the results obtained by Fredj (1974) for the vertical distribution of 1244 species of benthic invertebrates, from two interrogations of the data bank:

- number of species collected below a given depth, whatever the range of their vertical distribution:
- number of species living strictly below a given depth.

Table 2. Percentage of Species

Collected below	Endemic	Boreal	Non-endemic	Total number	Living strictly below	Endemic	Boreal	Non-endemic	Total number
0	100	100	100	100	0	100	100	100	100
50	46.0	74.2	67.3	63.5	50	25.1	24.1	22.9	23.3
100	25.1	55.0	48.4	44.3	100	11.8	15.6	15.2	14.6
150	20.8	46.0	40.1	36.7	150	8.5	11.9	10.7	10.3
200	15.2	42.2	34.9	31.4	200	6.6	10.1	9.0	8.6
300	10.9	34.6	28.0	25.0	300	4.7	7.2	6.4	6.1
500	6.2	24.4	20.0	17.5	500	2.8	2.8	2.4	2.5
1000	4.3	12.6	9.6	8.6	1000	0.9	0.7	0.6	0.7
2000	0.9	4.2	3.1	2.7	2000	0.9	0.3	0.2	0.3

The progress of knowledge since 1974 and the addition of phyla which were not taken in account at the time do not drastically change these figures. For sponges, for instance, 61 species were known below 100 meters and only 34 below 200 meters; new collections made by Vacelet (1969) increased those figures to 150 and 76 species respectively, including 17 species new to science and 4 new to the Mediterranean. Our knowledge of the sponges living below 200 meters has thus been increased by 2.2, but they represent only 15% of the total number of mediterranean sponges, which is lower than the average. There are still species to be discovered in the deep, as indicated by the recent cruise Escarmed in the Ionian Sea where unidentified hexactinellids have been discovered on deep cliffs below 2 000 meters (Zibrowius, pers. comm.).

The most important changes concern species living strictly below 2 000 meters: the ratio appears in 1983 to be slightly greater than 1%, to be compared with the figure of 0.3% in table 2. This is mainly due to the fact that sedentary polychaetes were not taken in account by Fredj (1974): some species recently described are confined strictly below 2 000 meters or even deeper. For instance, *Uschakovius enigmaticus* has been collected at 3 174 meters in the Matapan trench (Laubier, 1973), and *Lacydonia laureci* from 4 690 meters in the same area appears at present to be the deepest species from the Mediterranean Sea (Laubier, 1975).

From the 3 000 species considered, more than one third are confined to shallow waters (0 to 50 meters), more than two thirds do not live below 200 meters, and more than four fifth below 500 meters. These proportions differ significantly between the three biogeographic stocks. Endemics are generally shallow water species, while the boreal species increase with depth up to 20% of the total number of species living between 0 and 1000 meters.

Dealing with the species confined strictly below a given depth, a quarter of the total lives below 50 meters, hardly 10% below 200 meters, 2.5% below 500 meters and less than 1% below 2 000 meters. At present, the strictly deep fauna appears very reduced, although our knowledge has increased rather fast over the last twenty years.

The ratio between species living strictly below a given depth and species collected below a given depth is an estimation of the eurybathic degree of the fauna at the same depth. This ratio decreases gradually from the surface to the deepest parts. Up to 2 000 meters, species inhabiting a given range of depth come more and more from superficial layers: for instance, nearly 40% of the species living between 0 and 50 meters have a vertical distribution more or less extensive in depth. Below 500 meters, the fauna comprises six times more species coming from upper layers than strictly deep species. This eurybathic distribution coincides with the extra-mediterranean distribution.

The mediterranean endemic species offer a peculiar vertical distribution. More than half the number of endemic species live between 0 and 50 meters, hardly a quarter below 50 meters, 12% below 100 meters, 7% below 200 meters, 3% below 500 meters and a little more than 1% below 1 000 meters. This ratio is nearly the same below 2 000 meters, mainly due to the results of cruises Polymed I and II, both devoted to the study of the benthos below 2 000 meters. A comparison of these data with data on the total fauna demonstrates that as the depth increases, the originality of the Mediterranean Sea decreases.

The same analysis conducted at a generic level shows clearer results: with a few exceptions, all endemic genera live in shallow waters. In addition, more than 75% of the deep endemic species are known by a small number of individuals, from a single station and a few samples, as shown on table 3 resulting from the analysis of some 1 250 species.

Finally, with the exception of Uschakovius enigmaticus, for the time being ranked within the ampharetids (sedentary polychaetes), there is no mediterranean species which cannot be associated with a north-atlantic allied species, and no endemic genera. For this reason, the deep mediterranean endemic species set up numerous problems of taxonomy and distribution.

The analysis of the distribution of deep sea species demonstrates the strong north-atlantic affinity of deep mediterranean species (nearly 70% are in common with the boreal province), as well as their wide geographical distribution. Further, nearly 35% of these species are known on both sides of the North Atlantic.

The number of these species increases gradually with depth and generally speaking, the depper they occur in the Mediterranean, the wider is their extra-mediterranean distribution.

The situation is very different when the same analysis is conducted within the Mediterranean. For the whole fauna, the general situation is that of eastward depletion of species (92% are known from the western basin versus 54% in the eastern basin). This depletion is even more strengthened for deep sea species: 97% are known in the western part, versus 33% in the Adriatic Sea and a little more than 20% in the eastern Mediterranean.

This is the occasion to discuss the results from the Pola cruises. For species living in both western and eastern Mediterranean, the records by Marenzeller for the eastern basin are systematically the deepest. This is the case for the following species of echinoderms recorded by Marenzeller in the eastern Mediterranean: Luidia sarsi from 755 to 1 292 meters, Echinus acutus from 1 196 to 1 298 meters, and the endemic species Leptometra phalangium by 1 292 meters.

Table 3. List of endemic species strictly confined below 2 000 meters (from Fredj, 1974)

Species	Date of description	number of stations	number of individuals	number of recapture	Northeast Atlantic allied species
Hedingia mediterranea	1914	1	1	1	Trochostoma arcticum
Kolga ludwigi	1893	3	3	1	Kolga obsoleta K.hyalina(2)
Acanthicolepis cousteani	1961(1)	1	1	1	Acanthicolepis asperrima
Harmothoe lunulata					Harmothoe lunulata(2)
f. fauveli	1960(1)	1	++	1	
Lagisca drachi	1961(1)	1	3	1	Lagisca extenuata(2)
Macellicephala annae	1971(1)	1	1	1	Macellicephala hadalis
Macellicephala laubieri	1971(1)	4	27	1	Macellicephala grimaldii
Simrothiella (=Kruppomenia)					
minima	1903	2	7	1	Kruppomenia borealis
Limopsis pygmaea	1836	1	?	1	Limopsis minuta(2)
Sepiola steenstrupiana	1912	+	+	+(3)	Sepiola spp.(2)
Munidopsis marionis	1882	1	1	2(4)	Munidopsis sp.
Plesionika gigliolii	1903	+	+	+	Plesionika edwardsii(2)
Richardina fredericii	1903	4	5	2(4)	Richardina spinicincta
Siboglinum carpinei	1972(1)	+	6	1	Siboglinum ekmani

These surprising distributions can be explained either by a systematic mistake on the depth measurement, or by a deepening of the species in the eastern Mediterranean. This question of the Pola results needs to be cleared up in the future.

Qualitative Aspects of the Deep Mediterranean Benthos and Faunistic Remarks

From a qualitative point of view, the first level of faunistic change occurs around 200 meters, which coincides with the beginning of the homothermic layer in the western basin. Of some 3 000 species, nearly 700 species have been collected at least once below this depth. The following discussion is based on these 700 species.

Sponges. The question of the deep sponges has been raised several times, and Vacelet (1969) underlined the fact that "peu de régions au monde sont aussi peu connues... que les profondeurs de la Méditerranée". The recent improvements in our knowledge have mainly concerned the depths between 100 and 150 meters.

From a total number of some 502 species, 19 species live below 500 meters and only 2 are confined strictly to this range. Eurybathy is the rule, and applies to the deepest mediterranean sponge, Hamacantha facula, whose distribution extends from 146 to 2 280 meters. The occurrence of hexactinellids on rocky cliffs by 2 000 meters during the recent cruise Escarmed appears to be an important finding as far as the deep mediterranean fauna is concerned.

Cnidaria. There are only a few things to say about the hydrozoans from the deep Mediterranean: they are generally eurybathic species, with the single exception of Egmundella valdiviae collected at 2 300 meters off Nice and previously known from Indonesian waters (Marinopoulos, 1981; Vervoort, 1966).

The situation is different for gorgonians and scleractinians, which, independently from purely faunistic aspects, are of biocenotic, paleontologic and paleoecological interest.

The gorgonians have been recently reviewed by Carpine and Grasshoff (1975). Among 21 mediterranean species, 5 species are confined below 200 meters, Callogorgia verticillata, the single primnoid mediterranean gorgonian, occurring off Corsica at 123 meters.

The more interesting deep gorgonian Isidella elongata is widely spread in the western basin, and is sufficiently abundant to characterize a typical facies of the biocenosis of the deep oozes (Pérès and Picard, 1964). I. elongata has been recorded by Ruggieri (1967) in the very first layers of the Pliocene, lying directly on a transgression level: this fact has been used as an argument by the partisans of evaporite deposition in a deep basin. In spite of several records in the scientific literature, I. elongata is absent from the Atlantic, and replaced by Acanella eburnea. The vertical distribution of I. elongata extends from 170 to 1 000 meters, as extreme limits. The more shallow records are scattered and weak specimens (Carpine, pers. comm. and Rossi, 1949-50), while the deeper records are located on the algerian coast and appear uncommon in the other parts of the Mediterranean (Dieuzeide and Roland, 1957). Very dense "meadows" occur between 400 and 650 meters, and the only paleoecological conclusion of this distribution is the importance of the Pliocene transgression (Maurin, 1968; Carpine, 1970; Fredj, 1964; Vaissière and Fredj, 1964).

The scleractinian fauna shows clearly the recent changes of the deep benthos, with the elimination of species living in the Mediterranean during the last glaciation and the quantitative fluctuations which affect the existing species.

An "atlantic" fauna with a few species lived in the deep Mediterranean between 1 000 and 2 500 meters at least during the last glaciation. This fauna contained species such as Caryophyllia atlantica, C. sarsiae and C. ambrosia, now found as fossils, and Desmophyllum cristagalli of which fossil colonies are found from 2 000 to 2 500 meters and which still lives, with a much reduced extension, in waters from 200 to 500 meters.

This species impoverishment due to climatic fluctuations during the quaternary goes with quantitative variations still affecting the deep scleractinian fauna. The so-called white corals (mainly Madrepora oculata and Lophelia pertusa) represent the single, well-defined deep hard substrate community (from 200 to 500 meters). They are presently in recession, from the number of dead colonies which seems out of proportion with the living colonies, at least in certain regions (Zibrowius, 1980).

Echinoderms. The major originality of the deep Mediterranean as far as echinoderms are concerned is the lack of pedunculate crinoids and elasipod holothurians, with the exception of Kolga ludwigi which has never been sampled since the Pola collections by Marenzeller.

The ophiuroids, which are usually very abundant in the deep ocean (Gage and Tyler, 1982), do not go below 700 meters in the Mediterranean. The asterid Plutonaster bifrons is the single species confined below 2 000 meters (from 2 020 to 2 525 meters) while its atlantic distribution extends from 100 to 2 500 meters. The deepest mediterranean record for an echinoderm species is the eurybathic holothurian Pseudostichopus occultatus collected at 3 624 meters in the Tyrrhenian Sea.

Ascidians. The mediterranean ascidian fauna raises several taxonomical problems: from the 200 recorded species, probably no more than 160 species correspond to true or valid taxa. Apart from french and italian coasts, the rest of the Mediterranean is far from being well-explored.

As far as the deep Mediterranean is concerned, our knowledge comes from recent cruises (Monniot, 1969; Monniot and Monniot, 1974, 1975a,b). Among the eleven species living below 200 meters, several species justify special mention:

- Agnesiopsis translucida collected at great depths off the coast of Provence, is very common in the Atlantic from 500 to 5 000 meters.
- Dicopia sp. described from the south of Malta at 500 meters, are juveniles of 3 mm long belonging to the small family octacnemids strictly bathyal and abyssal. Adults of the same genus from the Atlantic measure between 5 and 6 cm. Due to the shortness of the larval life, it is not possible to consider the possibility of their transportation from the Atlantic through the Gibraltar strait. Thus, adults may be found in this part of the Mediterranean.
- Finally, Molgula pumila, M. tethys and Polycarpa candata are all three strictly deep species, two of them being recently described species.

Crustaceans. Compared with the nearly neighbouring deep Atlantic, the mediterranean crustacean fauna lacks several groups, such as the genus Munidopsis, which is represented in the Mediterranean by a dwarf and very rare species, M. marionis, also living in shallow waters.

On the contrary polychelids are commonly caught in the Mediterranean. Surprisingly enough, there are only two adult species, Polycheles typhlops (300- 2 025 meters) and Stereomastis sculpta (1 900-2 400 meters), while four different larval types are known from the deep Mediterranean:

- Eryoneicus puritanii (600-1 900 meters), generally considered as the larva of P. typhlops.
- Eryoneicus faxoni (2 800 meters), E. kempi (2 800 meters) and E. spinoculatus (600 meters); none of them can be accepted as larval stages of Stereomastis sculpta.

Two other adult species of the family polychelids must occur in the deep Mediterranean and have still to be discovered. They perhaps belong to the genus Willemoesia, occuring in the eastern Atlantic.

The polychelids are a good example of blind deep benthic decapods. The study of deep crabs shows that 10% of the species are blind, 10% have ordinary eyes while 80% have eyes twice the ordinary size such as the Mediterranean Geryon.

The polychelids originate from the eryonids, very common in shallow waters in the Triassic and Jurassic periods. The polychelids are in a sense one of the rare groups of "living fossils" from the Mediterranean. This led to the following remarks:

- One of the striking aspects of the deep mediterranean fauna is the fact that most species belong to the older groups within each phylum (crinoids, aphroditoids, protobranchs, filibranchs, polychelids), while a much smaller number of species contains very specialised organisms (Anamathia, Geryon and Munidopsis...).

  This is not specific to the Mediterranean, and has been underlined by Ekman (1953) for the world abyssal fauna.

- On the other hand, while the major part of the deep mediterranean species lives in superficial depths in other areas (Leptometra celtica, Echinocucumis typica, Plutonaster bifrons, Amphilepis norvegica,Macellicephala mirabilis, Nephthys ciliata, Dentalium agile, Malletia cuneata, Anamathia rissoana, Calocaris macandreae for instance), the polychelids belong to the small group of species genuinely coming from the deep, both in the Mediterranean as well as in the Atlantic.

Conclusion

The major points of this qualitative analysis of the deep mediterranean benthic fauna can be summarized as follows:

- The atlantic origin of the deep mediterranean fauna is beyong doubt. North european or widely distributed species predominate, most of them living in superficial waters outside the Mediterranean. The lessepsian migration through the Suez canal has not yet contributed to the deep fauna of the eastern Mediterranean.
- Compared with the nearby ocean, the deep fauna, if not drastically reduced as considered fifty years ago, is relatively poor. Even if recent investigations led to the discovery of new species in the deepest trenches, large taxonomic groups typically abyssal or bathyal are still lacking, as well as the cold stenothermic species which elsewhere represent the main part of the fauna.
- On the other hand, the originality of the mediterranean fauna decreases gradually with increasing depth. With a single exception, all deep species can be related to allied atlantic species. If one admits that there is some parallelism between the taxonomic rank of an endemic unit and the time spent by this unit in

its new environment, the almost complete absence of endemic genera and family clearly demonstrates the youth of the deep mediterranean fauna, youth increasing with depth.

What are the reasons of the lack of originality and the poverty of the fauna ? We can assume that they rest in the recent history of the fauna and the very peculiar physico-chemical conditions of the environment.

The deep homothermy with a relatively high temperature represents a strong barrier for a large number of species, but if these conditions had prevailed since the end of the messinian crisis and the first settlement in the deep basins, then the degree of endemism should be similar to that of the continental shelf fauna.

In fact, it seems highly probable that since the Pliocene and at different times, cold atlantic waters invaded the Mediterranean, which did not always display the type of stratification we observe now. It may be the case at the beginning of the Pliocene (the sediments of this period contain psychrophilic ostracods which in the Atlantic tolerate temperatures between 4 and 6°C (Benson, 1972) and certain glaciation maxima if one agrees with the hypothesis of inverted currents at Gibraltar as proposed by Mars (1963).

The wurmian regression left a rich fossil fauna among which <u>Chlamys septemradiata</u> and <u>Cyprina (Arctica) islandica</u> are revealing examples.

<u>Chlamys septemradiata</u> settled in the western basin, but some remains have been found in the Aegean Sea. The species has completely disappeared from the Mediterranean, except from the Alboran Sea where living specimens have been recorded by Picard (pers. comm.). In the Atlantic, <u>C. septemradiata</u> tolerates salinities around 36.5 o/oo. If it does not spread east of the Alboran Sea, it is because of the high mediterranean salinity.

<u>Cyprina islandica</u> which is very frequent in fossil beds, reproduces on the american coast of the Atlantic when the temperature reaches 13.5°C during the summer: this is similar to the mediterranean conditions, and it is the salinity which eliminated <u>C. islandica</u> from the Mediterranean.

The recent disappearance of the deep scleractinians (31 000 to 28 000 years), even if it is related to other physical factors, demonstrates that the deep mediterranean fauna has suffered a lot from the quaternary climatic changes.

The qualitative impoverishment from West to East, stronger for the deep fauna than for the whole benthic fauna, corresponds not only to the main direction of colonisation nor to some kind of proportionality with the number of data available. It also reflects changes in the deep circulation of water masses, which have periodically led to the establishment of anoxic conditions as proved by the sapropelic deposits of the Levant basin.

The occurrence of the polychelids together with some other strictly deep species demonstrates that the Gibraltar sill has not always been an impassable barrier.

Is this sufficient to state that the deep mediterranean fauna is similar to the abyssal fauna of the world ocean ? In fact, the youth of the settlement of the fauna together with the peculiarities of the present physical environment have prevented the appearance of a sufficient number of well adapted species. However, in the present state of our knowledge, it is important to underline the occurrence of a small group of authentic deep species found during the investigations of the last decade.

ECOLOGY AND BIOLOGY OF THE DEEP MEDITERRANEAN SPECIES

Widely-distributed eurybathic species constitute the major part of the deep mediterranean fauna. They have mainly been studied by taxonomists, and very few things are known about their nutrition and reproductive biology.

The hard substrates characterized by the occurrence of large colonies of the ahermatypic scleractinians (white corals) and the soft sediments with "meadows" of Funiculina quadrangularis and, in deeper waters, Isidella elongata have in common a high percentage of suspension feeders. This type of feeding is typical for all dense communities, from the lower part of the continental shelf (Leptometra phalangium and Gryphus vitreus facies) to a depth of 1 000 to 1 500 meters, throughout the above biocenoses. Suspension feeders occur in deeper waters, such as the asteroid Brisingella coronata, which has been collected up to 2 900 meters by the Washington in the Tyrrhenian Sea.

The biological activity is probably responsible for the resuspension of organic particles after their deposition on the sediment. In the Isidella elongata "meadows" (and this is probably the same with Funiculina quadrangularis), the presence of numerous large penaeid prawns traditionally caught by spanish fishermen using otter trawls such as Aristeus antennatus and Aristeomorpha foliacea could well be the main cause for the establishment of the facies. The penaeid prawns, while searching for their food in the first millimeters of the sediment, remove the organic particles and make them available for the suspension feeding gorgonian (Vaissière and Fredj, 1964).

Apart from suspension feeders which appear to dominate on the continental margins, organisms belonging to higher trophic levels (carnivorous and scavenging species) have been collected below 3 000 meters: decapods, asterids, fishes. They gradually decrease and disappear with increasing depth. The benthic fishes, although less diverse and abundant than the bathypelagic fishes, are not uncommon. They mainly belong to the sharks and rays, the chimaeriforms, anguilliforms, beryciforms and lophiiforms. Submarine pictures of large sharks attracted by baits have been obtained at 900 meters. The traditional macrourids, which are absent from another young deep basin, the Norwegian Sea, seem to be the more frequent group of carnivorous animals. Below this depth, other families of fishes join the macrourids, such as alepocephalids. Alepocephalus rostratus, Benthosaurus grallator, Nezumia aequalis, N. sclerorhynchus and Chalinura europea and some others occur below 2 000 meters. Chalinura europea, caught at 2 904 meters, is probably the deepest record for a fish in the Mediterranean (Tortonese, 1970).

To our present knowledge, the giant scavenging amphipods very common in the world ocean, are missing in the deep Mediterranean.

The feeding type of the single mediterranean elasipod Kolga ludwigi known from three specimens sampled at three different stations, and never seen since is completely unknown. The almost complete absence of elasipods and the small number of hexactinellids again demonstrate the peculiarity of the Mediterranean, which lacks large deposit feeders as well as deep suspension feeders.

Data on the reproductive biology of deep mediterranean species are even more scattered. In the Atlantic ocean, the reproduction cycle of deep sea ophiuroids shows evidence of a seasonal rhythm probably related to organic particle deposition (Gage and Tyler, 1982). However, the gonads are active all year round, and the yearly recruitment observed in the population is probably due to seasonal differences in larval survival rate. In deeper waters, no reproductive cycles have been observed.

In the Mediterranean, some preliminary studies of the reproductive cycle of species living at the edge of the continental shelf have been recently performed. The results are somewhat inconsistent.

The fissiparous asterid Sclerasterias richardi shows a yearly gonadic cycle with ripe ovocyte emission between mid-september and mid-october (Falconetti et al., 1976,1977; Febvre et al., 1981). At this depth the temperature of the water varies from 13.21°C (Spring) to 13.27°C (Summer) and 13.25°C (Fall), i.e. less than 0.1°C during the year, and the salinity range from 37.5 to 38 o/oo: these variations cannot account for the clearcut periodicity of reproduction. Other possible factors are the seasonal photoperiodic variations and especially the flux of organic particles. It must be underlined that sexual reproduction plays a minor role compared with the asexual reproduction occuring all the year round.

In the same area and at the same depths, the large brachiopod Gryphus vitreus, whose larva is still unknown in spite of the abundance of adult populations, apparently displays continuous gamete production.

It is also very possible that a number of eurybathic species could well reproduce in the upper layers, while individuals living in deeper waters would never reproduce. Detailed comparison studies of the same species at different depths are necessary to solve this problem.

Important larval ontogenetic vertical migrations have recently been demonstrated for abyssal gastropods of the northeast Atlantic (Bouchet and Fontes, 1981). It is also the case with deep mediterranean species: larvae of the decapod Geryon accomplish vertical migrations from 450 to 1 800 meters before they return to the depth. Among the deep sea fishes, some apods (Nettastoma melanurum, Nettodarus brevirostris) and macrourids do the same. On the contrary, Santucci (1932) claimed that the Eryoneicus larvae of polychelids migrate in great depths.

These few examples clearly show that the question of periodicity of the reproduction remains one of the major biological questions to be resolved in the Mediterranean as in the world deep ocean.

CONCLUSION AND PROSPECTS

This brief synthesis of our knowledge of the deep mediterranean benthos does not lead to definitive conclusions, but can help to define some of the research trends for the next future.

Firstly, the inventory of the deep fauna is far from being complete. From the arbitrary bathymetric limit of 200 meters to the deepest parts, the number of species recorded has increased by 30% within the last 25 years, and includes numerous species described from old records with a very small number of specimens.

Furthermore, the different parts of the deep Mediterranean have not been equally sampled: the western basin and the northern area of the eastern basin and the northern are relatively well explored, but wide areas such as the southern part of the Levant Sea have practically never been studied.

The transition zones such as Gibraltar strait, sicilo-tunisian sill, Messina strait, channels between the Aegean Sea and the hellenic islands, must be explored in detail as well as the Herodote plain off Egypt, mount Eratosthenes which culminates at 600 meters or the trench located East of Cyprus. Biogeographical gradients should be strong in some of these regions. The Gibraltar strait has a special importance for the understanding of the colonisation processes.

Generally speaking, morphological, taxonomical and ecological comparisons of deep allied species from the Mediterranean and the Atlantic are necessary.

Recent geological cruises and submersible dives have demonstrated the importance of hard substrates at great depths. Their exploration should be carried out using the manned submersibles. Comparative studies of white corals communities in the Mediterranean and in the Atlantic is probably necessary to assess the real conditions of the Mediterranean scleractinans.

From an ecological point of view, the structure of the deep ecosystem below 1 000 meters is poorly known. Quantitative estimates in terms of densities and especially in biomass, are lacking, more or less completely, even for the macrofauna. The composition of the different trophic levels is also almost unknown and certain ecological niches such as the scavenging amphipods of the world ocean seem absent.

A better knowledge of the flow of particulate organic matter reaching the bottom of the Mediterranean, and its time-space fluctuation must be achieved, as well as the importance of labile organic carbon compared to the total amount.

The biological features of deep sea fauna are completely unknown from a few hundreds of meters. Together with feeding and reproductive strategies, the metabolism of mediterranean species should be compared with other areas.

In situ experiments have recently started in the western Mediterranean for the estimation of the bioturbation rates in the canyon of Stoechades during the 1982 cruise of the submersible CYANA in the Mediterranean: this also must be pursued.

The study of the pollution of the deep sediments is in progress, but very few results are available apart from the exploration of an area of industrial waste dumping (Bourcier and Zibrowius, 1972).

Still, the major reason to study the deep mediterranean fauna lies in the fact that the Mediterranean is a unique example of a young deep sea, of which the recent paleoecological history starting at the messinian crisis, is rather well known. In this approach, a better knowledge of the deep fauna is useful not only for the biological oceanographers, but also to the geologists and paleontologists.

Better links between the different scientific interests would facilitate progress.

SUMMARY

The study of the deep-sea benthic fauna started in the Mediterranean. Forbe's dredgings in the Aegean Sea lead him to express the hypothesis of a deep "azoic" zone, and the discovery of a sessile fauna attached to the submarine cable joining Sardinia to Algeria by Milne-Edwards preceded the start of deep sea explorations. The period of the great oceanographic expeditions and the establishment of the first marine biological laboratories has been followed during the last twenty years by the first in situ observations and a clear increase in our knowledge.

Independent of a number of peculiar physiographic features (average depth 1 650 meters, maximum depth 5 093 meters), the mediterranean deep sea environment offers a series of original characteristics:

- deep homothermy at a relatively high temperature (nearly 13°C),
- deep circulation controlled by more or less sporadic seasonal movements,
- fast sedimentation rate with a biogenic dominance,
- recent geological discoveries show that the Mediterranean Sea can be considered as a remarkable natural laboratory to study the deep sea colonisation process since a known event (messinian crisis) relatively recent on a geological scale.

Due to insufficient data, the only macrofauna has been considered in the study.

A quantitative analysis is mainly based upon the data maintained in the Mediterranean fauna data bank developed at the University of Nice. The major results are the following:

1. The Mediterranean benthic macrofauna numbers some 6 000 species from which 3 000 have been considered to identify the main lines of the vertical distribution of species.

2. The recent and fast increase of our knowledge on the deep sea fauna is principally due to the multiplication of samplings, the use of observation techniques, new sampling gear and improvement of sorting operations.

3. Analysis of vertical distributions shows that more than one third of the species has not been recorded below 50 meters, more

than two-thirds below 200 meters and more than four-fifths below 500 meters. The strictly deep Mediterranean fauna appears to be very reduced. The more significant changes during the last ten years apply to the fauna living strictly below 2 000 meters (1% of the total fauna) including a small group of true "abyssal" species.

4. The deep fauna is mainly composed of very eurybathic species which have a very wide extra-mediterranean geographic distribution. Below 1 000 meters, 20% of the species are known in northern Europe waters where they generally occur in shallow areas.

5. Endemic species distribution displays several peculiarities: the deep endemic species, related to north Atlantic species, sets several problems regarding their systematic relationships and distribution. Generic endemism is almost absent. The specificity of the Mediterranean decreases when depth increases.

From a qualitative point of view, an important faunistic change occurs from 200 meters depth. The list of species collected at least once below 200 meters leads to several remarks:

1. Cold stenothermic species are lacking, and several typically deep sea taxonomical groups are very rare, such as the hexactinellid (sponges) and the elasipods (holothurians), which also play an important ecological role.

2. Within the Cnidaria, problems of the Isidella elongata facies and the deep coral community are underlined. Due to the number of species and abundance of colonies, the deep corals seem to regress since the last glaciation.

3. One of the striking features of the mediterranean fauna is that it is mainly composed of the older groups of the analysed phyla while a more reduced fraction comprise the more developed types of their group.

4. The polychelids (decapods) are the only "living fossils" of the Mediterranean. They belong to the small group of strictly deep species, and demonstrate that the Gibraltar sill has not always been a zoogeographical barrier.

The deep homothermy, together with an average high salinity compared with the Atlantic, and above all the consequences of the quaternary climatic fluctuations account for the relatively poorness and weak originality of the fauna. The quaternary climatic changes can also partially explain the eastward qualitative impoverishment which appears more important for the deep fauna than for the whole fauna.

Up to 1 000 meters depth, suspension feeding species are

numerous. Although decreasing in number, species belonging to higher levels (carnivorous and scavengers) have been collected deeper than 3 000 meters.

The reproductive biology of the deep mediterranean species through some preliminary data appears as one of the major biological problems of the mediterranean fauna.

As a conclusion, some future research areas are suggested:

- The continuation of the exploration of the lesser known areas.
- The description of deep hard bottom communities.
- The composition and function of the deep sea mediterranean soft bottom ecosystem below 1 000 to 1 500 meters depth.
- In situ experiments, better knowledge of the meiofauna and level of pollutants in the sediments.

A better knowledge of the deep sea mediterranean fauna is not only important for the biological sciences but can also contribute to the development of other scientific areas.

REFERENCES

Alinat, J., and Cousteau, J.-Y., 1962, Accidents de terrain en mer de Ligurie, in: "Océanographie géologie et géophysique de la Méditerranée occidentale, Villefranche-sur-Mer, 4 au 8 avril 1961," Centre National de la Recherche Scientifique, ed., Paris.

Alinat, J., Cousteau, J.-Y., Giermann, G., Leenhardt, O., Perrien, C., and Pierrot, S., 1969, Lever de la carte bathymétrique de la mer Ligure, Bull. Inst. Océanogr. Monaco, 68, 1395:1.

Allain, C., Furnestin, M-L., Maurin, C., 1966, Essai de synthèse sur l'océanographie physique et biologique dans la zone d'influence du courant atlantique en Méditerranée du Sud et du Levant, Comm. int. Explor. sc. Mer Médit., ed., Paris.

Antipa, G., 1941, Marea Neagra. 1: Oceanografia, bionomia si biologia generalà a Marii Negre, Publi. Fdn V. Adamachi Acad. roman., 10, 55:1.

Bellaiche, G., Orsolini, P., Petit-Perrin, B., Berthon, J-L., Ravenne, C., Coutellier, V., Droz, L., Aloisi, J-C, Got, H., Mear, Y., Monaco, A., Auzende, J-D, Beuzart, P., et Monti, S., 1983, Morphologie au sea-beam de l'éventail sous-marin profond du Rhône (Rhône deep-sea fan) et de son canyon afférent, Comptes rendus séances Acad. sci. série II, 296, 7:579.

Benson, R.H., 1972, Ostracodes as indicators of threshold depth in the Mediterranean during the Pliocene, in: "The Mediterranean Sea: a natural sedimentation laboratory," D.J. Stanley, ed., Dowden, Hutchison and Ross,Stroudsburg.

Bouchet, P., et Fontes, J-C, 1981, Migrations verticales des larves de gastéropodes abyssaux: arguments nouveaux dûs à l'analyse isotopique de la coquille larvaire et postlarvaire, Comptes rendus des séances l'Acad. série III, 292:1005.

Bourcart, J., 1960, Carte topographique du fond de la Méditerranée occidentale, Bull. Inst. Océanogr. Monaco, 57, (1163):20.

Bourcier, M., et Zibrowius, H., 1972 (1973), Les "boues rouges" déversées dans la canyon de la Canidaigne (région de Marseille). Observations en soucoupe plongeante SP 350 (juin 1971) et résultats de dragges, Téthys, 4:811.

Brouardel, J., et Fage, L., 1954, Variation de la teneur en oxygène de l'eau au proche voisinage des sédiments, Deep-sea res., 1:641.

Bruun, A.F., 1957, Deep-sea and abyssal depths, Memoir geological Soc. America, 67 (Treatise on marine ecology and paleoecology), 1:641.

Carpenter, W.B., 1872, Report on scientific researches carried on during the months of August, September and October 1871 in H.M. Surveying Ship "Shearwater", Proc. Royal Soc. London, 20:535.

Carpine, C., 1970, Une expérience de chalutage profond (recherche de la "Cravelle" engloutie au large de Nice), Bull. Inst. Océanogr. Monaco, 69:1408.

Carpine, C., et Grasshoff, M., 1975, Les gorgonaires de la Méditerranée,Bull. Inst. Océanogr. Monaco, 71:1430.

Chamley, H., 1972, Sur la sédimentation argileuse profonde en Méditerranée, in: "The Mediterranean Sea: a natural sedimentation laboratory," D.J. Stanley, ed., Dowden, Hutchinson and Ross, Stroudsburg.

De Buen, R., 1934, Liste chronologique des croisières océanographiques. (Comm. int. explor. sc. Mer Médit. Manuel pour les observations océanographiques à la mer, II).

De Maio, A., Bregant, D., and Sansone, E., 1968, Oceanographic data of the R.V. "Bannoch" collected during the international NATO cruise in the Tyrrhenian Sea (16 September-24 October 1963), Cah. océanogr. 20:1.

Dieuzeide, R., and Roland, J., 1957 (1958), Prospections des fonds chalutables des côtes algériennes. Recherches de nouvelles zones (années 1956-1957), Bull. Sta. Aquic. Pêche Castiglione, 9:9.

Ekman, S., 1953, "Zoogeography of the sea," Sidgwick and Jackson, London.

Emelyanov, E.M., 1972, Principal types of recent bottom sediments in the Mediterranean Sea: their mineralogy and geochemistry, in: "The Mediterranean Sea: a natural sedimentation laboratory," D.J. Stanley, ed., Dowden, Hutchinson and Ross, Stroudsburg.

Emelyanov, E.M., and Shimkus, K.M., 1972, Suspended matter in the Mediterranean Sea, in: "The Mediterranean Sea: a natural sedimentation laboratory", D.J. Stanley, ed., Dowden, Hutchinson and Ross, Stroudsburg.

Emery, K.O., Heezen, B.C. and Allan, T.D., 1966, Bathymetry of the eastern Mediterranean Sea, Deep-sea res., 13:173.
Falconetti, C., 1980, Bionomie benthique des fonds situés à la limite du plateau continental du banc du Magaud (Iles d'Hyères) et de la région de Calvi (Corse), Thèse sc. nat., Nice.
Falconetti, C., Fredj-Reygrobellet, D. and Fredj, G., 1976, Sexualité et fissipareté concomitantes chez l'Astérie Sclerasterias richardi, Mar. Biol. 34:247.
Falconetti, C., Fredj-Reygrobellet, D. and Fredj, G., 1977, Induction de l'émission des gamètes et premiers stades du développement larvaire chez l'Astérie fissipare Sclerasterias richardi, Mar. Biol., 39:171.
Febvre, M., Fredj-Reygrobellet, D., and Fredj, G., 1981, Reproduction sexuée d'une Astérie fissipare, Sclerasterias richardi (Perrier, 1982). Int. invert. reproduction, 3:193.
Fredj, G., 1964, Contributions á l'étude bionomique de la Méditerranée occidentale (Côte du Var et des Alpes-Maritimes, côte occidentale de Corse). Fascicule 2: La région de Saint-Tropez: du cap Taillat au cap de Saint-Tropez (région A 1). Bull. Inst. Océanogr. Monaco, 63:1311.
Fredj, G., 1972, Stockage et exploitation des données en écologie marine. A: Un fichier sur ordinateur des Invertebrés macrobenthiques. Mém. Inst.Océanogr. Monaco, 4:61.
Fredj, G., 1974, Stockage et exploitation des données en écologie marine. C: Considérations biogéographiques sur les peuplement benthique de la Méditerranée, Mém. Inst. Océanogr. Monaco, 7:88.
Fredj, G., 1983, Interrogation conversationnelle et nouveaux développements de la banque de données sur les Invertebrés macrobenthiques méditerranéens. Rapp. Comm. int. Mer Médit., 28: 311.
Fredj, G., and Giermann, G., 1982, Observations en soucoupe plongeante SP 300 des peuplements d'Errina aspera (L.) (Stylosterina) du détroit de Messine, Téthys, 10:280.
Forbes, E., 1843, Report on the Mollusca and Radiata of the Aegean Sea and on their distribution considered as dearing on geology. Report of the British association for the advancement of science:178.
Gage, J.D., and Tyler, P.A., 1982, Growth and reproduction of the deep-sea brittlestar Ophiomusium lymani Wyville Thomson, Oceanol. Acta, 5:73.
Giglioli, E.H., 1884, La scoperta di una fauna abissale nel Mediterraneo, International geographical congress, 3:165.
Gostan, J., 1961, Contribution à l'étude hydrologique de la Mer Ligure, Bull. Inst. Océanogr. Monaco, 58:47.
Herman, Y., 1972, Quaternary eastern Mediterranean sediments: micropaleontology and climatic records, in: "The Mediterranean sea: a natural sedimentation laboratory," D.J. Stanley, ed., Dowden, Butchinson and Ross, Stroudsburg.

Hersey, J.B., 1965, Sedimentary basins of the Mediterranean Sea. Colston papers, (Submarine geology and geophysics):75.

Holthuis, L.B., 1977, The Mediterranean Decapod and Stomatopod Crustacea in A. Risso's published works and manuscripts, Ann. Mus. Hist. nat. Nice, 5:37.

Kossina, E., 1921, Die Tiefen des Weltmeeres, Ver. eff. Inst. Meeres. Berlin, 9:70.

Laban, A., Pérès, J-M,and Picard, J., 1963, La photographie sous-marine profonde et son exploitation scientifique, Bull. Inst. Océanogr., Monaco, 60:1258.

Lacombe, H., et Tchernia, P., 1972, Caractères hydrologiques et circulation des eaux en Méditerranée, in: "The Mediterranean sea: a natural sedimentation laboratory,"D.J. Stanley, ed., Dowden, Hutchinson et Ross, Inc., Stroudsburg.

Laubier, L., 1961, Acanthicolepis cousteani n. sp., un Aphroditien de profondeur récolté par la soucoupe plongeante, Bull. Inst. Océanogr., Monaco, 58:3.

Laubier, L., 1973, Uschakovius enigmaticus gen, sp. n, Ampharetidae (Annélides Polychètes sédentaires) aberrant de Méditerranée orientale profonde, C.R. hebd. Acad. Sc., (D) 277:2723.

Laubier, L., 1975, Lacydonia Laureci sp. n. Annelide polychète nouvelle de l'étage abyssal de Méditerranée orientale, Vie Milieu (1), 25:75.

Logan, A., 1979, The recent Brachiopoda of the Mediterranean Sea, Bull. Inst. Oceanogr., Monaco, 72:1434.

Marinopoulos, J., 1981, Contribution à la connaissance des hydraires profonds de la Mer Méditerranée. Rapp. Comm. int. Mer Médit., 27:2.

Mars, P., 1963, Les faunes et la stratigraphie du Quaternaire méditerranéen, Rec. Trav. Sta. mar. Endoume, 43:61.

Maurin, C., 1968, Ecologie ichthyologique des fonds chalutables atlantiques (de la baie ibéro-marocaine à la Mauritanie) et de la Méditerranée occidentale, Rev. Trav. Inst. Pêch. marit., 32:1.

Merriman, D., 1968, Speculation on life at the depths: a XIXth century prelude, Bull. Inst. Océanogr., Monaco, no. spécial 2:377.

Miller, A.R., Tchernia, P., and Charnock, H., 1970, Mediterranean Sea atlas of temperature salinity, oxygen. Profiles and data from cruises of R.V. Atlantis and R.V. Chain, with distribution of nutrient chemical properties by D.A. McGill, Woods Hole, Woods Hole Oceanographic Institution. (The Woods Hole Oceanographic Institution atlas series:3).

Milne-Edwards, A., 1861, Observations sur l'existence de divers mollusques et zoophytes à de très grandes profondeurs dans la mer Méditerranée, Ann. Sc. Nat. Zoologie, (4) 15:149.

Monniot, C., 1969, Ascidies récoltées par la THALASSA sur la pente du plateau continental du Golfe de Gascogne (3-12 août1967), Bull. Mus. nation. Hist. Nat., 2e série, 41:155.

Monniot, C., et Monniot, F., 1974, Ascidies abyssales de l'Atlantique récoltées par le JEAN CHARCOT (campagnes NORATLANTE, WALDA, POLYGAS A), Bull. Mus. Nation. Hist. Nat., 3e série, 226:721.

Monniot, C., et Monniot, F., 1975, a. Ascidies abyssales de Méditerranée récoltées par le JEAN CHARCOT (campagnes POLYMEDE), Bull. Mus. Nation. Hist. Nat., 251:1353.

Monniot, F., et Monniot, C., 1975, b. Sept espèces d'Ascidies profondes de Méditerranée, Bull. Mus. Nation. Hist. Nat., 330:1117.

Monod, Th., and Hureau, J-C, 1977, Introduction, Ann. Mus. d'Hist. Nat. Nice, 5:5.

Nairn, A.E.M., Kanes, W.H., and Stehli, F.G., 1978, The ocean basis and margins. 4B: the western Mediterranean, Plenum Press, New York and London.

Nielsen, J.N., 1912, Hydrography of the Mediterranean and adjacent waters. Report on the Danish oceanographical expeditions 1908-1910 to the Mediterranean and adjacent seas, 1,2: hydrography:77.

Ovchinnikov, I.M., et Plakhin, E.A., 1965, Revue historique des recherches hydrologiques en Méditerranée, in: "Traits fondamentaux de la structure géologique, du régime hydrologique et de la biologie de la Méditerranée," ed. Nauka, Moscou.

Pérès, J.P., 1961, "Oceanographie biologique et biologie marine. 1: la vie benthique," Presses universitaires de France, Paris.

Pérès, J.M., et Picard, J., 1964, Nouveau manuel de bionomie benthique de la Mer Méditerranée, Rec. Trav. Stat. Mar. Endoume, 47:3.

Perrier, E., 1886, "Les explorations sous-marines," Librairie Hachette et Cie, Paris.

Petit, G., et Laubier, L., 1962, Les canyons de la côte catalane. Aperçu de nos connaissances et programmes de recherches, in: "Oceanographie géologique et géophysique de la Méditerranée occidentale," C.N.R.S., ed., Paris.

Picotti, M., 1968, L'océanographie méditerranéenne au 18e et 19e siècles, Bull. Inst. Océanogr. Monaco, (Congrès international de l'histoire de l'océanographie), 1:317.

Pruvot, G., 1894, Essai sur la topographie et la constitution des fonds sous-marins de la région de Banyuls, de la plaine du Roussillon au Golfe de Rosas, Arch. Zool. expér. gén., 3:599.

Reyss, D., 1961, Lagisca drachi n. sp., Polychète Aphroditidae récoltée dans le rech Lacaze-Duthiers, Vie Milieu, 12:473.

Reyss, D., 1970, Bionomie benthique de 2 canyons sous-marins de la mer catalane: le rech du Cap et le rech Lacaze-Duthiers, Thèse Sc. nat., Univ. Paris.

Reyss, D., 1971 (1972), Résultats scientifiques de la campagne POLYMEDE II. Polychètes Aphroditidae de profondeur en Méditerranée. Remarques systématiques et biogéographiques, Vie Milieu, 22:243.

Richards, F.A., 1957, Oxygen in the ocean, Mem. geol. Soc. America, 67 (Treatise on marine ecology and paleoecology), 1:185.
Risso, A., 1810, "Ichthyologie de Nice ou histoire naturelle des poissons du département des Alpes Maritimes," F. Schoell, ed., Paris.
Risso, A., 1816, "Histoire naturelles des crustacés des environs de Nice," Librairie-grecque-latine-allemande, Paris.
Risso, A., 1820, Mémoire sur un nouveau genre de poisson nommé Alépocéphale vivant dans les grandes profondeurs de la mer de Nice, Mem. Real. Accad. Sc. Torino, 25:270.
Risso, A., 1826, "Histoire naturelle des principales productions de l'Europe méridionale et particulièrement de celles des environs de Nice et des Alpes Maritimes", Levrault, Paris, Strasbourg,
Risso, A., 1827, Observations sur quelques nouvelles espèces de Crustacés de la mer de Nice, Nova Acta Acad. Caes. Leopol.-Car. germ. Nat. cur., 13:817.
Rossi, L., 1949-50, Celenterati del golfo di Rapallo (Riviera Ligure), Boll. Ist. Zool. Univ. Torino, 2:193.
Rouch, J., 1946, "La Méditerranée," Flammarion, Paris.
Rouch, J., 1952, Recherches océanographiques en Méditerranée avant 1914, Bull. trim. Am. Mus. Océanogr. Monaco, 21:5.
Ruffo, S., 1892, The Amphipoda of the Mediterranean. Part I: Ganamaridae (Acanthonotozomatidae to Gammaridae), Mém. Inst. Océanogr. Monaco, 13:364.
Ruggieri, G., 1967, The Miocene and later evolution of the Mediterranean Sea, (Aspects of Tethyan biogeography), Systemat. assoc. publ.
Ryan, W., Olausson, E., and Fairbridge, R.W., 1966, Mediterranean Sea, in: "The encyclopedia of oceanography," R.W. Fairbridge, Reinhold publ. Comp., New York.
Ryan, W.B.F., Stanley, D.J., Hersey, J.B., Faihlquist, D.A., and Allan, T.D., 1970, The tectonics and geology of the Mediterranean Sea, in: The Sea 4: New concepts of sea floor evolution (II): Regional observations concepts," A.E. Maxwell, ed., Wiley Interscience, New York.
Santucci, R., 1932, Per la conoscenza del Polycheles tythlops Heller del Mediterraneo, Bull. Mus. Lab. Zool. Anat. Comp. Univ. Genova, 12:56.
Selli, R., and Fabbri, A., 1971, Tyrrhenian: a Pliocene deep sea, R.C. Accad. Lincei, 8:580.
Sevastyanov, V.F., 1968, Redistribution of chemical elements as a consequence of oxidation-deoxidation processes in sediments of the Mediterranean Sea, Litologie i Poleznye Iskopayeme (lithology and mineral resources), 1:3.
Stanley, D.J., 1972, "The Mediterranean Sea: a natural sedimentation laboratory," Dowden, Hutchinson et Ross, Stroudsburg.
Stocks, T., 1938, Morphologie des Atlantischen Ozeans. Statistik der Tiefenstufen der Atlantischen Ozeans. Deutsche Atlantische Expedition, Meteor. 1925-1927, Wiss. Ergeb., 3:35.

Tyler, P.A., and Gage, J-D, 1979, Reproductive ecology of deep-sea Ophiuroids from the Rochall through, in: "Cyclic phenomena in marine plants and animals," E. Naylor and R.G. Hartnoll, eds., Pergamon Press, Oxford, New York.
Tortonese, E., 1970, "Osteichthyes (Pesci Ossei)," Fauna d'Italia, Calderini, Bologna.
Vacelet, J., 1969, Eponges de la roche du large et de l'étage bathyal de Méditerranée (récoltes de la soucoupe plongeante COUSTEAU et dragages), Mem. Mus. Hist. Nat., 59:145.
Vaissière, R., et Fredj, G., 1964, Contributions à l'étude bionomique de la Méditerranée occidentale (Côte du Var et des Alpes Maritimes - Côte occidentale de Corse). 5: Etude photographique préliminaire de l'étage bathyal dans la région de St. Tropez (ensemble A), Bull. Inst. Océan. Monaco, 64:1323.
Vamvakas, C., 1970, Peuplements benthiques des substrats meubles du sud de la Mer Egée, Téthys, 2:89.
Vervoort, W., 1966, Bathyal and abyssal Hydroids, Galathea report:97.
Wüst, G., 1961, On the vertical circulation of the Mediterranean Sea, J. Geoph. Res., 66:3261.
Zibrowius, H., 1977, Review of Serpulidae (Polychaeta) from depths exceeding 2 000 m, in: "Essays on Polychaetous Annelids in memory of Dr Olga HARTMAN," D.J. Reish et K. Fauchald, eds., The Allan Hancak Foundation, University of South California, Los Angeles.
Zibrowius, H., 1980, Les Scléractinaires de la Méditerranée et de l'Atlantique nord-oriental, Mém. Inst. Océanog. Monaco, 11:284.

THE EASTERN MEDITERRANEAN SHELF ECOSYSTEM IN GLOBAL CONNEXION INCLUDING SOME BIOLOGICAL AND GEOLOGICAL IMPLICATIONS

G. Høpner Petersen

Zoological Museum, University of Copenhagen
Universitetsparken 15, DK-2100 Copenhagen Ø, Denmark

INTRODUCTION

The Mediterranean is not only a sea between continents, it is also a sea between oceans. In attempts to formulate general rules in marine biology, the Mediterranean is an important test area.

In my comparative studies on energy flow in subarctic, temperate and tropical ecosystems, a quantitative model for the eastern Mediterranean shelf was constructed. The Mediterranean model supported the general trend I had previously found, but it also underlined the lack of information from many components of the Mediterranean ecosystems.

The results from these comparative studies on energy flow showed that the benthic pathway is favoured in subarctic environments whereas the pelagic pathway is favoured in tropical environments. I define a subarctic sea to be a sea in which sea ice is formed every winter, and thus include not only the seas in the North Atlantic and North Pacific, but also the northern part of the Baltic and of the Caspian Sea. A tropical sea has a constant high temperature and a constant thermocline. The energy flow in the eastern Mediterranean resembles that of a tropical sea, while that in the northern Adriatic resembles the northern European seas.

METHODS

The ecosystems selected are defined as water columns from different shallow, marine continental shelf regions with a sediment bottom at, or just below, the base of the photic zone (about 50 m depth). Additional selection criteria were: the entire primary production should be pelagic, and hence there should be no benthic plants; decomposition in the aphotic layer should be negligible; the ecosystem should not be influenced by pollution, freshwater upwelling or exchange of organic material with other ecosystems (Petersen and Curtis, 1980). The study was later extended to freshwater lakes which fulfilled the same requirements. It was also extended to ecosystems over deeper water, i.e. the continental slope, where I suggested that the soundscattering layer or "false bottom" replaces the benthic fauna (Petersen, in press). A further important requirement was that comparable production measurements were obtainable for the more important ecosystem steps from incident solar energy to fish yield to man. Fig. 1 shows a model of a fully-developed aquatic ecosystem with the major steps and flows; the particular steps selected for this study are indicated.

The production measurements are given in kilocalories per square meter per year. The incident solar energy was usually measured directly by photometer. The pelagic primary production was in most cases measured by the 14-C method. Zooplankton production was usually calculated from standard hauls, and zoobenthos was estimated from quantitative bottom samples. Fish production is generally taken as the sustainable fish yield to man, estimated from fish statistics. The conservative conversion factors are: 1 g wet organic weight = 0.25 g dry weight, and 1 g carbon in organic matter = 10 kcal.

RESULTS

The trend for both the marine shelf and the freshwater lakes is that energy mainly flows through the benthic link in the subarctic and through the pelagic link in the tropics and subtropics (Figs 2-4). In the high arctic, where sea ice is present also during the summer, a kryopelagic ice biota is developed, which can be considered as an "inverted bottom community" with relatively little production in the water mass column itself. The food chain is, for instance, ice algae - amphipods - polar cod - ringed seal - polar bear. The high arctic ecosystem will not be considered further mainly because of lack of data. For the slope ecosystem over greater depths, more data must be obtained before a partition of the energy flow between a "sinking, benthic" link and an "in-situ pelagic" link can be made. However, some baleen whales may be examples of vertebrate, top predators which feed on "the false bottom".

The quantitative trend that the benthic link dominates in the subarctic, and the pelagic link in the tropics is also supported by the evolution of vertebrate, top predators. The subarctic has many predators that feed on benthos, for instance, walrus, bearded seal, the common eider, the king eider, several species of flatfishes, wolffishes, codfish, scorpionfish, etc. Such species are rare or absent in tropical waters. On the other hand common, vertebrate,

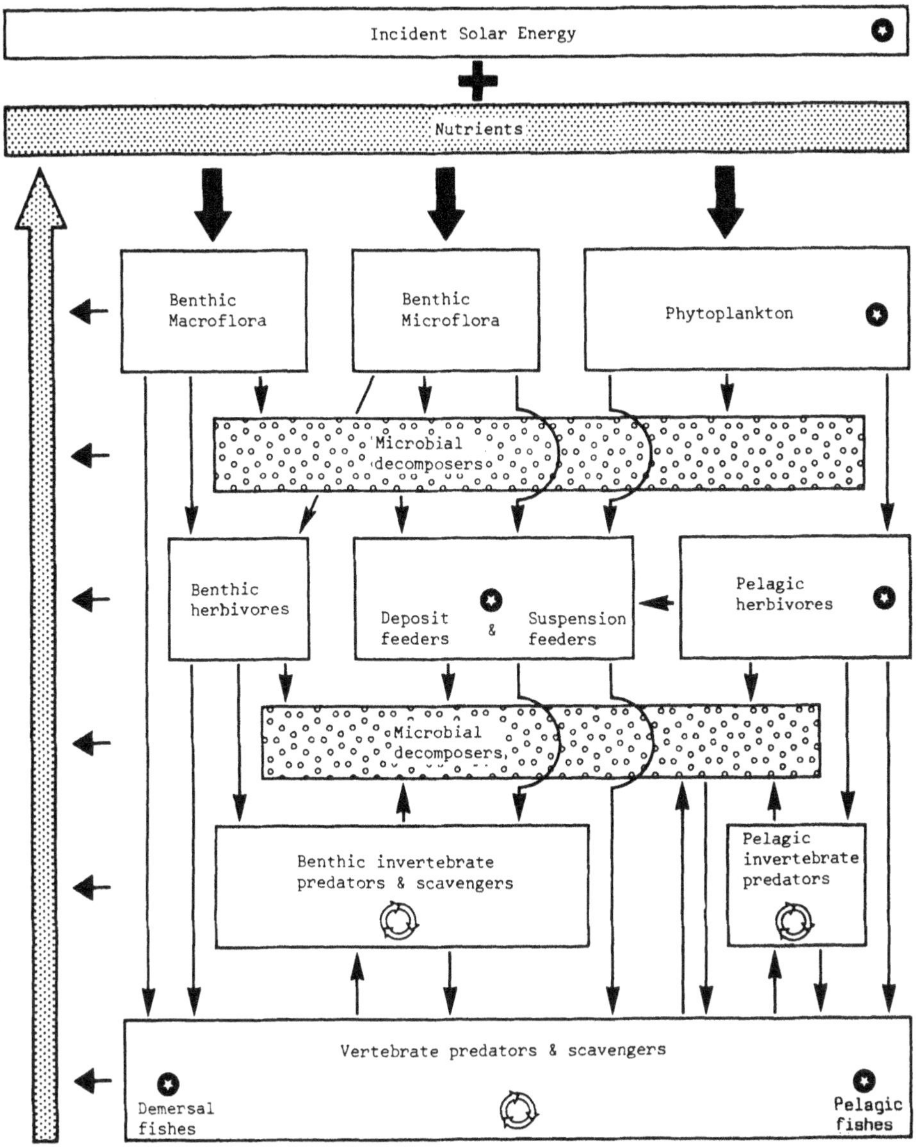

Fig. 1. A simplified model of major components in a marine ecosystem. The three large black arrows signify the input of organic energy, while the horizontal arrows depict the release of nutrients. A circle of arrows represents recycling within a component through predation and/or cannibalism. The remaining arrows indicate the important transfer pathways. The ecosystem components selected for the energy flow models in Figs 2,3 and 4 are marked by a white star on a black circle (Petersen and Curtis, 1980).

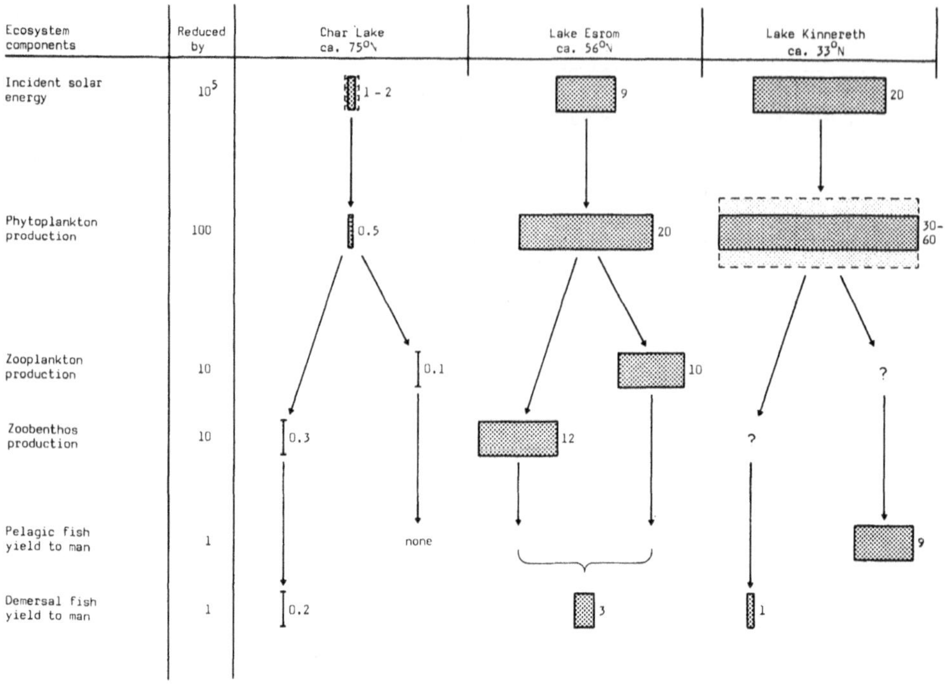

Fig. 2. Energy flow models for lakes from arctic, temperate and subtropical regions. All values expressed as kcal/m²/yr.

top predators in tropical waters include dolphins, pelicans, "sardine-like" fish, "mackerel-like" fish, tuna, etc., which all utilize a pelagic food source. This kind of top predators is rare or absent in the arctic.

Hypothesis to Explain the Observed Differences

The explanation for a different path of energy flow in subarctic and tropical waters must be sought in basic energy budgets that balance differently under different conditions. However, it is difficult to get reliable information, and convincing models, from the sources available today. Even important physiological experiments are in contradiction with each other. The following assumptions are preliminary, but hopefully they may lead to further research on the weaker parts of the argument.

1. Photosynthesis is only weakly-dependent on temperature (Fig. 5). It is mainly dependent on light conditions and nutrient supply.

2. Metabolism depends mainly on temperature. The physiological experiments are ambiguous. There may be some temperature adaptation, but in general, the metabolic rate in tropical

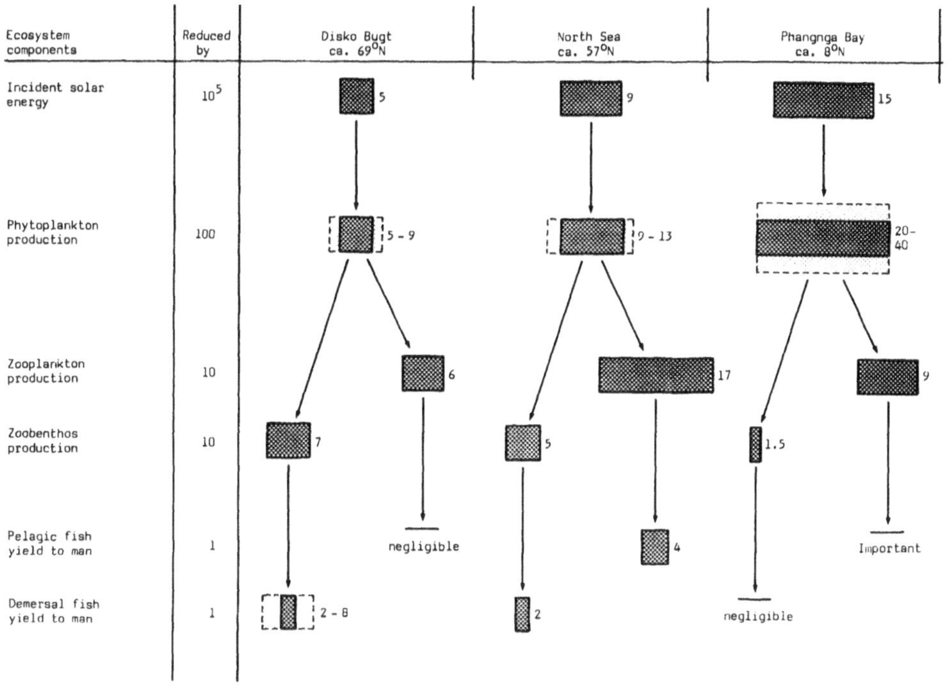

Fig. 3. Energyflow models for shelf ecosystems from subarctic, temperate, and tropical regions. All values expressed as $kcal/m^2/yr$.

waters seems to be from 4 to 6 times that in arctic waters. For a discussion see Clarke (1980). However, it is obvious that subarctic animals are slow and the tropical animals are fast. Moreover, arctic organisms are generally larger than their tropical counterparts. Both observations would mean higher metabolism in tropical waters (Fig. 5).

3. In cold waters many organisms have developed life-forms, life-strategies and modes of life that allow them passively to collect the food and oxygen they require. Thus the filter-feeding invertebrates seem able to utilize the hydrodynamic forces created around their body and siphon by the many different movements of tides, waves, currents, etc. Also sedimentation and buoyancy changes may be used to secure a passive food supply at low energy cost (Petersen, 1966; Petersen, 1978; Vogel, 1981).

Consequences of the different energy budgets

From the above assumptions one can tentatively predict 1) how

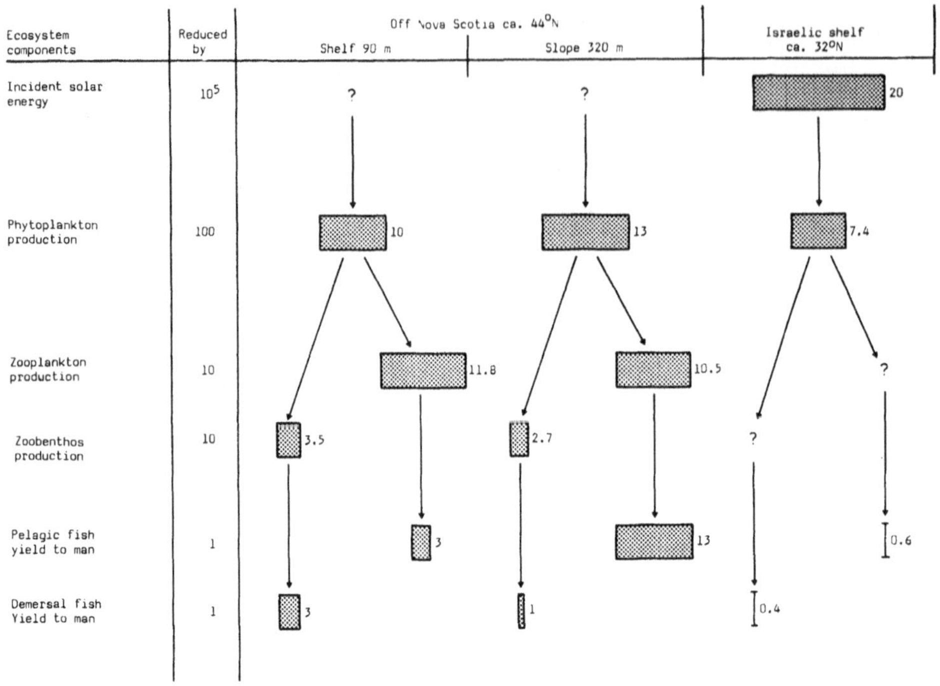

Fig. 4. Energyflow models for a shelf-slope locality (off Nova Scotia) and for the Israeli shelf in the eastern Mediterranean. All values expressed as $kcal/m^2/yr$.

ecosystems could be designed in different environments and 2) which solutions would be impossible. Thus, both in the ecosystem structure and in the survival and evolution of species one may talk of "energetically-possible niches" and "energetically-impossible niches". Different species may try to utilize the same "energetically-possible niches". Each species, each trophic level, each energetic niche is the result of complicated cost-benefit analyses. Several solutions may be possible, as can be seen in the different seas and oceans. A few solutions will be outlined. The subarctic and the tropical situations are taken as fixed points. The temperate, boreal and subtropical situations are intermediate, while the ice biotas (under shelf ice, near ice margins and in the kryopelagic ecosystems) are extreme situations in polar seas. In a typical subarctic situation, e.g., at 70°N, the incident solar energy varies from a polar night lasting 6 weeks to a polar day of a similar length (Fig. 6). This energy (both the heat energy and the photosynthetically-fixed energy) propagates through the ecosystem with a considerable time lag. The sun returns in January and solar energy increases considerably during the spring, but sea ice prevents this energy from reaching the water mass until late April, when sea ice usually disappears. The heating (and melting) create a strongly-stable water column through which heat penetrates only slowly. At a depth of 50 m the water begins

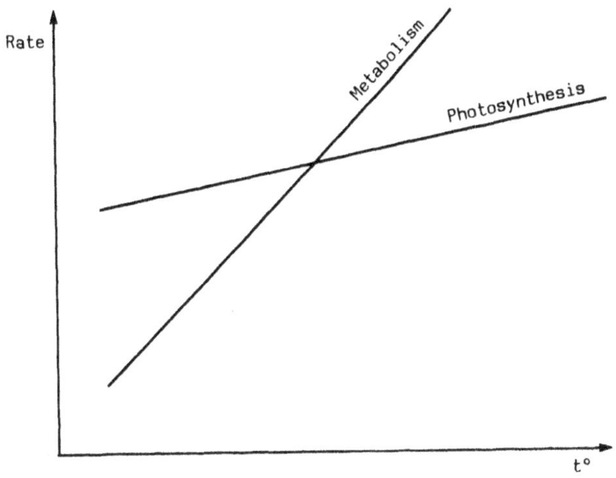

Fig. 5. The influence of temperature on the rates of photosynthesis and metabolism. The units are arbitrary.

to warm up in August and reaches a maximum of about 4°C in October, just before the vertical mixing in the autumn destroys the stabilized layer. When sunlight reaches the water, photosynthesis starts immediately, producing the spring diatom bloom.

The spring diatom bloom is grazed by the pelagic larvae of many benthic species, whose spawnings are synchronized with the diatom bloom. Subarctic plankton becomes dominated by meroplankton during spring and summer. Holozooplankton do not utilize the spring diatom bloom immediately because of the time required for them to build up a sufficiently large population. Subarctic benthos mainly consists of large old animals, and the most important species for biomass and production have planktotrophic larvae. The rule that the number of species with planktotrophic larvae decreases as one goes from tropics to arctic waters is correct; however, it is important to remember that biomass and production in the subarctic are dominated by the few species that do have planktotrophic larvae. Subarctic benthos has a relatively low metabolism in the cold, bottom water, but can utilize hydrodynamic forces for passive food collecting and thus conserve energy for growth and reproduction, even if the concentration of algae in the water mass is low. The subarctic, benthic biofilter consists of shallow burrowers on sediment bottoms and of a luxurious growth of sponges, hydroids, corals, tunicates, barnacles, etc. on rocky bottoms. Over deeper water the soundscattering layer or "false bottom" can be envisaged as a huge dipnet that is moved up and down by buoyancy changes.

In the typical equatorial situation (Fig. 7), day and night vary regularly all year round, the water temperature is always about 25-30°C and the thermocline is below the phototrophic layer. In

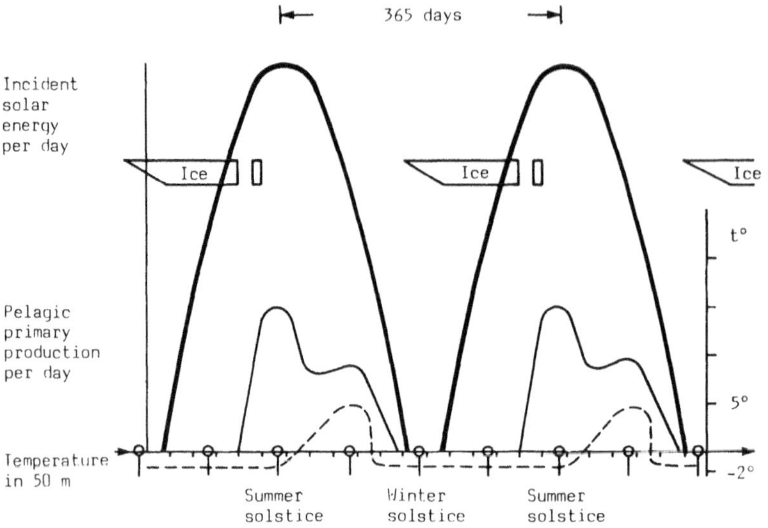

Fig. 6. The subarctic sea. The variation in incident solar energy (thick line) during a two-year period, showing the subsequent variation in pelagic primary production (thin line), sea-ice cover and temperature (broken line) at about 50 meters depth. The variation in solar energy is just as stable and just as old as the tropical variation, if current geophysical models are correct.

this situation the algal metabolism is probably the most serious decomposer important component of the degradation of photosynthetically-fixed energy. Hence in tropical waters it is important for the herbivores to be as close to the plants as possible, that is, to have zooxanthellae, or to be pelagic and fast. The high metabolism forced upon the animals by the high temperature is best utilized by fast active food collecting. The metabolic losses are, however, great, and not much is left over for bottom animals, sea birds and fish yield to man. Organic matter seems to accumulate at the surface of tropical water, where it nourishes the neuston, Halobatidae, etc., whereas on sandy beaches active species like Donax, Emerita, etc., are found. In tropical water it is also profitable to live in holes or caves, to be a borer in wood or soft rocks, or to be a deep burrower, all life forms known as kryptobiology. Life strategies such as those of symbiotic algae, neuston, and cryptobionts are rare or absent in subarctic waters.

The change in the life form "from boring to burrowing" as one goes from tropical to subarctic waters indicates when it can be profitable to change from active pumping to passive filter-feeding. Woodborers are very common in driftwood in tropical waters while they are absent from arctic waters even if driftwood is abundant.

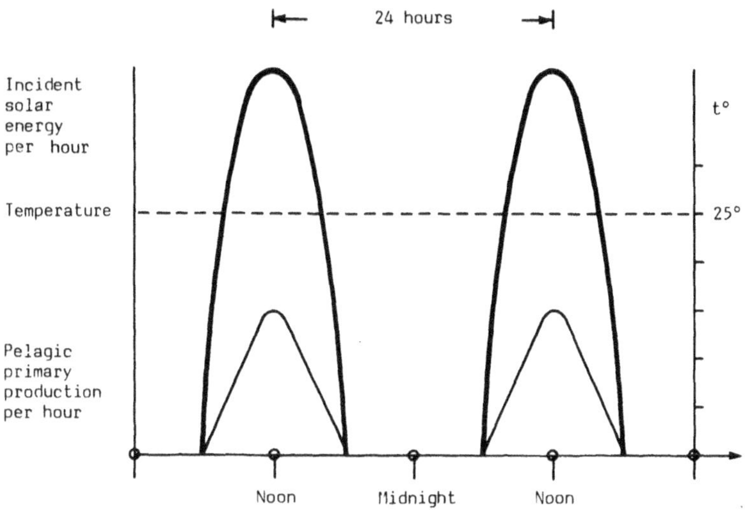

Fig. 7. The tropical sea. The variation in incident solar energy (thick line) during a two-day period, showing the almost constant high temperature (broken line) and negligible time lag for photosynthesis (thin line).

In temperate waters the bivalve Mya truncata is a deep burrower, while it is a shallow burrower in the subarctic. The genus Hiatella in temperate waters is represented by species which prefer caves and holes. In the subarctic it is represented by Hiatella byssifera, which lives freely on the bottom, attached only by the byssus, and is an important prey for the eiders. Thus, "to be a borer or a deep burrower" becomes an energetically impossible niche in the subarctic, obviously because this lifeform cannot control the hydrodynamic forces involved.

In tropical seas the very active crabs, hermitcrabs and other Decapoda are the dominant crustaceans, while amphipods and isopods are important crustaceans in polar seas. Similarly, the errant polychaetes dominate in tropical seas and sedentary polychaetes dominate arctic seas (Rasmussen, 1973:458, Fig. 119).

It is obvious that each energy niche is occupied by suitable life forms and that the same energetically-profitable niches are explored by several taxonomic groups that have evolved similar life forms. However, a full description of such niches and their corresponding life forms is still impossible.

Figs. 8 and 9 give the major patterns in the subarctic and tropical marine ecosystems, including some features from the tidal zone. I have tried to get information from other seas, for instance the Caspian Sea, which may have a gradient from subarctic to subtropical conditions and contains a rather unique fauna. It was

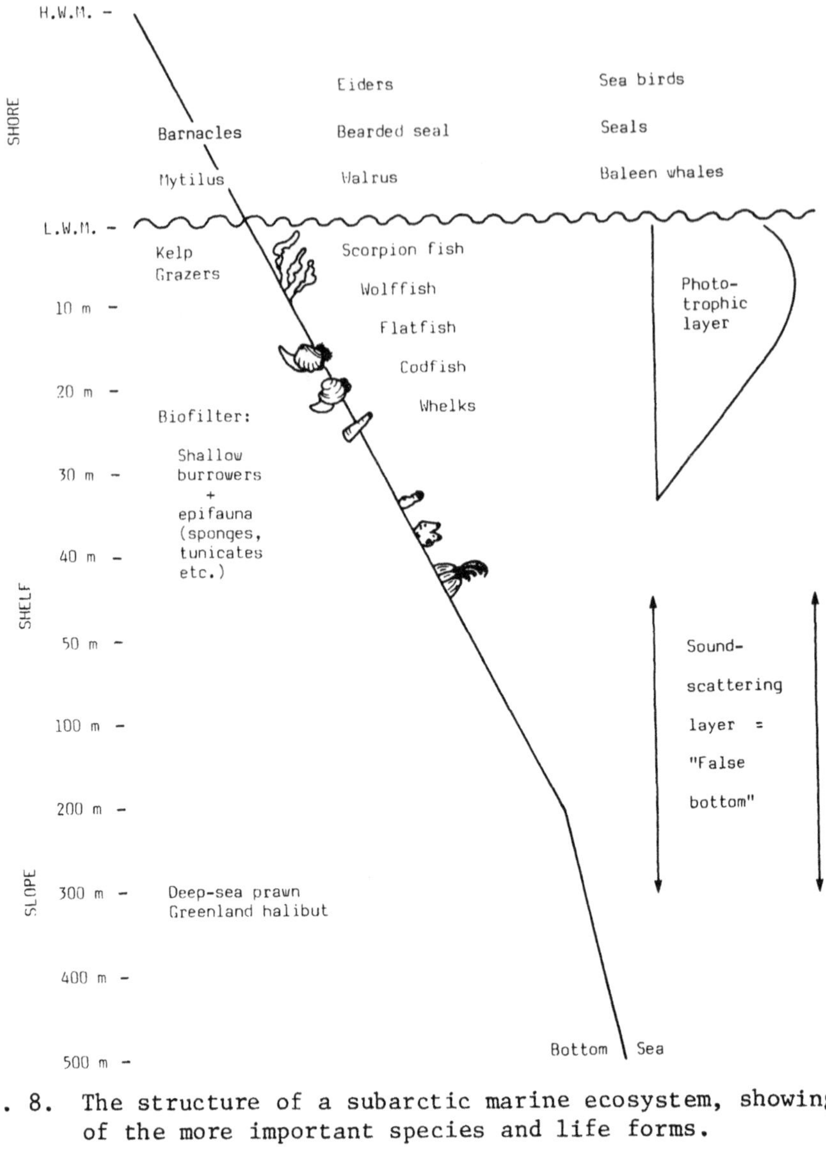

Fig. 8. The structure of a subarctic marine ecosystem, showing some of the more important species and life forms.

difficult to obtain quantitative data for a complete ecosystem model, but I did not find serious problems in the data that were available, and I was able to confirm that the ideas presented here have a general validity in different seas and for different faunas. The difference in number of species between the subarctic and the tropics cannot be included in Figs 8 and 9. The general statement that tropical ecosystems are more diverse than those in the arctic is too simple. For many taxonomic groups the opposite is true. A more general rule seems to be that the critical, minimum population size for the

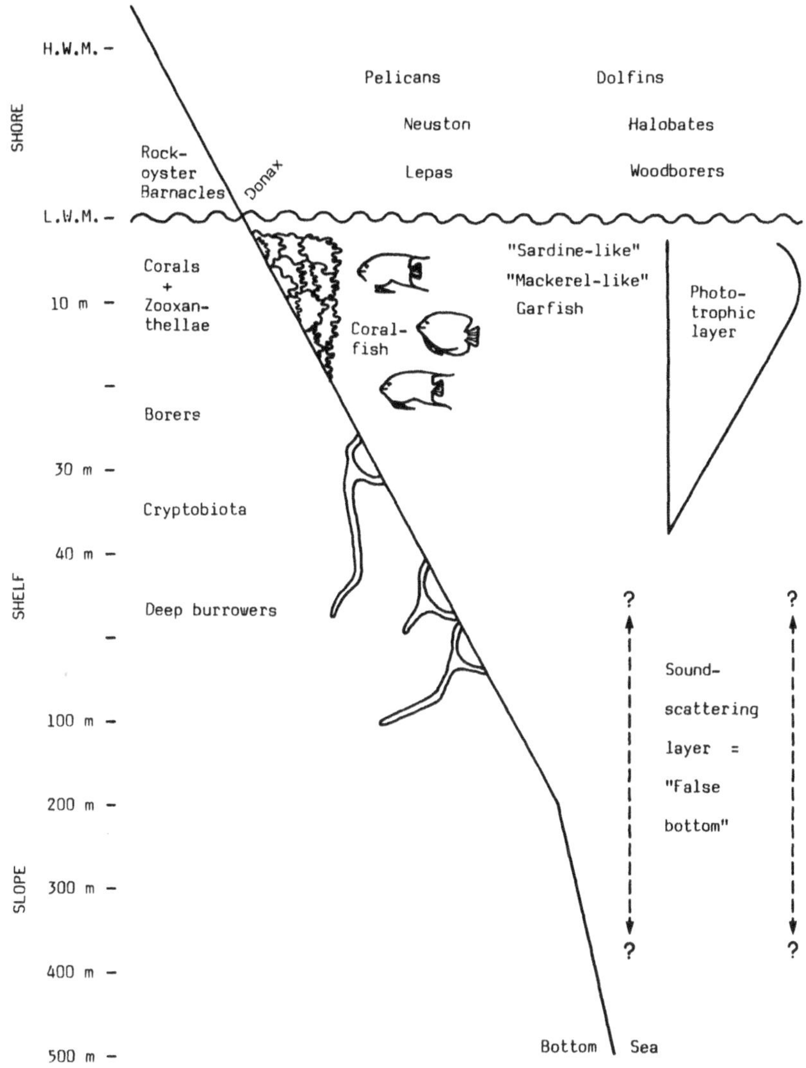

Fig. 9. The structure of a tropical ecosystem, showing some of the more important species and life forms.

survival of the species is larger where there are large variations in energy flow.

Consequences for Mediterranean Research

The Mediterranean has often been an obstacle for attempts to arrive at general marine biological rules. For example the results derived from numerous works on rocky shore zonation and wadden seas are of little use in the Mediterranean because of the absence of tides except for small areas off Gabès and Venice. The idea of "parallel level-bottom communities" (Thorson, 1951) failed in tropical waters, but it also fails in the Mediterranean for the same reason, namely, that the infauna is too scattered, even at times almost absent (Gilat, 1963; Gorenshtain-Galil, 1978; Micallef and Evans, 1968; Pérès, 1967). Quantitative grabs are insufficient tools for collecting animals. The fast epipelagic animals of the warm waters must be caught by trawls or dredges. The benthos simply does not occupy the same ecological niches as in colder waters. An exception is the northern Adriatic, which in this respect also shows its relationship to northern European seas. Thus it is influenced by cold, brackish water during winter and spring. Vatova (1949) erected, e.g., the "zoocoenoses" of Venus gallina and of Syndosmya alba in the northern Adriatic and these were subsequently considered to be parallel to the "Venus community" and the "Syndosmya community" of the Kattegat and North Sea. The resemblance between these communities is more likely due to the fact that some of the Mediterranean species can penetrate to the north in intermediate depths of about 5 to 30 meters, where neither winter cooling nor deep cold water are important. In Danish shallow water, where the animals must be able to survive sea ice cover, some Mediterranean species (e.g., Cardium glaucum) coexist with some northern species (e.g., Macoma baltica) which makes the "parallelism" for shallow water bottoms less obvious. The problem becomes even more complicated when the difference between neritic and oceanic species is considered (Petersen, 1968).

The Mediterranean benthos is treated extensively by Pérès, 1967,1982) and Gamulin-Brida (1967). These authors have erected an impressive system of zonation and biocoenoses. The system seems, however, difficult to use outside the Mediterranean, just as systems from other seas can be difficult to apply to conditions in the Mediterranean. Similarly, ideas on diversity and stability developed from work on plankton ecosystems in the Mediterranean region (see Margalef, 1968) are difficult to apply to subarctic plankton with its considerable impact of meroplankton in the short summer season.

A widely-used biogeographical method for comparing different faunas was developed using Mediterranean molluscs (Philippi, 1844), and the method has been used to divide the Mediterranean into several biogeographical regions (Pérès, 1967). In the North Sea the bivalve genera with a Mediterranean-Lusitanean origin have a simple taxonomy while those with a North Pacific origin have a complicated taxonomy. This was supposed to be a result of the different geographical separations during the Quaternary ice-ages (Petersen, 1977).

For fishery management the models offer energetic explanations for the failure of trawl fisheries in warm waters and the risks

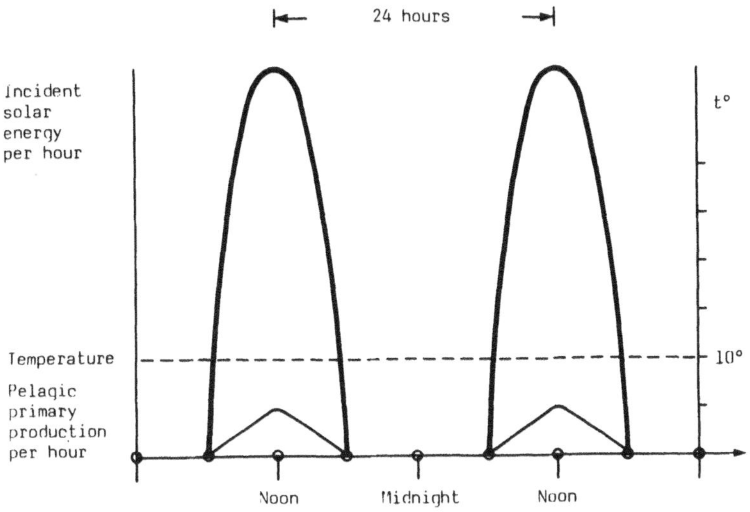

Fig. 10. Hypothetical situation for a tropical mountain lake with high incident solar energy (thick line) over a short period, low constant temperature (broken line), and low pelagic primary production (thin line) without a timelag. This situation probably occurs in tropical oligotrophic mountain lakes, but sufficient data on energy flow are lacking.

for pelagic fisheries in cold water. Such considerations can be applied to the different regions of the Mediterranean as well.

Some geological implications

The ideas represented in Fig. 8 and Fig. 9 are only at their initial stage and also have to be tested against palaeontological observations. On the other hand, they may be helpful for future studies on palaeoproduction. However, when studying fossils it should be remembered that deposition of calcium carbonate is related to high temperature and high metabolism and sometimes seems to be a burden or a waste product for tropical organisms. The deposition of chalk is not necessarily related to a high yield of organic matter.

The implications for palaeoclimatology depend on which parameter is supposed to be the most important, the temperature level or the variation in energy input. If the temperature level as such is the decisive parameter for the difference between tropical and subarctic waters, then the model can be used for considerations on general global heating and cooling. However, the argument of the present paper is that the structure of the subarctic ecosystem is

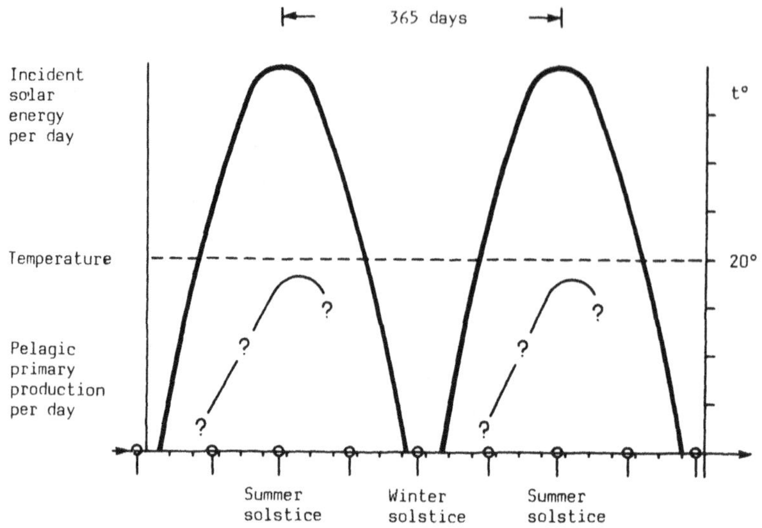

Fig. 11. Hypothetical situation for a warm polar sea, which should be the case in, e.g., Palaeocene times. The temperature (broken line) is high due to high heat flow from the core or to an enormous heat transport with ocean and wind currents from the equatorial belt. Pelagic primary production (thin line) could occur during summer and be grazed by migratory species. Most probably a warm polar sea will be a desert due to rapid mineralisation. However, sessile plants might depend on heterotrophy during winter, which may be the case with some recent subarctic laminarians. The benthic fauna might be similar to the present-day fauna near thermal vents. The life forms necessary for exploration of the possible energetic positions in a warm sea with polar darkness will be very different from those known today. These possible existence of such situations is not supported by geological evidence, fossils or palaeohydrography.

a consequence of the variation in the incident solar energy, with a period of 365 days and with a considerable time lag in the energy flow, while that of the tropical ecosystem is due to the 24-hour variation in incident solar energy and much smaller time lag. Figs 7 and 8 give a graphical representation of the different situations. Figs 10 and 11 give the hypothetical situations of equatorial solar energy and low temperature, and of subarctic light energy and high temperature. Furthermore, it is true that fossils indicating subarctic environments (walrus, eider, flatfish, Buccinum snails) are no more than 20-30 million years old (high-arctic not more than few mill. years old), and tropical fossils are more than 100 million years old. It follows that the earth's axis of inclination has changed considerably during the Tertiary. This supports the conclusions from fossil finds on Ellsmere island but is in conflict with the geophysical models for earth rotation and for plate tectonics (Donn, 1982; Petersen, in press). These are, of course, far-reaching deductions based on a handful of quantitative models from different seas, but they are also supported by studies of climate and geomorphology (Büdel, 1977), mainly done in the Mediterranean region and compared with similar studies in Greenland (personal communication from Ole Humlum, Arctic Station Godhavn).

The deposits of salt tell the dramatic history of the Mediterranean. However, the Mediterranean has also been exposed to climatic changes during the Quaternary, and several times it has been colder than it is today. Can Mediterranean fossils demonstrate the palaeoproduction and palaeostructure of the marine ecosystem ? Do the extinction, evolution and immigration of species fit into a picture of energetic possibilities changing according to the climate? These questions and others still remain unanswered.

SUMMARY

The general trend for aquatic ecosystems from different climatic zones that benthic production dominates in the subarctic-arctic and pelagic production dominates in the subtropical-tropical is valid also for the Mediterranean. The exaplanation for this difference is sought in the different cycles of the solar energy and in the timelags for their propagation through the ecosystem, and not only in the temperature difference. Tentative explanations for "energetically-possible niches" and "energetically-impossible niches" are given. Designs of subarctic and tropical ecosystems are shown, based on cost-benefit considerations for photosynthesis, metabolism and passive food collecting aided by hydrodynamic forces. The importance for palaeoecology is stressed and the conflict between biology and geophysics concerning tilt of the earth's axis is demonstrated by considerations of life conditions in a warm, dark, polar sea.

REFERENCES

Büdel, J., 1977, "Klima-Geomorphologie," Gebrüder Borntraeger, Berlin, Stuttgart.

Clarke, A., 1980, A reappraisal of the concept of metabolic cold adaptation in polar marine invertebrates. Biol. J. Linnean Soc., 14:77.

Donn, L., 1982, The enigma of high-latitude palaeoclimate, Palaeogeography, Palaeoclimatology, Palaeoecology, 40:199.

Gamulin-Brida, H., 1967, The Benthic Fauna of the Adriatic Sea, Oceanogr. Mar. Biol. Ann. Rev., 5:435.

Gilat, E., 1963, Methods of study in marine benthonic ecology. Com. int. Explor. sci. Mer Medit., Colloque Com. Benthos: Méthodes quant. Et. Benthos échel. dimension, benthontes. Marseille, Novembre 1963:7.

Gorenshtain-Galil, B., 1978, Macrobenthal Communities of Palmahim-Nizanim Area (Israel), Tel Aviv University, unpublished thesis.

Margalef, R., 1968, "Perspectives in Ecological Theory," The University of Chicago Press, Chicago.

Micallef, H. and Evans, F., 1968, "The Marine Fauna of Malta", Malta University Press.

Pérès, J.M., 1967, The Mediterranean Benthos, Oceanogr. Mar. Biol. Ann. Rev., 5:449.

Pérès, J.M., 1982, Zonation and Organismic Assemblages, Marine Ecology, 5:9.

Petersen,G.H.,1966, Balanus balanoides (L.) (Cirripedia) life cycle and growth in Greenland, Meddelelser om Grønland, 159:114.

Petersen,G.H.,1968,"Marine Lamellibranchiata. The Zoology of the Faroes,"III part I.

Petersen,G.H.,1977, The Density, Biomass and Origin of the Bivalves of the Central North Sea, Meddelelser fra Danmarks Fiskeri - og Havundersøgelser. N.S.7:221.

Petersen,G.H.,1978, Life cycles and population dynamics of marine benthic bivalves from the Disko Bugt area of West Greenland, Ophelia, 17:95.

Petersen,G.H.,(in press), Energy flow in comparable aquatic ecosystems from different climatic zones, Intern. Council for the Exploration of the Sea. Symposium on Biological Productivity in the Temperate Zone of the North Atlantic (Kiel, 2-5 March, 1982).

Petersen,G.H.,(in press), Energy flow budgets in aquatic ecosystems, and the consequences on a conflict between biology and geophysics about earth axis tilt. II Nordic Symposium on Climatic Changes, Stockholm, May 1983.

Petersen,G.H.,and Curtis, M.A., 1980, Differences in energy flow through major components of subarctic, temperate and tropical marine shelf ecosystems, Dana, 1:53.

Philippi, R.A., 1844, "Fauna Molluscorum, viventium et in tellure tertiaria fossilium," Regni utriusque Siciliae. S.E. Anton Verlag, Halis Saxonum.

Rasmussen, E., 1973, Systematics and ecology of the Isefjord Marine Fauna (Denmark), Ophelia, 11:1.

Thorson, G., 1951, Animal communities of the level sea bottom, Année Biologique, 27:481.

Vatova, A., 1949, La fauna bentonica dell' Alto e Medio Adriatico, Nova Thalassia, 1(3):1.

Vogel, S., 1981, "Life in moving fluids, the physical biology of flow," Willard Grant Press, Boston, Mass.

EFFECTS OF POLLUTION AND MAN-MADE MODIFICATIONS ON MARINE BENTHIC COMMUNITIES IN THE MEDITERRANEAN: A REVIEW

Gérard Bellan

Station marine d'Endoume, 13007 Marseille, France

GENERALITIES

Introduction

When in 1970, at an F.A.O. International Conference on Pollution of the Sea, Bellan and Pérès (1972) tried to assess certain general aspects of pollution in the Mediterranean, they found only two geographical sectors where pollution studies had been conducted in a comprehensive manner, from the point of view of sites studied and importance of pollution: the regions of Marseille on the French coast and Trieste in the upper Adriatic. Everywhere else, the data were limited in scope, as shown at the FAST ("Acqua per il domani") conference held the previous year in Milan, and in which one of the authors had participated (Pérès and Bellan, 1970). However, one of their principal conclusions was that it would be easy to evaluate the actual state of pollution in the Mediterranean by utilizing certain judiciously-selected benthic populations.

It is in the spirit of this proposition that I will treat the action of pollution and, more generally, the action of human activity on Mediterranean benthic populations.

Since 1970, considerable efforts have been made in certain sectors of the Mediterranean, and notably in the northern part of the different basins of this sea, to acquire a multidisciplinary knowledge of benthic communities affected by human activity. These efforts permit a rational evaluation of the mechanism of action of anthropogenic perturbation.

It is of course out of the question to prepare a complete inventory, which would be only a catalogue of the numerous studies conducted and of little utility.

General Data on Mediterranean Communities

I would like to recapitulate certain essential facts which show the originality and the complexity of action of deleterious factors linked to human activity on Mediterranean benthic communities. These communities are diversified, principally on account of the infinite variety of environmental conditions found in the Mediterranean - climatological, bathymetric, hydrodynamic, substrate, oligotrophic, local contributions of terrestrial matter etc., all of which have contributed to the richness and the originality of Mediterranean benthic studies (Pérès and Picard, 1964).

Depending on whether they are established on soft-bottomed or on rocky substrates, benthic communities are subjected to well-characterized and dominant environmental factors. Man-made modifications result in immediate repercussions. The communities established on soft bottomed substrates are dependent essentially on the nature of the sediment and the processes of sedimentation, all of which are largely conditioned by the quality and quantity of sediment influx, and by hydrodynamism, both of which could be modified directly or indirectly by human activity. The communities established on hard-bottomed substrates are conditioned by a group of factors of similar importance. This has resulted in a tremendous diversity in these communities, at least at the fine level and in a greater difficulty in studying and understanding the mechanisms which govern their distribution and their evolution.

The conditioning factors are:

- hydrodynamism, which on a jagged coastline can alter within the space of a few meters.
- nature of the substrate: rock quality gradient, cavernous nature etc.
- illumination, which does not depend solely on depth.

These factors do not evolve independently of each other. Pollution, sedimentary influx and direct human activity all interfere, and could modify the community in a radical manner.

Before I conclude this portion on the difficulties of dealing with hard bottomed substrate communities, I would like to recall that in the superficial levels (up to around 30 meters depth at least), the two kingdoms of the living world, the vegetal kingdom and the animal kingdom coexist in a competitive manner and colonise these substrates, and there exists a certain displacement of their spatial distribution in response to the environmental factors.

Brief Recapitulation of Perturbation by man in the Mediterranean

Pollution. The northern part of the western basin of the Mediterranean and the upper Adriatic receive more than half of the total load of pollutants entering the Mediterranean, on account of the highly-populated and heavily-industrialized cities in this area which, without exception, pollute the marine environment. All kinds of pollutants of diverse origins, ranging from domestic to industrial, are present in the sea and in the sediment. Toxic amounts of detergents have been analyzed in the mud 80 km off the Rhône River. Studies on other major rivers yield almost similar results. PCB, DDT and detergents are present in sediments at a depth of 2700 m in the centre of the western Mediterranean (Arnoux et al., 1981; in press).

Developments. Since the most ancient times, man has modified the coastline. Ports, relatively small originally, have become bigger. More generally, shoreline developments have considerably distorted the coastal edge: 174 km out of 281 km of the total coastline of the Department of the Bouches du Rhône in France and 5715 ha of the bottom between 0 and 20 m out of 38 847 ha, to cite an example provided by Meinesz et al. (1982) who provided similar data for the Departments of the Var of the Alpes Maritimes.

These modifications provoke considerable secondary effects on hydrodynamism and sedimentation (Picard and Bourcier, 1975). The regularization of the flow of the Rhône also modified the sedimentation processes in the Gulfs of Fos and Lion in a dramatic manner.

ACTION OF ANTHROPOGENIC PERTURBATION ON BENTHIC COMMUNITIES OF SOFT BOTTOMED SUBSTRATES

This action is, by far, the best documented. It has been particularly studied on the Provencal coasts, on the Italian and Greek coasts and in the bay of Izmir. Its manifestation depends on a certain number of mechanisms described in 1980(a) by Bellan et al. These mechanisms are given below, and are augmented by data subsequently gathered in the Marseille region and compared with selected examples from the Mediterranean.

Immediate Contact with the Source of Pollution Determining Concentric Areas of Pollution

As a result of a specific source of strong pollution (sewage in a large city such as Marseille, (Bellan, 1967a,b), the animal population present on the polluted,soft-bottomed substrate is distributed in a perfectly regular manner as a function of pollutant burden of the waters which drain into the area, or of sedimentation of organic matter.

Three concentric areas can be demonstrated: faunistic and geochemical data are provided in Bellan et al., 1980a; Bellan et al., 1980b and Arnoux et al., 1980 (Fig. 1, Table 1). It is unnecessary to recall these works here; however, much of the data presented here, by way of example, has been taken from the work of Zarkanellas and Bogdanos (1977) or Zarkanellas (1979).

Zone I or the zone of maximal pollution. This zone is deprived of all animal macroscopic life and, at times, of the meiobenthos.

Zone II or the polluted zone. One passes rapidly (at the most at around 10 meters) from Zone I into Zone II or the polluted zone, the sediment of which is typically a black and noxious mud which receives the maximum pollution dose. The population in this zone is characterised by a very small number of polychaete species. The quantitative evolution of these species can be extremely rapid.

From 1967 onwards, Bellan refused to envisage the existence of an organised biocenosis structure, principally an account of the apparent absence of a clear relation between these species and owing to the anarchic development of their populations.

Zone III or subnormal zone. The transition between zone II and III is almost as abrupt as that between zones I and II. The subnormal zone is characterised by a notable species enrichment, but also conserves a clear dominance of polychaetes: 45 to 55% of the species and 70 to 80% of the individuals. The species of the preceeding zone (Zone II) have practically disappeared. No species can be described as typical of the subnormal zone. The species characteristic of the defined biocenocis are always present (5 to 15% of the species) but they comprise only 3 to 10% of the individuals.

Particularly abundant:

- species with a broad ecological distribution.
- fine sediment and muddy species demanding or tolerating the presence of fine or very fine particles.

Some species such as *Myrtea spinifera*, *Thyasira flexuosa*, *Para-*

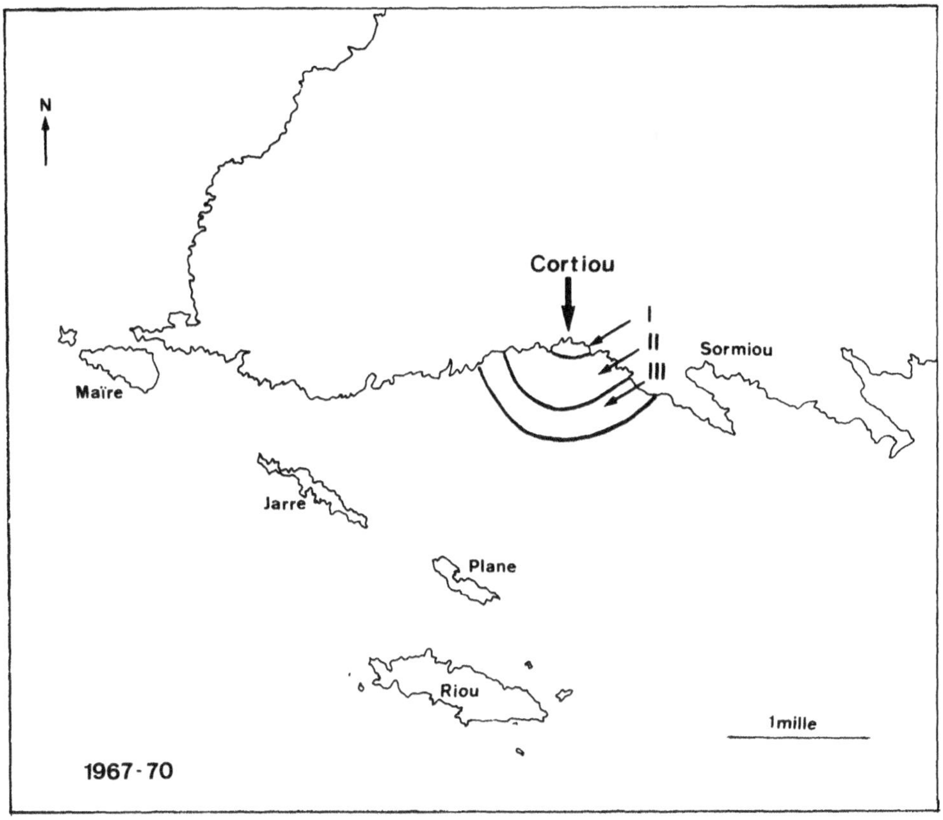

Fig. 1. Distribution of maximal pollution (I), polluted (II) and subnormal (III) zones off the Marseille-Cortiou sewer outfall (in 1966).

lacydonia paradoxa, Dorvillea kefersteini and other diverse forms such as Nematonereis unicornis, Spio multioculata, Corbula gibba show a notable quantitative development in the subnormal zone. In conclusion, one of the major characteristics of the population of this subnormal zone is its extreme monotony: the reclassification of a certain percentage of the species yields more than half of the total number of individuals.

In this subnormal zone, intermediate values of pollutant burden are found.

In fact, this zone can perhaps be divided into two subzones: a subnormal subzone in the strict sense, an "internal" subzone corresponding perfectly to the criteria indicated (Table 1).

Besides, there exists an external subnormal subzone where the pollutant burden is more feeble, where the "front line of pollution"

Table 1. Pollutant amounts in sediment classified on the basis of biological zonation

		Polluted zone	Subnormal zone	
			internal	external
Organic carbon	g/100 g	2-15	1-2	0.4-1
Hydrocarbons (total)	mg/100 g	100-1600	20-100	6.5-35
PCB	$ng \times g^{-1}$	1000-1600	50-500	5-110
DDT	$ng \times g^{-1}$	50-1600	10-50	2-40
Mercury	$\mu g \times g^{-1}$	2-21	0.5-1	0.3-1
Zinc	$\mu g \times g^{-1}$	300-2550	150-200	140-240
Lead	$\mu g \times g^{-1}$	100-1250	40-100	21-57
Copper	$\mu g \times g^{-1}$	100-680	20-100	12-54
Chromium	$\mu g \times g^{-1}$	50-240	25-55	22-30
Arsenic	$\mu g \times g^{-1}$	10-30	5-10	4-7

(Arnoux et al., 1980) corresponds to the limit between these two subzones; the populations are least altered, notably at the level of the biocenosis. This could correspond to an "ecotone zone", intervening before the normal zone where the action of pollution cannot be felt. It must be recorded that depending on this conception, the sampling conducted in 1965 and 1966 by Bellan (1967a) and that he considered as belonging to the "normal" zone, belongs to this external sub-normal subzone, as noted by Salen-Picard (1982) whereas Bellan and Bellan-Santini (1972) had simply considered this under the form of a "more or less marked transition zone".

Certain complementary remarks on these zones. Development both in space and in time can be noted within these zones and at their limits. The understanding of this development permits visualisation of the ultimate form of the zones described.

Zone I, the zone of maximal pollution can develop very little, at least over a long period.

Zone II, the polluted zone is subjected to heavy perturbation, provoking important faunal disequilibria, with spectacular decreases in species abundance, and, in fact, their total disappearance. Romano (1973) described the phenomenon of alteration of extension and regression between the zone of maximal pollution and the polluted zone and, above all, between the polluted zone and the subnormal zone. He demonstrated that in winter the faunal elements of the subnormal zone could temporarily occupy the external subzone of the polluted zone but could subsequently disappear very rapidly. Conversely, in summer, the species of the polluted zone had the tendency to multiply in the subnormal zone at its internal limit. There was thus a centrifugal zonal migration in summer and a centripetal one in winter. Zarkanellas recently (1979) presented a remarkable study on the effects of the fall of dissolved oxygen in summer in the Bay of Elefsis, provoking the complete disappearance of macrobenthic fauna in the centre of the bay (Fig. 2).

This zonation is, in reality only the reflection of a general world-wide evolutionary pattern of the macrobenthic biocenosis on the soft bottomed substrate (Reish, 1959; Pérès and Bellan, 1973; Pearson and Rosenberg, 1978). This pattern has recently been (1983) studied by Salen-Picard who considered it applicable, in a similar manner, to the initial population and the cause of the degradation. The influence of a perturbation factor, in this case of anthropogenic origin, on the initial state of a biocenosis has been considered. In this biocenosis, which is rich in individuals and in species linked essentially to the structure of the sediment, the following are noticed in a manner successive or otherwise:

a) a regression of the species strictly linked to the original conditions of the environment;

b) whereas this phenomena renders the development of tolerant species more pronounced, certain of them, considered as pollution indicators (Stora, 1982) tend to monopolize available space. A limited increase of diversity could be observed (Zarkanellas and Kattoulas, 1981). The biocenosis structure remains recognisable even if degraded (subnormal zone);

c) a destruction of the biocenosis is recorded; certain species exist and develop, apparently independently of each other. Species diversity decreases and becomes minimal (polluted zone);

d) the macrobenthos disappears (zone of maximal pollution).

The phenomena described are not restricted to the Marseille region. They have been demonstrated in other sectors of the Mediterranean and specifically on the coasts of the Italian peninsula (Cognetti, 1970;1978; 1982), in the sector of Trieste (Chirardelli et Orel, 1968; Ghirardelli et al., 1973; 1974; 1975), in the bay of Elefsis (Zarkanellas, 1979), in the upper part of the Saronikos Gulf (Zarkanellas and Bogdanos, 1977) and in the Gulf of Izmir (Kocatas, 1981).

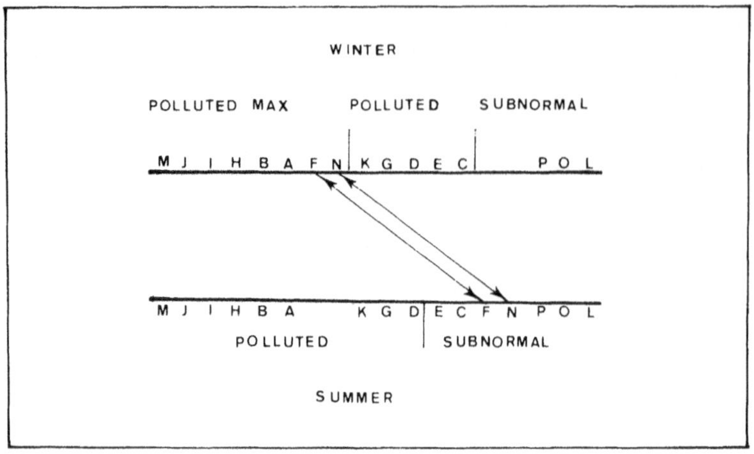

Fig. 2. Displacement of stations from one to the other according to season, in the Elefsis Bay (according to Zarkanellas, 1979).

The Areas of Decantation

The influx of organic matter is due to local hydrodynamic action; the most polluted populations thus correspond more or less to preexisting "decantation areas" and are more-or-less distant from the pollution source.

By "decantation areas" are meant the under sea surfaces where the decrease of hydrodynamic action permits a locally-enhanced settling of fine or light particles, specifically clay, rotting Posidonia fibers, vegetal or animal debris etc. By definition, these "decantation areas" are predestined to receive, and often accumulate, the influx of exogenous organic matter either from the surrounding natural environment or from human activity.

The first example that could be considered is that of the Gulf of La Napoule (Bellan, 1979). A preliminary study conducted in July 1973 on the commissioning of an under sea outlet established the fact that the pollutants ejected into the Gulf had the tendency to accumulate there which led to a general deterioration of the population, notably in the decantation area situated in the north west of the Gulf and referred to as the subnormal zone. Four years after the commissioning of the outlet, an improvement of the situation could be seen in the east of the Gulf and a deterioration in the western part (Fig. 2), notably in the western decantation area, the population of which had developed towards that typical of the polluted zone. Between this area and the opening of the outlet pipe, there existed populations perfectly representative of the subnormal zone (zone III).

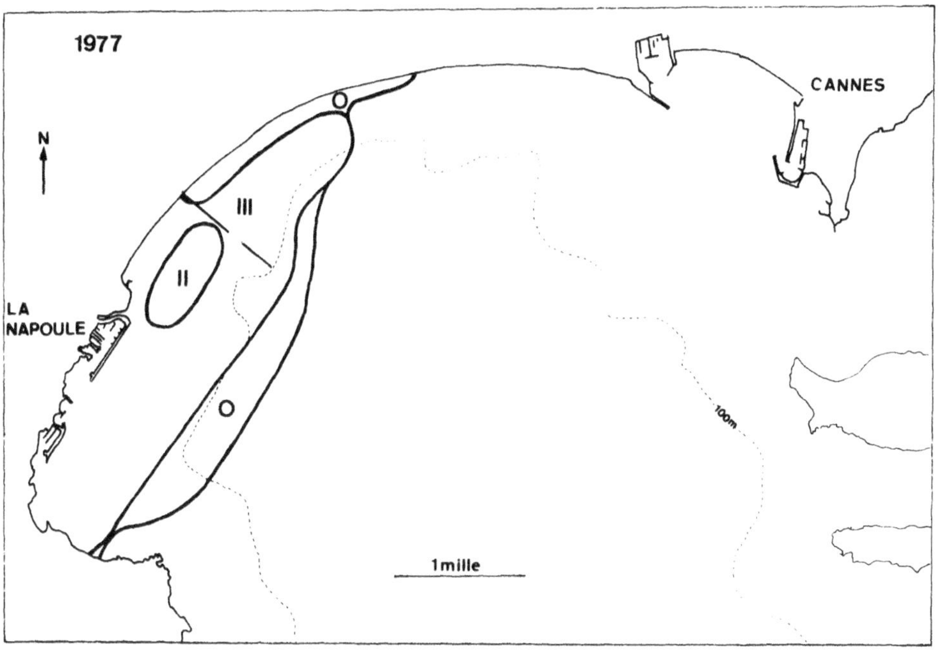

Fig. 3. Distribution of polluted (II), subnormal (III) and slightly or potentially contaminated (0) zones in the bay of La Napoule (in 1977).

As a second example the occurrence, south-west of Maïre island, close to Marseille, of a strong core of pollution within the interior of the preexisting decantation area could be cited. This decantation area, rich in *Posidonia* fibers was nourished, in 1971, by influx from the north (Marseille-Prado road) during the mistral (north-west wind) (Picard, 1978). Since 1973, the sector had already reached, by extension towards the west, the subnormal zone arising from the outlet of the Marseille-Cortiou drain (Fig. 4). Within the same decantation area of *Posidonia* fibers the development was noted of a localised core already described in 1974, consisting of a "metastasis" of the polluted zone (zone II). The sedimentary material consisted of a relatively thick fraction. This was due, in fact, to the previous existence of a deposit of *Posidonia* fibers and to the decantation of diverse debris from a raft of floating detritus which remained within the polluted core because of the flow of a moderate east wind (Bellan et al., 1980a). These decantation areas are undoubtedly frequent on the Mediterranean coasts, and have been studied in the Marseille region. Traces have also been found in other areas: off the Tiber (Gambi and Giangrande, 1983), in Augusta Bay in Sicily (Cantone and Pilato, 1981; Di Geronimo, in press), the Trieste region (Ghirardelli et al., 1974) and in the Mar Grande of Tarente (Tursi et al., 1978; Tursi et al., 1981).

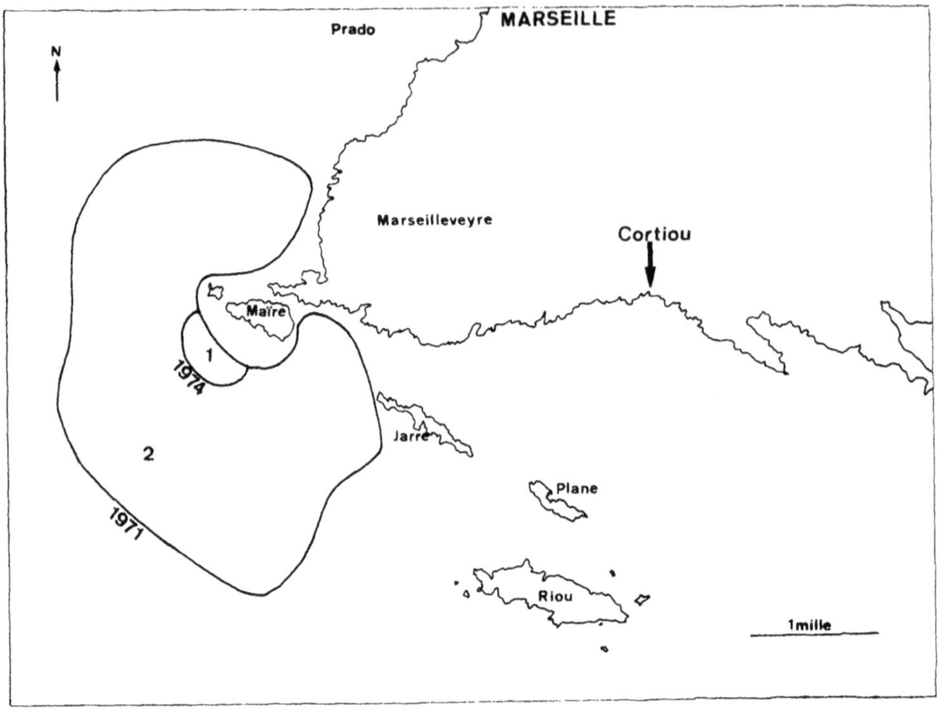

Fig. 4 Maïre island deposition area: 1) deposition area in 1971; 2) polluted zone settlement in 1974.

These decantation areas serve very often as a "bridge head" to the advance of polluted populations. In them, sedimentation accompanied by excessive influx of organic matter and pollutants, can be produced, as demonstrated by Bourcier (1980), by progressive and successive contamination on a grand scale of neighbouring coastal basins, with the appearance of species of the subnormal zone (Thyasira flexuosa, Myrtea spinifera, Corbula gibba) and exceptionally of the polluted zone (Nereis caudata).

In the region of the capes separating these bays, the populations in optimal hydrodynamic conditions (including non-influx of excessive sedimentation) are protected, but in the interior of the bays the processes of alteration are accelerated. These contaminations are amplified when natural climatic phenomena, on the grand scale, are superimposed on human activity (Bourcier, 1980).

Arrangement of Populations in "mosaic elements"

The dispersion of organic matter and of pollutants from weak, but multiple sources, can be hindered by the existence of artificial obstacles such as breakwaters and sea walls protecting the harbours. The most polluted populations are composed of "mosaic elements" in the regions where these obstacles form a screen to water circulation.

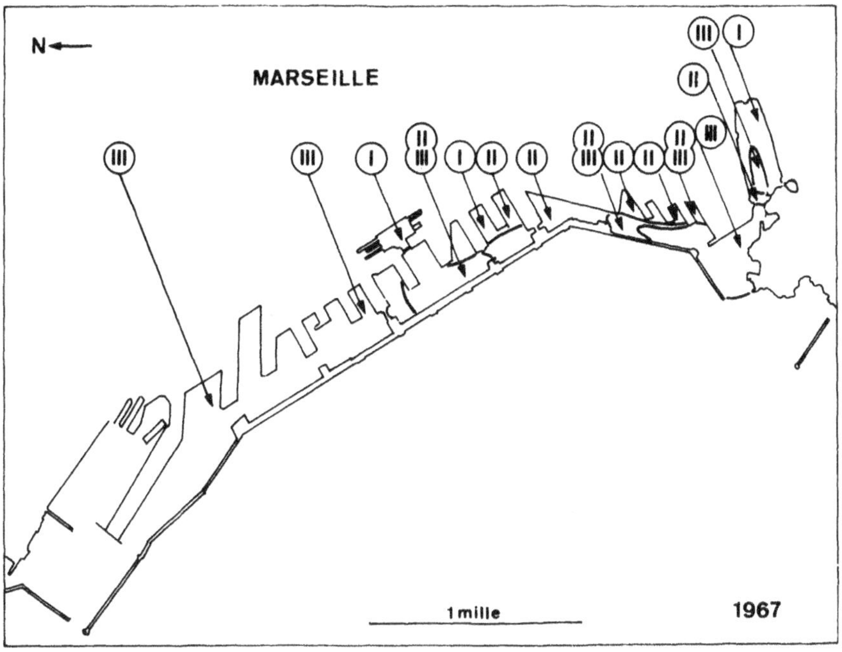

Fig. 5 "Mosaic" distribution in Marseille Harbour in 1966: maximal pollution (I), polluted (II) and subnormal (III) zones, (II/III) zones of "mixing" assemblages.

Such phenomena seem to be the rule in large harbours such as that of Marseille (Fig. 5) (Bellan, 1967a,b), Genova (Fig. 6) (Bellan, 1976) and Long Beach-Los Angeles, California (Reish, 1959). The populations present in these large harbours appear entirely analogous; they are identical in detail (polluted zone) or at least homologous (subnormal zone) as Bellan (1976) has demonstrated. In these large harbours (fig. 5) different population zones are found which have been described with respect to the first zone of maximal azoic pollution, a classical polluted zone and a subnormal zone with Venerupis aureus, a species characteristic of muddy sand in sheltered water biocenosis, Corbula gibba in the superficial levels and a glimpse of Thyasira flexuosa in the greater depths. There is a large structural analogy between the populations of large harbours and those in the vicinity of a large outlet. All the same, this structural similarity of the populations is opposed by a double dissimilarity at the level of their arrangement.

First, it is this that, in the harbours, characterises the spatial distribution of populations, which are not distributed in concentric circles. They are disposed in "mosaic elements" corresponding to a basin in the interval between two breakwaters, or to a zone protected by a sea wall, for example. Each site is a

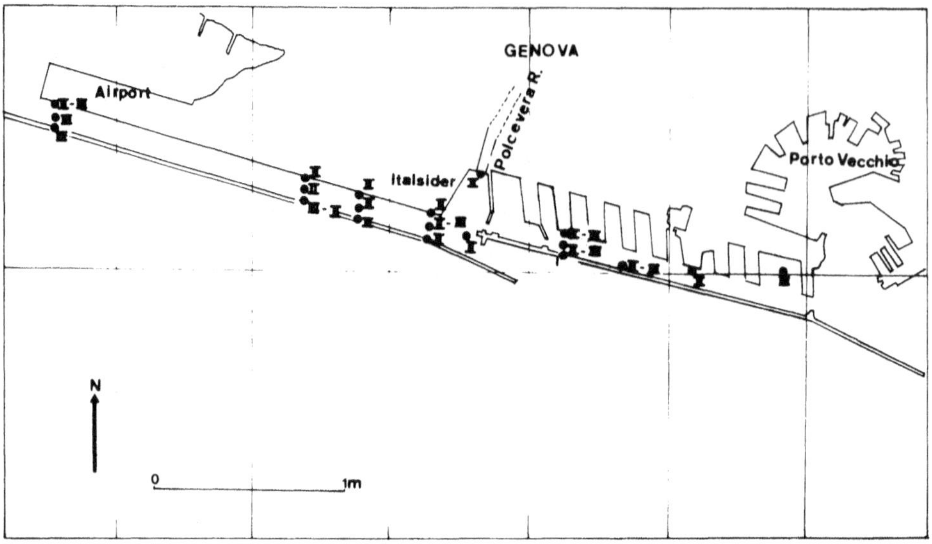

Fig. 6. "Mosaic" distribution in Genova Harbour in 1975.

particular case, depending on pollution inflows and on the importance of the obstacles which oppose the circulation of water and on "cleaning" by normal hydrodynamics. Although zones I, II and III succeed each other regularly in the sewage outlet area, we see, in the harbours, an abrupt passage from one zone to another. One can thus pass, almost without transition, from the azoic area (Zone I) to populations typical of zone III.

The second dissimilarity concerns the disposition of harbour populations in comparison with what is observed in the open sea. The arrangement in "mossaic" patterns is not simply spatial. Frequent superimpositions of elements of the polluted zone are observed in populations referable to the subnormal zone. In the outlet area such phenomena have been observed but always on a reduced scale: either a space scale or a time scale. The case is different in harbours where zones of mixing occupy large surfaces with an area comparable to that occupied by the distinct assemblages. All the same it seems that these mixed populations persist at least on the yearly scale and evolve very little quantitatively (on account of the species recruitment). Finally, it is useful to specify that these zones of mixing of populations II and III are not the simple juxtaposition of very small areas of populations that could be individually ascribed to the polluted zone and to the subnormal zone but a real mixture as Bellan et al. (1980a) have pointed out. The local and permanent coexistence of population of Venerupis aureus, Corbula gibba and Scolelepis fuliginosa and Capitella capitata are certainly due to the fact that these species respond rapidly (within 2 to 4 weeks) to erratic variations of environmental conditions (influx of pollutant, prolonged stagnation of water, disturbance during storms, etc.).

It is necessary, nevertheless, to state that the phenomena described in large harbours, abundantly provided with breakwaters, are not always the rule. Thus in the harbour basins of the Fos complex, between 1972 and 1978 Stora and Romano (1980) described communities that could belong to the subnormal zone but which are not seen to exhibit the complexity that we have just described.

Particular Problems

Extension of degraded populations by the blocking up of excess of organic matter. When there is a loading in the sediments of an excess of organic matter, the recycling of which appears blocked by the intervention of chemical substances adsorbed by clay before setting,the benthic populations reached by pollution occupy sometimes very enlarged areas, which correspond to the extension of the masses of polluted water as has been described in Bellan et al. (1980a) and Picard (1978) (Fig. 7). Since its commissioning in 1986 and until 1970, the influence of the Marseille-Cortiou sewer has not surpassed about 1.5 km radius (Fig. 1) a zone centering on the exit of the sewer. Since 1970, the diversion in summer of the little river Huveaune into the sewer of Cortiou results in the inflow of chemical industry pollutants which mix with the domestic inputs of the Marseille complex. All of this comes into contact with the suspended clay in the sea water. It has been established that after 1970, there was an abrupt acceleration, and considerable extension of the subnormal zone (sediments becoming black and noxious and modifications of benthic populations). This progression of the external limit of the subnormal zone (Fig. 7, state in 1979) was all the more rapid on reaching the preexisting decantation area. This extension is, since 1979, much stronger towards the open sea by the mere fact of the contact (and the greater dilution) of neritic polluted waters with waters of the general east-west circulation of the north western basin of the Mediterranean. It follows a still higher rhythm towards the east of the La Ciotat Gulf aread and could serve there as an example for the second mechanism. Some similar phenomena have been described in the region of Trieste by Ghirardelli et al. (1974) as also in the Saronikos Gulf (Zarkanellas and Bogdanos, 1977) and in the Izmir Bay (Kocatas, 1981).

Seasonal destruction of population by ambient oxygen deficiencies. In the highly polluted Elefsis Bay (in Greece) Zarkanellas (1979) observed the summer destruction of the population, to the impoverishment of this subnormal zone, "enriched" locally by the presence of *Scolelepis fuliginosa*. This seasonal destruction over a considerable area (several km^2) is in direct relation to the fall of ambient dissolved oxygen levels during the summer months.

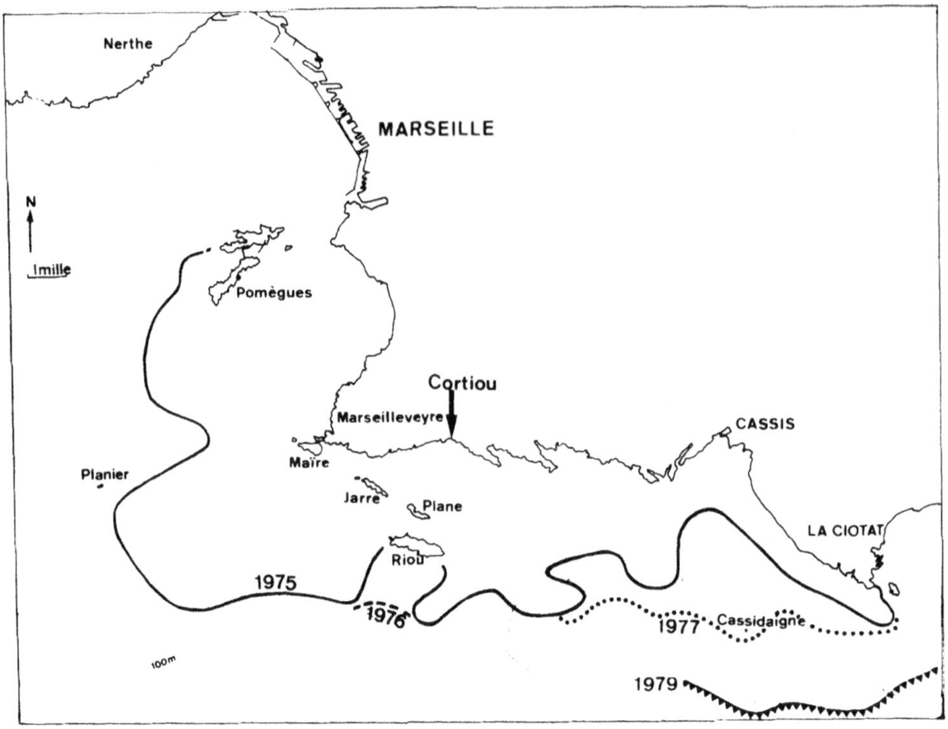

Fig. 7. Extension of the disturbed zone (subnormal zone) off Cortiou (1975-1979).

In winter the situation is restored and a population (poor) reappears. A zonal migration is observed (see Fig. 2).

This cyclic evolution is much more important than that observed sporadically in the Cortiou area. Kocatas (1981) provided similar data for the Izmir Bay.

The hydrodynamic phenomena reversing summer stagnation of the waters, and thus oxygen impoverishment, are the direct cause. More dramatically, and for the same fundamental reason, the excess of organic matter which enters the Elefsis Bay is homologous to that described in the preceding paragraph in the Cortiou area.

Degradation of populations owing to erratic influx of fresh water. The Berre lagoon situated around 30 km from Marseille has an area of around 15500 ha. Until March 1966, this lagoon was typically marine and the base community could have been classified as a muddy sand biocenosis in sheltered waters (Febvre, 1968). The diversion of the canal of the hydroelectric plant at St. Chamas carrying on an average, $3.6 \cdot 10^9 m^3$ of fresh water per year, that is to say, four times the volume of the Berre lagoon, provoked a major ecological catastrophe from the hydrobiological (Minas, 1973), geochemical (Arnoux et al., 1980), and bionomic (Bellan, 1972; Bellan and Stora, 1976; Stora, 1982) points of view. The processes can thus be recapitulated: formation of a halocline, provoking an oxycline which determines anoxia below a depth of 6 m. This anoxia totally blocks the recycling of organic matter (introduced into the lagoon or developing at superficial levels) in the sediments, which maintains the anoxia and results in the disappearance of all macrobenthic life over a considerable area (Fig. 8). The pollution or the excess of organic matter in the lagoon cannot be considered as an essential cause of this effect. It is known (Arnoux et al., 1980; Stora and Arnoux, 1983) that it is in the south-west of the lagoon ("coin salé") that the least unbalanced populations and large concentrations of pollutants and organic matter in the sediments coincide, but the salinities are also the least unstable and oxygenation sufficient (Bellan and Stora, 1976; Stora, 1982). Muddy sand in the sheltered water biocenosis (SVMC) give way, progressively, to a settlement that could be referred to as a euryhaline eurythermal biocenosis of brackish coasters (LEE). In the south, the salt water coming from the Gulf of Fos permits the maintenance, near Martigues, in the navigation channel dredged at -9m, of a degraded SVMC, giving way progressively to a SVMC/LEE population and then to a degraded

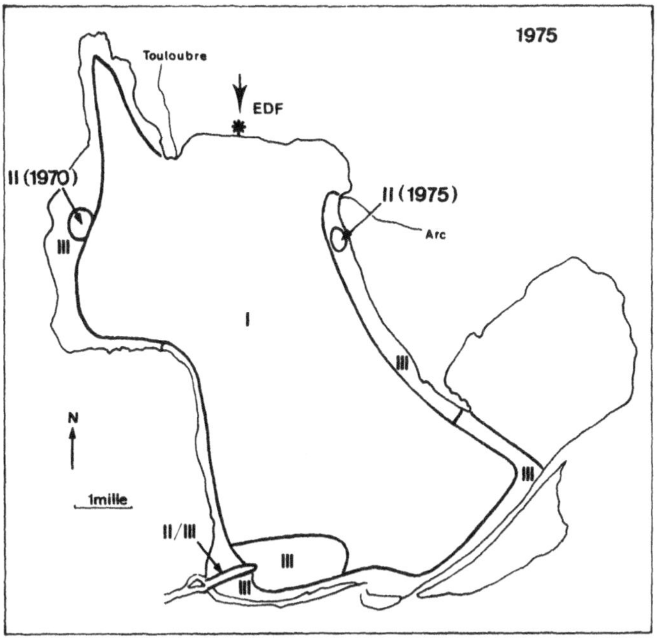

Fig. 8. Extension of settlement in Berre Lagoon in 1975. II. Sporadic appearance of polluted zone settlement (1970-1975). III. Subnormal zone.

LEE population on moving towards the center of the lagoon. These populations as a whole, can be compared to the subnormal zone. With the exception of the canal and of the "salty corner" which it encircles in the south-west, no species can develop beyong a depth of 5 m in the rest of the lagoon.

Perturbing action not directly linked to pollution. Apart from these phenomena linked directly to pollution, community alterations resulting from diverse perturbation of anthropogenic origin should also be emphasized. The phenomena observed completely fit into the general pattern proposed.

Similarly Salen-Picard (1981), studying the evolution in the Gulf of Fos of a dredged deposit assemblage, which could certainly be classified in the subnormal zone recorded the:

- abundance of *Thyasira flexuosa*, *Corbula gibba*, *Lumbrineris latreilli*.
- development of species of ecological significance.

Picard and Bourcier (1975) studied the impact of diverse human interventions, notably the enlargement of a port, on benthic populations in the La Ciotat area. The sedimentary instability linked to serious modifications of hydrodynamism in the La Ciotat Bay, provoked a profound alteration of the benthos, the reversal of which took ten years, but which nevertheless had a tremendous effect in the sense of displacement on the communities described.

Human perturbations in the bathyal region. Very little is known about the impact of human activity on the bathyal. Bourcier (1969) demonstrated, as a function of rhythm of sedimentation, the differing action of "red mud" (alkaline waste from the treatment of non toxic bauxite) on populations of the Canyon of the Cassidaigne (Marseille region). The hypersedimentation in the Canyon axis hinders the survival of the macrofauna.

Since it is known that the most diverse pollutants are found in the deepest mud (Arnoux et al., 1981; in press) one has the right to be disturbed about the future of the bathyal and abyssal Mediterranean.

ACTION OF ANTHROPOGENIC PERTURBATION ON ROCKY SUBSTRATE COMMUNITIES AND ON PHANEROGAM BEDS

A knowledge of communities established on rocky substrates and on phanerogams beds is also of interest since existing information appears inadequate.

The inventory of vegetal and animal groups is, certainly, relatively advanced, but studies on communities in general should be supplemented.

In reality, algologists on one side and zoologists on the other have conducted largely uncoordinated research on the phytocenosis and the zoocenosis. There are perceptible divergences in understanding of the phenomena observed and described. This has permitted Belsher and Boudouresque (1976) to present a healthy critical communication on the controversy and on the manner in which the algae (and I can say more generally the vegetation) have been utilised by certain bionomists (essentially zoologists).

I undertake the responsibility of stating that the points of view of both parties are not as far apart as it seems. I recall also having insisted on the difficulties that studies of the communities established on rocky substrates present.

Pollution and Shallow Water Rocky Substrate Assemblages

For several decades, it was accepted without serious verification that there existed a "nitrophilic community" in the surroundings of harbour complexes in general and in the open sea at the exit of domestic sewers. (The second category was only vaguely referred to). This nitrophilic community was poor in species, both algae and animals. The only consistent work in the domain of algology was that of Feldman (1937) and Feldman and Mazoyer (1941). It was necessary to wait for the work of Bellan-Santini (1962; 1966; 1969) to acquire a sure knowledge on the impact of pollution on the assemblages of the shallow infralittoral fringe. It is necessary, with Belsher and Boudouresque (1976), to regret that this author could only present a fraction of her work (perhaps not completely negligible) on the flora of the assemblages studied. Among the most significant data recorded by Bellan-Santini on the communities of the polluted environment, I will cite:

a) the confirmation of the appreciable reduction of the number of animal species.

b) the confirmation of a stock of animal species indicative of pollution,

c) the presence of certain sciaphilous species which however are photophilous in the polluted environment,

d) the separation of a stock of "biofouling" species present on artificial substrates as well as in the harbour environment and open sea and completely independent of pollution,

e) the progressive disappearance of certain floristic and faunistic elements with pollution and the precise correlation of certain disappearances with specific pollutant levels (Arnoux and Bellan-Santini, 1972) (Fig. 9).

In the same way, the disappearance of certain shallow water species of Cystoseira, present specially in zones of elevated hydrodynamism (Cystoseira stricta being of this type), had been subsequently verified on the Mediterranean Coasts as a whole, as well as in Spain (Gomez-Garreta and Ribera Siguan, in press), in meridional Italy (Gamulin-Brida et al., 1967), in the Upper Adriatic (Gamulin-Brida et al., 1967) and in the Gulf of Izmir (Kocatas, 1978).

Considering all the work that has been conducted, it appears that the zonation on rocky substrates is not as clear as that described on the soft-bottomed surface (but it seems to deal more simply with an artifact due to difficulties of methodology and cartography). It has been proposed, in particular, that the reported zone deprived of all vegetal and animal life does not really exist for the rocky substrate. It has not been found in the Vieux Port of Marseille studied by Leung Tack (1971). The author records only an impoverishment in the region most distant from the entry of the Port. He also emphasizes the seasonal variations of the epibionts and certain endobionts.

Again in the Marseille region, the work of Bitar (1983) on shallow water communities subjected to the influence of the large sewer of Marseille-Cortiou and that of Desrosiers (1977), Bellan-Santini and Desrosiers (1977) and Desrosiers et al. (1982) on identical communities subject to multiple industrial pollution of the Gulf of Fos amply confirms the impoverishment linked to pollution; provides precise data on faunistic replacements; demonstrates the quantitative increase of certain tolerant species and the alteration of the animal cycles; and finally emphasizes the destruction of the base concretion.

Certain local hydrodynamic phenomena often disturb an evolutionary pattern that is never simple. The present of capes and points reduces the impact of pollution, while the presence of indentations and coves enhances it as has shown by Bitar (1983) as well as by Belsher (1974).

It is evident that hydrodynamism can play an important part with respect to the communities established on a rocky substrate in the polluted environment. The relatively epibiont character of the communities and the possibility of over-oxygenation linked to the transport of water masses add up that to the fact that these are less subject to aggressive pressure than more "endobiontic" communities present on and in the badly drained sedimentary substrate.

At the level of particular taxonomic groups, certain specialists have perfected "Indices of Pollution". Limiting myself to the

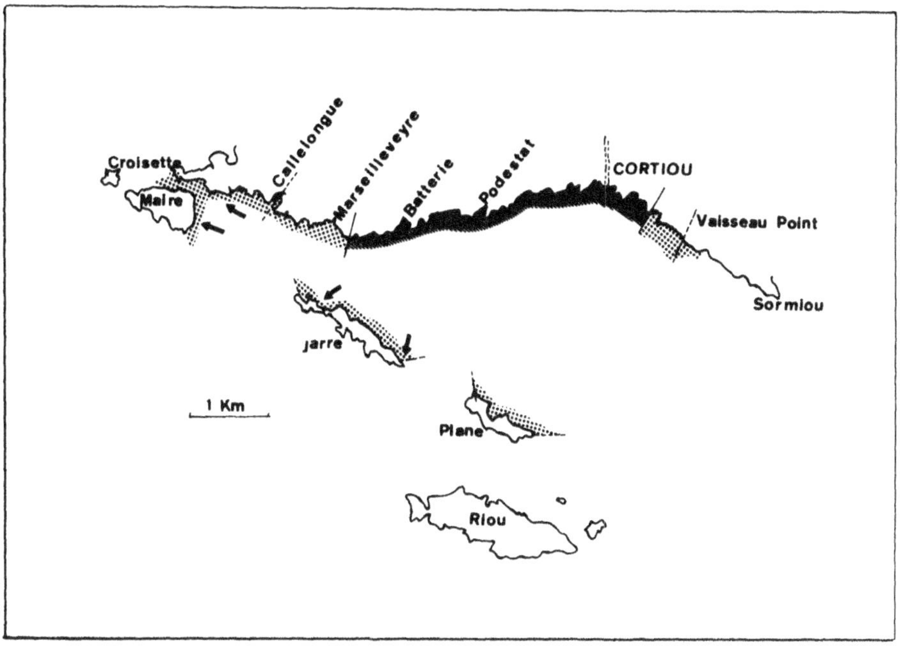

Fig. 9. Zone of disappearance of Cystoseira stricta assemblage off Cortiou according to a gradient of pollution. Slanting lines: zone without C. stricta, dotted lines: zones with sporadic presence of C. stricta.

Marseille region, I will cite the Index of pollution of Belsher (1979) taking into a ccount the concentration (Ψ) of diverse algal groups:

$$I.P. = \frac{\Psi \text{ Bangiophycae}}{\Psi \text{ Phaeophycae} + \Psi \text{ Cryptonemiales} + \Psi \text{ Ceramiales}} \times 100$$

which increases with proximity to the Cortiou sewer. Bellan (1980) proposed an Index of biological detection which is based on the premise of sensitive modifications of the polychaete populations as a whole and is defined as the ratio of the dominance of "biological detectors of polluted waters" (Platynereis dumerili, Theostoma oerstedi, Dorvillea rudolphi,...) to "pure water" ones (Syllis spp. Amphiglena mediterranea). This ratio increases with intensity of pollution and is always higher than 1 in a polluted environment. Bellan-Santini (1981) established a similar index for Amphipoda. Harmelin et al. (1981) utilised the echinoderms in a particularly interesting manner. There appeared to exist a clearly decreasing gradient with respect to the number of species of macroechinoderms in the zone of industrial pollution (Gulf of Fos) whereas the zone of preponderantly domestic pollution (Cortiou) is characterised by an appreciable increase of the abundance of Paracentrotus lividus alone, the size of which, on the contrary, decreases. These data amply confirm the previous observations of Ghirardelli et al. (1973) in the region of Trieste and these of Zavodnik (1977). Bellan-Santini (1981) emphasized the clear dominance of "suspension feeders" to the detriment of grazers in areas rich in organic matter.

The work of Verlaque (1977) on the phytobenthic population in the neighbourhood of the electric power plant at Ponteau (Gulf of Fos) provided decisive data on the action of this very specific type of pollution. The thermal stress induces a specific richness of flora and a dense vegetation cover at a certain distance from the outlet, an illustration of the diagram of Pearson and Rosenberg (1978) and of the localised increase of the "Index of diversity". All the same Verlaque (1977) pointed out a classical gradient of macrofaunal impoverishment on approaching the exit of the outlet pipe. In the Mediterranean, the examples cited, based on the excellent data available in the Marseille region, are not appreciably different from those known in other places, as I have already stated. The work realised on the coasts of Spain (Gomez Garreta and Ribeira Siguan, in press) and France (Marcot-Coquegniot, 1983; Meinesz et al., 1982) can be recapitulated. The contributions of Meinesz et al. (1982) have advanced our knowledge of the action of the alterations resulting from shore line developments on the French coasts of Mediterranean. Elsewhere, we can note the works of Pignatti and De Cristini (1976), Specchi and Orel (1966), Ghirardelli et al. (1973), Gamulin-Brida et al. (1967), Giaccone and Rizzi Longo (1974) along the Italian Coasts, of Zavodnik (1976, 1977a,b), Stirn (1973), Stirn et al. (1975), Munda (1974), Golubic (1970), Gamulin Brida et al. (1973), along Yugoslavian Coast, of Kocatas (1978) and of Basson et al. (1976) in Lebanon.

Among the most recent algological work, the detailed study of Gomez Garreta and Ribera Siguan (in press) established certain natural links with that of zoologists. Attention should be called, among the other significant data, to the progressive disappearance of <u>Cystoseira mediterranea</u>, the reduction of the Phaeophycea, the establishment of a facies of <u>Corallina elongata</u> and to the adaptation of the Ulvacea in contaminated sites. Bellan-Santini, as early as 1969, established similar conclusions based mostly on zoological data collected in the Marseille region.

Pollution and Assemblages of the Inferior Infralittoral and of the Circalittoral Zones

Although the shallow water communities (up to a depth of almost 5m) established on rocky substrates have been relatively well-studied, the same is not the case for deeper water communities, which require long dives in turbid water.

Bellan-Santini et al. (1980) and Hong (1980) studied, in a

detailed manner, the coralligenous concretions of the rocky substrate of the Marseille region, notably in the Gulf of Fos. With reference to homologous communities in pure waters, the following were noted:

a) a marked qualitative and quantitative impoverishment of the fauna,

b) the zoological groups, bryozoans, crustaceans and echinoderms suffer an important decrease in richness and abundance. Molluscs and polychaetes show no variations; hydroids exhibit an increase,

c) a decrease of the species diversity index,

d) considering functional groups, a diminution of the builders and of the constructed-habitat users, while the destructive organisms remain constant,

e) considering the ecological assemblages, a striking feature is a global diminution of the characteristic species whereas the stock of species with a broad ecological distribution are increased.

The Formation of Marine Phanerogams

I will restrict myself to recalling certain significant data:

a) In Mediterranean, as a whole, a regression of the beds of *Posidonia* is noticed, under the direct action of pollution or of any human activity (littoral developments, fishing and pleasure activities, etc.),

b) this regression is not necessarily linked to any natural deficiency of the beds, on arriving at their geographical limit,

c) this regression affects, as a whole, the floristic and faunistic fractions even if a specific fraction is temporarily favoured,

d) the regression of *Posidonia* beds often profit from the increase of surfaces occupied by *Cymodocea nodosa* lawns. It is known since the work of Molinier and Picard (1952) that the latter is the pioneer stage as well as the natural regression stage of the beds.

e) that the disappearance of the *Posidonia* leaves in places "dead mats", the fauna of which, initially studied by Harmelin (1964), seems to develop with the stage of degradation of the mat and with the level of environmental pollution.

f) that this "dead mat" is very often submerged by hypersedimentation which is frequently the rule in strongly polluted areas. This submersion leads to the destruction of the assemblages of the mat and of the installation of a very-much-altered muddy sand assemblage, in its sedimentary sheathing.

CONCLUDING REMARKS

Since the work of Bellan and Pérès (1972), the assessment of the impact of human activity on Mediterranean benthic communities has made astonishing progress in comparison with other geographical areas on a world-wide basis. The extent of this knowledge is certainly unequalled, whether considering population studies, diversity of the biotopes and deleterious factors or general precision of the studies. The smaller amount of work conducted in the south of the Mediterranean, although the effects of pollution are acutely felt there, must be regretted.

A certain number of points should be emphasized:

1. Adequate data on the action of the deleterious anthropogenic factors on soft bottomed communities are available. Interdisciplinary research involving hydrodynamic specialists, sedimentologists, geochemists and biologists should be systematically undertaken. In many areas, this research serves to complement earlier knowledge.

The problem is more complex when hard substrates are considered. On the one hand, owing to the impact of pollutants conveyed by water, phenomena such as sedimentation with the fixation of the pollutants and their concentration and retention in interstitial water intervene only exceptionally (in harbours). The impact on hard substrates could, perhaps, from the point of view of the mechanism, be intercalated between the impact on the plankton and that on the benthos of soft bottomed substrates. On the other hand, difficulties of methodology, linked essentially to the difficulties of studying the environment, still hinder the perfecting of evolutionary modules. Besides, it is unclear that the levels actually attained could be surpassed to a great extent. Nevertheless, one requirement remains: that the data acquired by algologists and by zoologists (the latter having often incorporated data essentially dealing with algal populations in their studies) must be considered. It is necessary that a strong effort be made to study phanerogam formations in a multidisciplinary fashion.

2. It appears that degradation processes and further evolution of the communities follow an analogous progression whatever be the nature of the perturbation: chemical or domestic pollution, hypersedimentation, erratic entry of fresh water, harbour developments etc. The intensity of the perturbation (and its duration) is more important than its nature.

3. The study of benthic communities present in the area subjected to human activity is not a goal itself. It permits assessment of the state of the marine environment, taking into consideration the "time" factor, as stated by Reish in 1959. It also permits the proposal of practical efficacious solutions to problems posed by nuisances. It is no longer considered that the ideal treatment of residuary water before ejection into the sea is dilaceration. Bellan (1967a), subsequent to research cited above, criticised this "treatment". The rich experience acquired on the effects of a complete stop or reduction of the pollutants on benthic communities permitted different scientists to formulate other propositions. Picard and Bourcier (1975) had also suggested the planning of work linked to harbour construction in such a manner that the maximal influence of perturbation is felt in the beginning of the work and that subsequent discharge of material is limited to a minimum. In a general manner, Mediterranean scientists are in agreement on the necessity of utilising the hydrodynamic structures which permit the better dispersion or dilution of the polluting elements.

The pollution source may be suppressed by halting the discharge of dredged materials, installing a purification station in the exit area of a sewage pipe draining into the sea, etc. In this case, the more superficial the perturbed area, the more rapidly and completely the diverse pollution zones are resorbed, owing to dispersion and cleaning by strong hydrodynamic action. It is also necessary to avoid all that could cause additive "synergistic effects": for example, the installation of an electric power plant in a zone strongly polluted by chemical discharge, the massive entry of a stream in a lagoon or bay subjected to the influx of organic matter, the discharge of particularly fine matter (clay) in the proximity of the exit of a sewer rich in detergents. All these examples are in fact derived directly from the knowledge acquired on Mediterranean benthic ecosystems.

4. These remarks are made in an optimistic spirit, but they should be qualified: only the reestablishment of a certain equilibrium, and not necessarily the reconstitution of the former state could be hoped for as long as the physical nature of the substrate (granulometric essentially) is not radically modified. It must also be remembered that the accumulation of non degradable pollutants is an irreversible phenomenon and that the disappearance of certain phylogenetically and ecologically fragile species leads to an impoverishment of our common natural heritage, an integral part of our common history.

Acknowledgments

This paper has been translated by Sheila S. Jacob, Ph.D.

SUMMARY

The consequences of human activity on benthic communities on hard or soft bottomed substrates have been studied in detail in the northern and central part of the different Mediterranean basins. An excellent resemblance exists between these communities, reinforced by the fact that their evolution is linked more to the intensity rather than the nature of perturbation.

The communities established on soft bottomed substrates are characterised by a general organisation in concentric circles, strictly correlated with the load of organic matter or sedimental pollutant. This simple zonation can sometimes be modified, either by the presence of "areas of deposition", located somewhat further away from the source of pollution and affected by local hydrodynamism, or by obstacles to water circulation, which lead to the formation of "mosaic assemblages".

The communities established on hard substrates are often more diversified and their arrangement depends, above all, on the local topography and on hydrodynamic factors which accentuate or reinforce the action of the pollutants.

A few suggestions for better protection of Mediterranean benthic communities are presented.

REFERENCES

Ardizzone, G., e Riggio, S., 1981, Similitudine e diversità nelle biocenosi benthoniche del porto di Palermo in relazione al substrato di sediamento, Quad. Lab. Tecnol. Pesca, 3:587.

Arnoux, A., et Bellan-Santini, D., 1972, Relations entre la pollution du secteur de Cortiou par les détergents anioniques et les modifications des peuplements à Cystoseira stricta, Téthys, 4:583.

Arnoux, A., Auclair, D., et Bellan, G., 1972, Etude de la pollution chimique des sédiments marins du secteur de Cortiou (Marseille): Relation avec les peuplements macrobenthiques (Résultats preliminaires), les Journées Etud. Pollutions, Athènes, C.I.E.S.M.:137.

Arnoux, A., Bellan, G., Jorajuria-Olivari, A., Monod, J.L., et Tatossian, J., 1980, Relations entre les peuplements benthiques et les caractéristiques granulométriques et chimiques des dépôts au large de l'émissaire de Marseille, Ves Journées Etud. Pollutions, Cagliari, C.I.E.S.M.:791.

Arnoud, A., Bellan-Santini, D., Monod, J.L., et Tatossian, J., 1981, Polluants organiques dans les sédiments prélevés entre la Provence et la Corse (Mission BIOMEDE I), Ves Journées Etud. Pollutions, Cagliari, C.I.E.S.M.:423.

Arnoud, A., Monod, J.L., Bouchard, P., et Airaudo, C.B., 1981, Evolution et bilan de la pollution des sédiments de l'étang de Berre, Ves Journées Etud. Pollutions, Cagliari, C.I.E.S.M.: 433.

Arnoud, A., Chamley, H., Bellan-Santini, D., Tatossian, J., et Diana, C., (in press), Etude minéralogique et chimique des sédiments profonds de la Méditerranée occidentale (Mission BIOMEDE I), VIes Journees Etud. Pollutions, Cannes, C.I.E.S.M.

Bakalem, A., et Romano, J.C., in press, Pollution et peuplements benthiques dans la région algéroise (baie d'Alger et baie de Bou Ismail), VIes Journées Etud. Pollutions, Cannes, C.I.E.S.M.

Basson, P.W., Hardy, J.T., and Lakkis, S., 1976, Ecology of marine macroalgae in relation to pollution along the coast of Lebanon, Acta Adriat. 98:301.

Bellan, G., 1967, Pollution et peuplements benthiques sur substrat meuble dans la région de Marseille. 1ère partie. Le secteur de Cortiou, Rev. intern. Océanogr. méd., 6/7:53.

Bellan, G., 1967b, Pollution et peuplements benthiques de substrat meuble dans la région de Marseille. 2ème partie. L'ensemble portuaire marseillais, Rev. intern. Océanogr. méd., 8:51.

Bellan G., 1972, Effects of an artificial stream on marine communities, Mar. Poll. Bull., 3:74.

Bellan, G., 1976, Recherches biologiques dans les grands ensembles portuaires: leurs enseignements, in: "Atti Trav. rod. intern. Difesa e Produttivita del Mare," Livorno.

Bellan, G., 1979, An attempted pollution abatement in the Gulf of la Napoule (Cannes, France), Mar. Poll. Bull., 10:1963.

Bellan, G., 1980, Annélides polychètes de substrats solides de trois milieux pollués sur les côtes de Provence (France): Cortiou, golfe de Fos, Vieux-Port de Marseille, Tethys 9:267.

Bellan, G., (in press), Bilan du détournement d'un petit fleuve côtier dans l'émissaire des eaux usées de Marseille, sur les groupements annélidiens établis sur substrats solides, VIes Journées Etud. Pollutions, Cannes, C.I.E.S.M.

Bellan, G., et Pérès, J.M., 1970, Etat général des pollutions sur les côtes méditerranéennesde France. Cong. intern. "Acqua per il Domani". IVe Inciesta intern. "La Difesa del Mare", Milano, 4/1969, Quad. Civ. Staz. Idrobiol. Milano, 1:36.

Bellan, G., et Bellan-Santini, D., 1972, Influence de la pollution sur les peuplements marins de la région de Marseille (Méditerranée nord-occidentale), in: "Pollution and Sea Life," A. Ruivo Ed., Fishing News Ltd., W. Byfleet, Surrey.

Bellan, G., et Pérès, J.M., 1972, La pollution dans le bassin méditerranéen (quelques aspects en Méditerranée nord-occidentale et en Haute-Adriatique: leurs enseignements), in: "Marine Pollution and Sea Life," M. Ruivo, ed., Fishing News Ltd., W. Byfleet, Surrey.

Bellan, G., Kaim-Malka, R.A., et Picard, J., 1975, Evolution récente des différentes auréoles de pollution marine des substrats

meubles liées au grand collecteur de Marseille-Cortiou, Bull. Ecol., 6:57.
Bellan, G. et Stora, G., 1976, Effets et conséquences d'un apport artificiel d'eaux douces sur les peuplements benthiques d'un étang marin. IIIes Journées Etud. Pollutions, Split, C.I.E.S.M.:147.
Bellan, G., and Reish, D.J., 1978, Technique of studying the modifications of biocoenosis, Rev. intern. Océanogr. méd., 50:5.
Bellan, G., Bellan-Santini, D., et Picard, J., 1980a, Mise en évidence de modèles éco-biologiques dans des zones soumises à perturbations par matières organiques, Oceanol. Acta, 3:383.
Bellan, G., Jorajuria-Olivari, A., et Picard, J., 1980b, Le peuplement des substrats meubles dans le couloir d'écoulement des eaux usées de la ville de Marseille, Ves Journées Etud. Pollutions, Cagliari, C.I.E.S.M.:649.
Bellan-Santini, D., 1962, Etude floristique et faunistique de quelques peuplements infralittoraux de substrat rocheux, Rec. Trav. St. mar. End., 26:237.
Bellan-Santini, D., 1966, Influence des eaux polluées sur la flore et la faune marines benthiques dans la région marseillaise, Sc. Techn. munici., 61:285.
Bellan-Santini, D., 1968, Influence de la pollution sur les peuplements benthiques, Rev. intern. Océanogr. méd., 10:27.
Bellan-Santini, D., 1969, Contribution a l'étude des peuplements infralittoraux sur substrats rocheux (Etude qualitative et quantitative de la frange supérieure), Rec. Trav. St. mar. Endoume, 47:5.
Bellan-Santini, D., 1981, Influence des pollutions sur le peuplement des amphipodes dans la biocoenose des algues photophiles, Téthys, 10:185.
Bellan-Santini, D., et Desrosiers, G., 1976, Distribution du benthos du substrat dur dans un golfe soumis à des multiples pollutions (golfe de Fos), IIIes Journées Etud. Pollutions, Split, C.I.E.S.M.:153.
Bellan-Santini, D., et Desrosiers, G., 1977, Action du rejet thermique de l'usine de Margigues-Ponteau sur les peuplements de substrats durs à dominance de zoobenthos dans la partie supérieure de l'étage Infralittoral, Téthys, 8:83.
Bellan-Santini, D., Harmelin, J.G., et Hong, J.S., 1980, Influence d'un gradient de pollution à dominance industrielle sur la faune des fonds de concrétionnement de type coralligène, Ves Journées Etud. Pollutions, Cagliari, C.I.E.S.M.:771.
Belsher, T., 1974, Séquence des effets d'un égout urbain, en fonction de l'éloignement de la source de pollution, sur les peuplements photophiles de mode battu (fraction algale); premiers résultats, Bull. Soc. Physiol. France, 19:158.
Belsher, T., 1975, Présence d'un peuplement macrophytobenthique dense au débouché en mer de l'égout de Cortiou (région de Marseille-France), Rapp. Comm. int. Mer Medit. 23:91.

Belsher, T., 1979, Essai de définition d'un Indice de Pollution, Rapp. Comm. int. Mer Médit., 25/26:187.

Belsher, T., et Boudouresque, C.F., 1976, L'impact de la pollution sur la fraction algale des peuplements benthiques de Méditerranee, in: "Atti Trav. rod. inter. Difesa et produttivita del Mare," Livorno.

Belsher, T., Cerceau, C., Feunteun, C., et Le Guen, S., 1977, Etude de similitude entre revelés situés dans une zone polluée (région de Marseille, France), Rapp. Comm. int. Mer Médit., 24:199.

Bitar, G., 1983, Influence d'un grand émissaire urbain sur la distribution du zoobenthos de substrat dur dans la région de Marseille (Méditerranée nord-occidentale), Téthys, 10:200.

Bogdanos, C., 1981, A preliminary survey of an area in Oreos Channel (Greece), Ves Journées Etud. Pollution, Cagliari, C.I.E.S.M.: 811.

Bonvicini-Pagliai, A.M., Crema, R., Joannille, E., and Vitali, R., 1980, Spatial and temporal patterns of sandy bottom communities near a coastal power plant, Ves Journées, Etud. Pollutions, Cagliari, C.I.E.S.M.:815.

Bourcier, M., 1969, Ecoulement des boues rouges dans le canyon de la Cassidaigne (décembre 1968), Téthys, 3:779.

Bourcier, M., 1980, Evolution des peuplements macrobenthiques entre la Ciotat et les Iles des Embiez (côtes de Provence). Processus de contamination du Benthos entre bassins côtiers voisins, Téthys, 9:197.

Bourcier, M., 1982, Evolution au cours des quinze dernières années des biocoenoses benthiques et de leurs faciès dans une baie méditerranéenne soumise a l'action lointaine de deux émissaires urbains, Téthys, 10:303.

Cantone, G., and Pilato, G., 1981, Effects of pollution on the polychaetous populations in the Roadstead of Augusta (Sicily), Ves Journées Etud. Pollutions, Cagliari, C.I.E.S.M.:857.

Cantone, G., Cormaci, M., Fassari, G., Furnari, G., Gallazzo, G., Pavone, P.,e Brigandi, S.,1980, Primi dati sul fouling del porto di Catania, Thalas. Salent., 10:44.

Cognetti, G., 1970, Influenza degli inquinamenti sulle popularzioni del benthos marino, Pubbl. Staz. Zool. Napoli, 38(suppl.):5.

Cognetti, G., 1978, On some aspects of the ecology of benthic littoral polychaetes, Boll. Zool., 45:145.

Cocnetti, G., 1982, Aspects de la colonisation des milieux imprevisibles, Symbioses, 14:205.

Crema, R., Bonvicini-Pagliai, A.M., 1980, The relationship between community and pollution abatement in the Gulf of Follonica (Tuscany), Mem. Biol. mar. Oceanogr., 10(suppl.):159.

Cristiani, G., 1980, Biomasse et répartition de l'herbier de Posidonia oceanica de la Côte Bleue (Bouches du Rhône, France) et pollution marine par les métaux lourds, Thèse 3ème cycle, Univ. Aix-Marseille Fac. Sc. Techn. St. Jérome.

Di Geronimo, I., in press, Influence de la pollution sur les peuplements a Mollusques de la baie d'Augusta (Sicile), VIes Journées Etud. Pollutions, Cannes, C.I.E.S.M.

Febvre, J., 1968, Etude bionomique des substrats meubles de l'Etang de Berre, Tethys, 1:421.

Feldman, N., 1937, Recherches sur la végétation marine de la Méditerranée. La côte des Albères, Revue Algol. 10:1.

Feldman N., et Mazoyer, C., 1941, Recherches sur les Céraminacées de la Méditerranée occidentale, Minerva, Alger:510.

Gambi, M.C., et Giangrande, A., 1983, Distribution des annélides polychètes dans les fonds meubles devant l'embouchure du Tibre (Mer Tyrrhénienne), Rapp. Comm. int. Mer Médit., 28:249.

Gamulin-Brida, H., 1972, Pollution de la mer et sa protection avec égard special à la mer Adriatique, in: "Atti 2ème Simp. naz. Conserv. natur.," Bari.

Gamulin-Brida, H., Giaccone, G., et Golubic, S., 1967, Contribution aux études des biocoenoses subtidales, Helgol. wiss Meeres., 15:429.

Gamulin-Brida, H., Pozar, A., Simunovic, A., et Span, A., 1973, Les conséquences biologiques de la pollution des eaux marines sur les biocoenoses benthiques. Groupement des biocoenoses benthiques de la mer Adriatique du point de vue de la pureté des eaux, in: "Atti 5ème Coll. intern. Oceanogr. med.," Messina.

Gherardi, M., e Lepore, E., 1976, Gli anellidi policheti del mar piccolo di Taranto, Rivis. Idrobiol., 15:151.

Ghirardelli, E., 1972, La protezione del mare e dell'Adriatico in particolare, Boll. Zool., 39:427.

Ghirardelli, E., et Orel, G., 1968, Conséquences de la pollution sur les peuplements du "Vallone de Muggia" près de Trieste, Rev. intern. Océanogr. méd., 10:111.

Ghirardelli, E., Orel, G., e Giaccone, G., 1973, Metodologie e ricerche sugli effeti biologici di un impianto di depurazione con condotta sottomarina di liquami domestici trattati nel golfo di Trieste, in: "Atti V Cong. Naz. Soc. It. Biol. Mar."

Ghirardelli, E., Giaccone, G., et Orel, G., 1974, Evolution des peuplements benthiques du golfe de Trieste, Rev. Intern. Océanogr. méd., 35:111.

Ghirardelli, E., Orel, G., e Giaccone, G., 1975, Esperienze sullo scarico a mare di Trieste: metodologie e ricerche par la valutazione degli effetti sul bentos, Ingegneria Ambientale, 4:413.

Giaccone, G., et Rizzi-Longo, L., 1974, Structure et évolution de la végétation marine dans les environnements pollués, Rev. intern. Oceanogr. méd., 34:67.

Golubic, S., 1970, Effects of organic pollution on benthic communities. Mar. Poll. Bull., 1:56.

Gomez-Garreta, A., et Ribera-Siguan, M.A., in press, Etude de l'impact de la construction du polygone industriel de Tarragone sur le macrophytobenthos. VIes Journées Etud. Pollutions, Cannes, C.I.E.S.M.

Griggs, G.B., Grimanis, A.P., and Vassilaki-Grimani, M., 1978, Bottom sediments in a polluted marine environment, upper Saronikos Gulf, Greece, Environ. Geology, 2:97.

Harmelin, J.G., 1964, Etude de l'endofaune des "mattes" d'herbier de Posidonia oceanica Delile, Rec. Trav. St. mar. Endoume, 35:43.

Harmelin, J.G., Bouchon, C., et Hong, J.S., 1981, Impact de la pollution sur la distribution des Echinodermes des substrats durs en Provence, Méditerranée nord-occidentale, Téthys, 10:13.

Hong, J.S., 1980, Etude faunistique d'un fond de concrétionnement de type coralligène soumis à un gradient de pollution en Méditerranée nord-occidentale (golfe de Fos). Thèse 3eme cycle, Univ. Aix-Marseille II.

Kocatas, A., 1978, Contribution a l'étude des peuplements des horizons supérieurs de substrat rocheux du golfe d'Izmir (Turquie). Ege Univ. Fen Fak. Monogr. Ser., 12:1.

Kocatas, A., 1981, Evolution cyclique du benthos dans les zones de pollution du golfe d'Izmir (Turquie). Ves Journées Etud. Pollutions, Cagliari, C.I.E.S.M.:643.

La Noce, T., Pagnotta, R., Pettine, M., and Puddu, A., 1980, Coastal water pollution around Tiber river mouth. A case study, Prog. Wat. Tech., 12:193.

Lepore, R., e Gherardi, M., 1976, Osservazioni sugli anellidi policheti de nuovo porto di Taranto, Oebalia:121.

Leung Tack Kit, D., 1971, Etude du milieu pollué: le Vieux-Port de Marseille. Influence des conditions physiques et chimiques sur la physionomie du peuplement du quai, Téthys, 3:767.

Maggi, P., 1973, Le problème de la disparition des herbiers à Posidonies dans le golfe de Giens (Var), Sc. Pêches, Bull. Inst. Pêches mar., 221:7.

Maggi, P., Gruet, Y., et Lassus, P., 1977, Influence de la pollution urbaine sur la vitalité des herbiers à Posidonies dans le golfe de Giens (Var); Sc. Pêches, Bull. Inst. Pêches mari., 269:5.

Marano, G., Vaccarella, R., Iaffaldano, D., e Bello, G., 1976, Bacino portuale di Bari: Communità'del bentos sessile (1), Oebalia, 2:55.

Marcot-Coqueugniot, J., 1983, Le peuplement algal du port de Port Vendres: Evolution et dérive de 1974 a 1981, Rapp. Comm. int. Mer Médit., 28:223.

Matarrese, A., Sciscioli, M., e Tursi, A., 1979, Aspetti biocenotici dell'inquinamento locale del mar grande di Tarento, in: "Atti 8ême Simp. naz. Inst., Univ. Bari, Conserv. nat" : 71.

Meinesz, A., Astier, J.M., Bodoy, A., Cristiani, G., et Lefevre, J.R., 1982, Impact de l'aménagement du domaine maritime sur l'Etage Infralittoral des Bouches du Rhône (France, Méditerranée occidentale), Vie Milieu, 32:115.

Minas, M., 1973, Sur la synthèse et la dégradation de la matière organique dans l'Ecosystème de l'Etang de Berre. Dynamique et bilans, Rapport avec le régime hydrodynamique. Thèse Doc. d'Etat, Univ. Aix-Marseille.

Molinier, R., et Picard, J., 1952, Recherches sur les herbiers de Phanérogrames marines sur le littoral méditerranéen français, Ann. Inst. Océanogr. Paris, 37:250.

Munda, I., 1974, Changes and succession in the benthic algal association of slightly polluted habitats, Rev. intern. Océanogr. méd., 34:37.

Munda, I., in press, The effects of different pollutants on Marine Benthic Algae, VIes Journées Etud. Pollutions, Cannes, C.I.E.S.M.

Pearson, T.H. and Rosenberg, R., 1978, Macrobenthic succession in relation to organic enrichment and pollution of the Marine Environment, Oceanogr. Mar. Biol. Ann. Rev., 16:229.

Pérès, J.M., 1967, The Mediterranean Benthos, Oceanogr. Mar. Biol. Ann. Rev., 5:449.

Pérès, J.M., 1978, Vulnérabilité des Ecosystèmes méditerranéens à la pollution, Ocean Management, 3:205.

Pérès, J.M., et Picard, J., 1964, Nouveau manuel de bionomie benthique de la mer Méditerranée, Rec. Trav. St. mar. Endoume, 7:3.

Pérès, J.M., et Bellan, G., 1972, Aperçu sur l'influence des pollutions sur les peuplements benthiques, in: "Marine Pollution and Sea Life," M. Ruivo Ed., Fishing News, W. Byfleet, Surrey:375.

Pérès, J.M. et Picard, J., 1975, Causes de la raréfaction et de la disparition des herbiers de Posidonia oceanica sur les côtes françaises de Méditerranée, Aquat. Bot., 1:133.

Picard, D.J., 1965, Recherches qualitatives sur les biocoenoses marines de substrats meubles dragables dans la région marseillaise, Rec. Trav. St. mar. Endoume, 52:1.

Picard, J., 1978, Impact sur le benthos marin de quelques grands types de nuissances liées a l'évolution des complexes urbains et industriels de la Provence occidentale, Oceanis, 4:214.

Picard, J., et Bourcier, M., 1975, Evolution sous influences humaines des peuplements benthiques des parages de la Ciotat entre 1954 et 1972, Téthys, 7:213.

Pignatti, J., e Cristini (de), M., 1976, Associazioni di alghe marine come indicatori di inquinamento delle acque nel Vallone di Muggia presso Trieste, Arch. Oceanogr. Limnol., 15 suppl.: 185.

Reish, D.J., 1959, An ecological study of pollution in Los Angeles-Long Beach Harbors, California, Occas. Rap. Allan Hancock Found., 22:1.

Relini, G., 1972, Degradazione dell'ambiente marino adopera di scarichi di corpi solidi: Alcune consequenze economiche, estetiche ed ecologiche, in:"Atti 2 Simp. naz. Conser. nat.," Bari.

Relini, G., 1980, Insediamento di organismi marini di substrato duro in ambienti portuali mediterranei, Mem. Biol. mar. oceanogr., 10:61.

Relini, G., Montanari, M., Viale, S., e Di Cintio, R., 1975, Considerazioni sulle caratteristiche biologiche ed abiologiche di zone del porto di Genova sottoposte ad intensi scarichi fognari, in: "Atti V Simp. naz. Conserv. nat," Bari.
Relini, G., Montanari, M., Viale, S., e Pisano, E., 1976, Prime fasi di insediamento su substrate duri immersi in acque del Porto di Genova a diverso grado di inquinamento, Arch. Oceanogr. Limnol., 18, suppl. 3:113.
Relini-Orsi, L., e Relini, G., 1971-1972, Recenti uperti ittiologici su Fondi fangosi Batiali del Mar Ligure, Quad. Civ. St. Idrobiol. Milano, 3-4:5.
Riggio, S., e Ardizzone, G., 1981, Eutrofizzazione e communità' bentoniche su substrati artificiali. Indagine preliminare sulle coste della Sicilia nord-occidentale, Quad. Lab. Tecnol. Pesca, 3:605.
Romano, J.C., 1973, Etude cyclique des peuplements marins de substrats meubles situés au débouché en mer du grand collecteur de Cortiou (Marseille), Thèse Doct. Océanogr., Univ. Aix-Marseille II.
Salen-Picard, C., 1981, Evolution d'un peuplement de Vase Terrigène Côtière soumis à des rejets de dragages dans le golfe de Fos, Téthys, 10:83.
Salen-Picard, C., 1982, Contribution a l'étude dynamique des peuplements marins de substrats meubles: les peuplements macrobenthiques circalittoraux soumis à l'envasement dans la région provençale, Thèse Doct. Etat-Sciences, Univ. Aix-Marseille II.
Salen-Picard, C., 1983, Schémas d'évolution d'une biocoenose macrobenthique de substrat meuble, C.R. Acad. Sc., Paris, Ser. III, 296:587.
San Martin, G., Vieitez, J.M., e Campoy, A., 1980, Contribucion al estudio de la fauna de annelidos poliquetos de las costa españolas: polihetos errantes recolectados en la bahia de Palma de Mallorca, Bol. Inst. Esp. Oceano., 6:289.
Sara, M., 1972, I popolamenti bentonici marini e gli effetti dell' inquinamento, in:"Atti II Simp. Naz. Conserv. Nat.," Bari.
Scalera-Liaci, L., 1978, Aspetti ecologici del porto di Manfredonia (Foggia), Oebelia, 4:1.
Specchi, M., e Orel, G., 1966, I popolamenti dei fondi e delle rive del Vallone di Muggia presso Trieste, Bol. Soc. Adriat. Sc. Trieste, 56:137.
Stirn, J., 1973, Ecological consequences of marine pollution, in: "Atti 5 Coll. intern. Oceanogr. med.," Messina:25.
Stirn, J., Avcin, A., Kersan, I., Marcotte, B.M., Meith-Avcin, N., Vriser, B., and Vutovic, S., 1975, Selected biological methods for assessment of marine pollution, in: "Marine Pollution and Marine Waste Disposal," Pearson and Frangipane Ed.
Stora, G., 1982, Recherches de bionomie descriptive et expérimentale (in vivo et in vitro) dans quelques biotopes littoraux soumis à des variations naturelles ou artificielles des conditions du milieu (notamment dans l'étang de Berre et le golfe de Fos),

Thèse Doct. Etat-Sc., Univ. Aix-Marseille II.
Stora, G., in press, Mise en évidence de la parturbation de peuplements benthiques littoraux soumis aux effets synergiques de différents facteurs altéragènes. VIes Journées Etud. Pollutions, Cannes, C.I.E.S.M.
Stora, G., et Romano, J.C., 1980, Influence de la construction et l'exploitation d'un complexe portuaire et industriel sur la distribution et l'évolution des peuplements benthiques littoraux d'un golfe méditerranéen (golfe de Fos), Prog. wat. Techn., 12:137.
Stora, G., and Arnoux, A., 1983, Effects of large freshwater diversions on benthos of a Mediterranean lagoon, Estuaries 6:11.
Tursi, A., Lepore, E., e Gherardi, M., 1977, Relazione fra inclinazione del substrato e insediamento degli organismi bentonici, in: "Atti Cong. Soc. ital. Biol. mar.,"
Tursi, A., Piscitelli, G., Matarrese, A., e Gherardi, M., 1978, Aspetti ecologici del Porto di Taranto (Mar Grande), Oebalia, 4:41.
Tursi, A., Matarrese, A., Scalera-Liaci, L., Gherardi, M., Lepore, E., Sciscioli, M., Piscitelli, G., e Ghieppa, M., 1980, Associazoni bentoniche del Mar Grande di Taranto: primi risultati di une analisi multivariata, Mem. Biol. marine, 10:331.
Tursi, A., Matarrese, A., Piscitelli, G., e Gherardi, M., 1981, Biocenosis del Mar Grande di Taranto, Quad. Lab. Tecnol. Pesca, 3 (1 suppl.):563.
Verlaque, M., 1977, Etude du peuplement phytobenthique au voisinage de la centrale thermique de Martigues-Ponteau (golfe de Fos, France, Méditerranée), Thèse 3ème cycle Océanogr., Univ. Aix-Marseille II.
Zarkanellas, A.J., 1977, Benthic studies in Kavala Gulf, Greece, Thalassogr., 1:266.
Zarkanellas, A.J., 1979, The effects of pollution induced oxygen deficiency on the benthos in Elefsis Bay, Greece, Mar. Environ. Res., 2:191.
Zarkanellas, A.J., and Bogdanos, C.D., Benthic studies of a polluted area in the upper Saronikos Gulf, Thalassogr., 1:155.
Zarkanellas, A.J., and Kattoulas, M.E., 1981, The ecology of Benthos in the Gulf of Thermaikos, Greece. 1. Environmental conditions and Benthic Biotic Indices, Marine Ecol., 3:21.
Zavodnik, D., 1976, Investigation on pollution-stressed littoral communities in the northern Adriatic, in: "Proc. inter. Symp. Mar. Poll. Res.," S.P. Meyers Ed.
Zavodnik, D., 1977a, Benthic communities in the Adriatic Sea: Reflects of pollution, Thal. jugosl., 13:413.
Zavodnik, N., 1977b, Long-term changes of marine vegetation near Rovinj (North-Adriatic), Rapp. Comm. int. Mer Medit., 24:181.
Zunarelli-Vandini, R., and Cognetti-Variale, A.M., 1981, Effets de décharges polluantes sur une communauté a Polychètes de fonds meubles littoraux, Cah. Biol. mar., 22:123.

THE EFFECTS OF THE GEOLOGICAL AND PHYSICO-CHEMICAL FACTORS ON THE DISTRIBUTION OF MARINE PLANTS AND ANIMALS IN THE MEDITERRANEAN

Mihai Băcescu

Musée d'Histoire Naturelle "Grigore Antipa"
1, Chaussée Kisselef, Bucuresti 79744, Romania

Considering the spectacular results of deep-sea geological drilling, as well as the new outlook on the expansion of the ocean bottoms and on continental drift, and especially on the evolution of palaeoclimates, one can trace the evolution of the oceans and therefore better understand the present evolution of marine flora and fauna.

The Mediterranean, with its mosaic of faunas and its varied world of plants and microorganisms raises difficult problems of biogeography. Their solution asks one to know among other things, the geological past of the sector and the inventory, as complete as possible, of the present and past fauna and flora.

As for the present status of the plant world, Boudouresque (1981) sums it up as follows: "The simple census of the species and genera of marine plants is very far from being taken in the Mediterranean; each year, many species new for the region or for science, are being discovered."

Such a statement is in my opinion much more valid for the animal world.

A brief review of the Mediterranean evolution after its separation from the great Tethys Ocean and the formation of the Atlantic Ocean (about 100 million years ago) offers us a first key to understanding the present fauna and flora of this sea and of its annexes.

During the Eocene, the marine climate of the Mediterranean sector continued to be tropical like that of the Tethys Sea; the animals that occurred in that sector were completely different from the present ones; coral reefs and their characteristic fishes prevailed. The palaeo-Mediterranean fauna of Bolca (northern Italy) is self-evident in this respect as well, as the crabs occurring near Vienna and Turkey. The coral reefs of the Eocene resemble the coral ecosystems of the Tethys Sea as they survive in the Caribbean Sea or in the Indian-western Pacific sector.

Beginning with the Miocene, the history of the Mediterranean became very turbulent with rising abyssal waters and rising bottoms that, once on the surface, dried and formed thick evaporitic and salt layers. Other bottoms rose creating sills of the Siculo-Tunisian type, which is at the origin of the separation of the Mediterranean into the two known biogeographical areas. Immense necks of land - from Sicily to the Azores - collapsed, leaving behind only a few islands; likewise, the Thyreneid in the middle and the Aegeid to the east. Once the Pleistocene began the glacial catastrophe took place.

At the end of the Messinian (which corresponds to the Meotian and Pontian) the tectonic stage followed, in which mountains and new deep basins appeared.

At the beginning of the Messinian salinity crisis (6.5-5 million years ago), the Mediterranean, separated from the Atlantic, received freshwater from the area of the Paratethys Sea (the Black Sea region).

During the Miocene, 23-20 million years ago (Sharf el Din and Karam, 1979), strong orogenic movements separated the Mediterranean both from the Indian Ocean and from the northern part of the ancient Tethys - the Paratethys basin. The sea water grew cold causing a first massive reduction of the Tethys thermophile fauna; it is only the Tethysian Foraminifera fauna that seems to have partially survived. However, 14-15 million years ago (upper Miocene), the Mediterranean fauna required its tropical character with a rich generic and specific diversity.

During the Pliocene, some other tropical survivors were replaced by northern, temperate-water general, a fact well-illustrated by the changing of the echinid and molusc populations: the genera Clypeaster and Diadema for example were followed by the genera Echinus, Sphaerechinus, Strongilocentrotus. Moreover, the Pleistocene ice age resulted also in the disappearance of many temperate-water organisms. Accordingly, the Pliocene and especially the Pontian, offer a second key to explain the composition of the present Mediterranean fauna, formed of the following categories, of various zoogeographical origins, located in various sectors descendants of the Tethys tropical water survivors; Pliocene boreo-Atlantic elements; post-glacial Pleistocene elements; glacial, boreo-arctic relicts; immigrants from the Red Sea; Pontic, Pleistocene and

Quaternary elements; endemics of various components that were mentioned above.

Certainly, I shall only mention that besides the movements of the under-ocean crust, which greatly influence the ecosystems, there also are sudden natural catastrophes that can destroy in a few seconds not only entire ecosystems, but their substrata as well. I would mention the earthquakes, the case of the Thyreneid or of the Aegeid, or the volcanic eruptions (as a result of an eruption, the Santorini island has become the archipelago that we see today).

But to these natural catastrophes, one should also add those caused today by man: radioactive pollution or pollutions by industrial wastes, hydrocarbons, coastline modification, etc.

The pollution of the Gulf of Genoa, of north-western Adriatic, of the north-eastern corner of the Marmara Sea and of western Black Sea are good examples. The dams of the rivers flowing into the Black Sea and the Asswan dam are further eloquent examples.

The regulation of the Nile's floods and the reduction of its alluvial deposits with their biogenic salts are affecting the ecosystem of the Levantine sector of eastern Mediterranean which is in the process of formation.

The influence of this new anthropogenic activity - the river dams and the location of new industries, for the moment with negative results for the ecosystems - must be kept under continuous study.

The Ligurian Sea (under the influence of ever-more-numerous industries along the Rhone), the predelta area of the Po in the northern Adriatic and that of the Levantine Egyptian coast, under the influence of the Suez Canal and the Nile basin - are all sectors to be closely investigated in monitoring the evolution of ecosystems that were thoroughly studied some time ago, that is those with excellent baseline data.

The geological data, correlated with the hydrobiological, palaeoclimatic and biological criteria, confirm the geographical division of the Mediterranean into a western basin and an eastern one isolated by the Siculo-Tunisian sill (400m) - a geographical, hydrological, and climatic frontier, as well as its separation from the Atlantic about 100 million years ago as an independent province separated by the Gibraltar Strait.

Among the specific features of the great Mediterranean province, I mention below the following:

- the Atlantic superficial current occurring all along the southern coast and affecting the shallow waters;

- a permanent 12°-$13.50^{o}C$ homothermy, established according to the meridians from depths of 100-400m to maximum depths, this homothermy allowed the extension of the vertical distribution of certain littoral species (Ianirella bonnieri, Desmosoma chelatum) (George and Menzies, 1968),
- the scarcity of biogenic salts in the water and a rich content of Ca in the sediments; being an isolated sea that evolved between the same limits the Mediterranean engendered a large number of endemics; 19 species of isopods, 44% among demo sponges, etc.,
- the absence of a true abyssal fauna (Tortonese, 1951) caused by the stagnation of waters exceeding 2000m in depth and the Gibraltar Strait that prevents the penetration of the rich abyssal stenothermic fauna of the Atlantic.

As showed by the Lamont Laboratory analyses (Menzies et al., 1973; Ninkovich and Heenzen, 1967) the black silt layers common on the eastern Mediterranean bottoms were and still are unfavourable to life and the appearance of these black sediments was repeated in various periods.

Today one could find there conditions similar to those of the Black Sea sediments. A completely azoic black silt layer was deposited 156,000 years ago during both the Wisconsin ice age and the last one (dated 9000 years ago).

The poverty of abyssal benthos in the post-Pliocene and recent periods, resulted because time was sufficient neither for a recolonization nor for an active genogenesis.

The reduced number of abyssal isopods in the Mediterranean do not belong to endemic genera as one might expect but to genera with a large subtropical distribution, perhaps of Tethysian origin, which disappeared from the Mediterranean because of the azoic bottoms (George and Menzies, 1968). A single species of Brisingeidae, for example, reaches 3000m deep.

In the western sector, the fauna is of a north-Atlantic or Ibero-Moroccan type while to the east, it is of an Indian-western Pacific type, partly of a probably older origin than that which penetrated through the recent Suez passage.

As the distinguished marine botanist Feldman (1937) showed, the Mediterranean may be considered - as far as algae are concerned - as a temperate domain in winter and a subtropical one in summer with 12-$14^{o}C$ annual temperature differences.

The eurythermic algae are generally located towards the coasts all through the year; the stenothermic ones are located deeper, where the annual temperature difference is not too strongly felt.

Given the water transparency, benthic diatoms occur up to 200m deep to the west of the Mediterranean. Strong currents in the littoral belt protect the algae against the deposition of mud which threatens them during the calm periods; it is also the currents that influence the duration of emersion of the algae.

Given the high sill of Gibraltar and the existence of the Atlantic current that continually brings Ibero-Moroccan elements, i.e. cold waters and subtropical waters at the same time, the western Mediterranean is characterized by a richer ocean plankton; a wide continental shelf; a weak salinity (36.5 o/oo in the Alboran basin, reduced by the freshwaters of the Rhone, and increasing to 37 o/oo on the North-African coasts) before mixing with the eastern Mediterranean waters in the Tyrrhenian Sea area.

For such reasons these two great sectors may be also separated into other zoogeographical areas, taking into account the climate, the hydrological conditions and the floristic and faunistic composition: the Alboran Sea with the Gulf of Lyon where the salinity of the Atlantic waters (originally 37 o/oo) decreases to 35 o/oo under the influence of the waters of the Ebro or of the Rhone, and the homothermy begins from 50-100m deep (Oliver, 1983); the North-African coasts rich in nutrient salts and with a 36.5-37.5 o/oo salinity etc.; the Sirte Gulf where the Atlantic fauna changes; the Adriatic, the Levant, the Aegean Sea, the Pontic domain - all well circumscribed as separated units, even by the composition of plankton.

I deal below with some of these sectors which are more isolated and less studied from the biogeographical viewpoint.

1. The Adriatic Sea

In contrast to all the biogeographic sectors of the Mediterranean, the Adriatic sector - especially in its northern part - is characterized by its special benthic flora, among which survived many boreo-Atlantic, North Atlantic and even Pontic temperate elements carried by the Sarmatian Sea (Paratethys) with which it was connected from the Pliocene.

Among these elements the following are worth mentioning; *Fucus*, *Catenella*, *Pilaiella*; associations of brackish or even freshwater phanerogams (*Najay*, *Utricularia*, *Potamogeton*, *Ceratophyllum*, *Zanichelia*, *Zostera*), to note but the phytobenthos and the abundance of halophile characea. The latter achieved, in NE Adriatic, a true "pseudo-coralligenous structure" (Lovric, 1981) that I would better designate as a "coralloid structure"; and this because they are - just like the "coralligenous structures" of other plants in the

Mediterranean - only structures that are similar (-oid) to the coral reef structures, but have nothing in common with the birth of corals, as the "genous" ending suggests. Accordingly, the term to be used for these structures is "coralloid" and not "coralligenous", an etymological nonsense.

Despite the fact that the large number of endemics mentioned for the Adriatic (25% for the benthic flora) - that given by Gamulin-Brida and Span in 1981 (100) as well as that noted by Giaccone in 1981 (169) - has been criticized, it shelters, however, more endemics than the Mediterranean (especially plants); dozens of species have been described bearing the specific name adriatica (Gamulin-Brida and Span, 1981).

As for the fauna of this sea, which according to Rield could amount to more than 10,000 species, the inheritance of Paratethys and the long freshwater contribution of the Pô, rich in nutritive salts and detritus, created in the north ecosystems similar enough to the ecosystems of NW Black Sea. I would mention the biocoenoses dominated by Lentidium mediterraneum (up to 130,000 specimens/m^2), Balanus improvisus and Cyclope neritea on sandy-silty bottoms, while in brackish waters, a large quantity of organic matter: Glycera convoluta, Abra alba, Spisula subtruncata, Nephthys hombergi.

Among animals, there are also more boreal elements as compared to the remaining part of the Mediterranean (40 species) and many Paratethysian species especially in northern Adriatic, including 8 species of endemic coral; Epizoanthus - with 5 species; many Turbellaria unknown in the Mediterranean (Childia groenlandica, Callinera bürgeri, Saccoglossus nereschouskü (Riedl); and finally, Paratethysian species and many boreal-celtic forms which have survived in the northern Adriatic since the Pleistocene or the ancient Quaternary as Taviani stated in his speech about Mrs. Gamulin-Brida and Span's report in 1981. I would also note the rich association of Ponto-Caspian forms or of forms related to them, coming to the northern Adriatic also through the south-eastern canal of Paratethys (Acipenser stellatus, black gobies, cobitids, copepods, Siriella j. jaltensis).

Although in the Adriatic the NE-SW migration came to an end at the beginning of the Quaternary, in the Aegean Sea it is still going on, via the Bosphorus, but with a reduced intensity.

Conclusively, the northern Adriatic Sea belongs to the Paratethys basin while the southern sector with its more thermophile species indicates a Tethysian heritage.

Consequently, the biota of northern Adriatic, considering its phytogeographic features, has a boreo-Atlantic aspect, while the organisms from the south (e.g. Dubrovnic archipelago) show Levantine and eastern Mediterranean affinities (Lovric, 1981); the south preserves Tethysian thermophile elements, or elements enriched with tropical elements, recalling also the eastern Mediterranean.

The geographic isolation of the Adriatic favours speciation beginning both with the ancient stock inherited from the Paratethys and with the species of the eastern and western basins which meet in the southern area.

Anyway, the vast pre-delta ecosystem of the Pô - that seems to be a broken part of the NW ecosystem of the Black Sea - clearly demonstrates its palaeogeographic affinity with the old Tethys Sea.

2. The Eastern Mediterranean

Although the Adriatic sector is well-settled as far as vegetation and fauna are concerned, the Eastern Mediterranean is a sector undergoing a strong faunistic movement and transformation.

It is characterized ecologically by 12.75-13.55^{o}C in the bottom waters and 15-16°C at the surface, in winter (but the annual average temperature is 20°C (Allain et al., 1966; Furnestin, 1979; Lakkis, 1981); salinity = 38.7-39.5 o/oo in winter decreasing to 28 o/oo during the floods of the Nile; the homothermy is established towards 400m, but it can begin even at 75m (Lakkis, 1981). The sediments are poor in nitrates, but contain 50% Ca and are rich in organic substances (particularly remains of foraminiferans).

From the biological point of view, the Eastern Mediterranean is distinguished by a great percentage of thermophile tropical and subtropical elements originating from three different sources: direct descendants of the Tethys Sea population which survived in various niches; immigrants from the Indian Ocean - through the subsequent pliocene communications - or from the Red Sea. As a matter of fact, the erythraeic elements are the main characteristic of this this sector.

Among the Tethys relicts are surely Cassiopea andromeda (Schäfer, 1955), many foraminiferans, pteropods (Cavolinia g. gibbosa) (Rampal, 1981),many copepods and a few Hydromedusae as well as the fishes Aphanius fasciatus, Pseudoscarus cretensis (Kosswig, 1948).

In its northern part many north-Atlantic and boreo-Atlantic forms survived - besides Ponto-Caspian elements that immigrated during the east-west Tortonian transgressions (or Badenian ones, according to Voicu, 1981). Present colonists which came through the Suez Canal (lessepsian immigrants) - e.g. Rhysosolenia alata indica, one dozen of certain lessepsian algae and 11 probable lessepsian algae of which 5 occur in the western Mediterranean (Por, 1978), many commensals of sponges (Synalpheus gambarelloides and Alpheus dentipes in Euspongia officinalis (as a matter of fact 7 sponge species the Alpheidae prevailed) the goby Cromogolins quadrivitatus recorded in the waters off Rhodes (Koukouras et al., 1979).

The abundance of sponges mentioned by Forbes since 1844, is also a special characteristic of eastern Mediterranean.

In the Libyan sector begins the replacement of Atlanto-Mediterranean temperate forms by subtropical ones. Just as in the Adriatic, the Mediterranean plankton of the eastern basin shows minimum values (Tortonese, 1951; Furnestin, 1979). Haritonides et al. (1981) identified 100 taxa in the flora of the northern Sporades islands and in the Gulf of Pagassitikos; in the former subtropical species dominated while in the latter, Atlanto-Mediterranean species.

Certainly it is difficult to establish positively the affinity of any one of these species for one or another of these categories and especially to distinguish an immigrant from a true relict (Holmquist, 1966) as far as one does not know well the fossil forms of eastern Mediterranean and of the Aegean Sea, and as one has not a precise faunistic index compiled before the opening of the Lessepsian Canal.

A good example showing the importance of fossils to identify the origin and period during which certain organisms occurred is provided by Vitreotubus digeronimoi, found in the Pleistocene layers of the lower bathyal floor in Sicily (Zibrowius and Bitar, 1981); this species still lives in the Azores and in the Indian Ocean; in 1982, the same authors found 5 more erythraeic polychaetes in the Mediterranean.

Among the more than 250 polychaetes recorded by Amoureux et al. (1978) in the Aegean Sea, 77 species also occur in the Red Sea and the rest of the Mediterranean and 72 also in the Red Sea and Indian Ocean, which indicates that most of them show erythraeic affinities whose certain origin is deduced only with difficulty.

The fauna of the Mediterranean sector shows a decrease in the number of species and in the number of individuals from the west to the east; in addition, many species occurring also in the western Mediterranean are affected here by dwarfism (Pérès and Picard, 1958).

A global poverty is seen in all the holobios here (for example, among the 30 species of Ampelisca from western Mediterranean, only one was recorded in eastern Mediterranean among the 51 cephalopods on the west, only 30 were also found in the east etc. Almost the same species of algae and phanerogams exist also in eastern

Mediterranean, but they have a poorer distribution and development than in the west. Here the phanerogam Halophila stipulacea Forsk, is abundant, an immigrant from the Red Sea as well as Cymodocea, considering the silty bottoms rich in Ca and organic substances. The ecosystem of Posidonia oceanica which forms one of the most characteristic ecosystems of the Mediterranean province is in regression in the eastern Mediterranean and does not occur below 35m deep.

The algae Sarconema filiforme, Solieria dura count among the Tethysian relicts; in exchange, Acanthophora seems to have entered the eastern Mediterranean through the ancient openings of the Suez Canal; moreover, a species of Sargassum and of Caulerpa prolifera were recorded before near Rhodes. Charylodes longicornis is considered by Tortonese (1959) as a Tertiary relict.

The lessepsian migration has been rigorously studied during recent decades (Steinitz, 1929; Ben Tuvia, 1971; Por, 1978), and it might be a serious reference but the percentage advanced by Por (1981), i.e. 15% erythraeic elements which came through the lessepsian migration, needs new confirmations. It is not known, for example, whether certain species, recorded only in recent years, did not exist there before the opening of the Suez Canal (the Case of Athanas or Diogenes and its parasites in Black Sea).

The same thing is valid for the newly described species, among which there are some with obvious Indo-western Pacific affinities (more than 20 Peracarida; Alpheus migrans and other 3 shrimps; 5 oligochaetes, many Turbellaria etc.); can they not be descendants of some taxa surviving from the Tethys or from the Pleistocene ?

Around the Eastern Mediterranean, in the phreatic fluvial-lacustrine domain and in the caves, many representatives of the Tertiary marine fauna survived, driven towards the phreatic waters, as a result of water cooling or by the regressions and transregressions of post-Tertiary seas, so that it is not difficult to suppose that a great percentage of other animals could survive directly in certain niches, at sea. For example, in the first case are known many freshwater cyprinodontids (Anatolichthys, Koswigichthys etc.) and many troglobitic crustaceans in Turkey (Kosswig, 1948); hydrocarians (Motas, 1962); Thermosbenaceae, Bathynellaceae, Thyphlocaris etc. in the Jordan Valley or Tiberiada lake (Por, 1975); Troglomysis, Spelaeomysis proteus in Yugoslavia or in southern Italy, etc.

The abyssal forms lack also on the azoic bottoms of that region, and even the bathypelagic forms are in a reduced number (Menzies, 1972); but, as George and Menzies (1968) wrote "for a better understanding of abyssal zoogeography and origins of Mediterranean deep sea fauna, more data should come from modern intensive surveys and expeditions".

The most interesting area of the Eastern Mediterranean is the Levantine sector, "a fascinating scene of the zoogeographic modifications of the present marine world" (Por, 1975). Many aspects of this sector have been studied by the researchers of Israel (particularly Por, 1975, 1978, 1981), But Tortonese (1951), Băcescu (1961), Furnestin (1979), Lakkis (1981) have also contributed. The last author mentions 263 planktonic algae 29 of which have an erythraeic origin and 300 zooplanktonic ones, 6 of which are Indo-Pacific recorded for the first time in the Mediterranean. Almost the whole plankton in this area is of Mediterranean temperate-warm and subtropical type; the floods of the Nile engender an important development of the phytoplankton along the Levantine coast.

Among the 284 fish species recorded along the coast of Israel, 30 are lessepsian immigrants (Ben Tuvia, 1971). Siriella thompsoni probably owes its presence between Cyprus and Crete to the lessepsian immigration (Băcescu, 1961).

Por (1975) puts forward the idea that a lessepsian province might even be envisaged having as northern limit the area where the winter temperature does not go below 16°C. It might shelter 10 to 15% lessepsian immigrants as compared to the natives and this percentage is growing, a fact proved by recent discoveries: 6 opistobranch (Barash and Danin, 1972); 12 species of cumaceans, 10 tanaidaceans, 8 mysids (Băcescu, 1977) etc.

In the NE waters off Libya prevail Atlanto-Mediterranean and Ibero-Moroccan species as well as endemics: Paramysis motasi, Heterotanais longidactyla, Heteromysis libyana (Băcescu, 1977); Leptanthura mülleri, Apanthura libyana (Negoescu, 1981) etc.

Certainly, the coral world - the most characteristic ecosystem of the warm Tethys - disappeared in many zones of its vast area because of the replacement of the warm climate by a temperate or even cold climate.

The thermophile fauna also largely disappeared or withdrew to the phreatic domain and to a few caves. Thus, many psammobitic representatives of the Tertiary fauna have been preserved; others have survived associated with sponges.

In the first category I would like to mention the hydracarians, the amphipods, the isopods, the mystacocarida (Derocheilocaris (occurring on the littoral sand of Israel), but especially the Thermosbenaceae and the Bathynelaceae. A single example would tell convincingly about the fauna of Tethys. Bogidiella or subgenus Medigidella - a subgenus very wide-spread in the caves around the Mediterranean - includes a species, i.e. B.(M.)arganoi (Rufford Vigna, 1973), also in the Gulf of Mexico; in 1983, Stock describes a new species B.(M.)sarawacensis of Kalimantan hypogée. Whence the conclusion that some Bogidiellidae occupied the inland waters before the Pangea split(Stock, 1983) (why wouldn't we consider that they came from the Tethys?).

Taking into account that the fauna of Levant differs much from that of western Mediterranean, just in the presence of thermophile elements, I consider that its poverty should not be explained through the impoverishment of the Atlanto-Ibero-Moroccan fauna, as it is usually asserted (Por, 1975), because this sector had not that fauna. The Tethysian fauna was impoverished by the climatic vicissitudes of the past and by the present winter cooling of the shallow waters up to 16-17°. Certainly, the coral reefs, with their rich fauna did not survive in that area; here and there isolated deep-sea species were preserved; abundant are the sponges in whose association surely survived ancient forms (perhaps new species of Synapseudes, Heteromysis (Băcescu, 1977), or of Synalpheus; S. hululensis recorded in the sponge Eusopongia officinalis of the Aegean Sea, (Koukouras, 1979a,b) also belongs to this category.

3. The Aegean Sea and the Pontic Domain

Colder waters; lower salinities towards the Marmara Sea and the Black Sea; influence of the Pontic fresh waters; absence of a thermophile fauna; richness of boreal - Atlantic and Ponto-Caspian elements that could be included in the term "Paratethysian descendants"; the azoic bottom established at depths shallower than the rest of the Mediterranean - here are a few characteristics of the Aegeic sector.

The fauna of the Aegean Sea as well as that of southern Turkey began to be actively studied in recent years.

Among the more than 200 species, 106 (54.92%) arrived on the coast of Israel and on that of Syria, only 40 reached the Anatolian coast and the waters off the islands Rhodes, Cyprus, Crete, Santorini (Demir, 1952; Geldiay and Kocatas, 1968a,b; Kocatas, 1981), while about 20 did not go beyong the waters off Egypt (e.g. 10 epiphytous algae - Aleem); in the Aegean Sea, the lessepsian immigrants do not extend beyond northern Crete (only 3 species), as the warm water fauna does not surpass the Andipsara-Euboea line (Por, 1978).

Por (1978) presents the following as characters identifying a lessepsian immigrant: stenohalinity, monophagy, the necessity of a certain substratum and reproduction through planktono-trophic larvae for the holoplankton.

Recently special attention has been paid to the parasites of the species which utilized the lessepsian migration; the fish

Aphanuis dispar entered the Mediterranean with its ectoparasite Ancyrocephalus salinus; 2 species of Siganus came with their monogenes of genus Pseudohalitrematoides.

Moreover, the parasites of the crustaceans Athanas nitescens and Diogenes pugilator (Indo-western Pacific origin) - Paratheges racovitzai, Anisarthrus pelseneeri, Pseudione diogeni, Septosaccus cucnoti, Liriopsis pygmaea (Godreanu and Godreanu, 1956) - were preserved; they are also Indo-Pacific elements, but older ones that survived from the ancient Tethys Sea.

A few, still-living tropical species or some others which have disappeared from the Aegean Sea, were found as fossils in the 3500 year-old hologenous sediments around Crete. The sponges Petrobiona or Merlia normani (Labourel Vacelet) species that do not live any more in that area, rapidly disappear as a consequence of the warm-cold alternation, just like Lithophylum tortuosum of eastern Mediterranean; other species become rarer, e.g. Coralium rubrum, certain molluscs, and even sponges.

In conclusion, I would like to underline that the Aegean Sea raises difficult biogeographic problems especially because the knowledge of the physico-chemical and dynamic structure of the Greek seas, as well as the inventory of the organisms are matters tackled only these last three decades (Băcescu, 1978) despite the studies carried out during the expeditions 140 years ago (Forbes, 1844) and despite the first precise data on the benthic populations obtained by the "Calypso" campaigns (1954-1955). Pérès and Picard (1958) state that "these campaigns are not sufficient to fill in the serious gaps in our knowledge of benthos in eastern Mediterranean (i.e. Aegean Sea) ....for comparison of the western and eastern basins."

These authors conclude that the populations greatly resemble the well-represented corraloid ecosystem, but the latter shows a facies of eastern type and descends deeper than in western Mediterranean (100-200 m as compared to 60-70 m), being poorer in associated forms (Pérès and Picard, 1955). The presence of the populations of Halophila stipulacea and of the alga Caulerpa prolifera - a eurytop and eurybathial alga descending to 146 m, the absence of Isidella elongata, so common in western Mediterranean are also characteristic of the Aegean ecosystem.

As for the Pontic sector, more than 50 relict species of Ponto-Caspain type were found not only around the Aegean Sea, but also in the rivers and lacustrian basins, to the far east of Europe and Asia (Mardukhai-Boltovskoi, 1964).

I quote after Băcescu (1966): "Dans nombre de rivières de l' Europe, d'Anatolie et jusqu'au lointain bassin de la mer d'Aral, ou dans le lac de Tibériade et dans l'Euphrate, on a trouvé et on continue toujours de trouver, des espéces caspiennes ou des espèces qui leur sont apparentées de près, par exemple: Cobitis (Sabanejewia) larvata De Fil. et C.(S.) conspersa Cant., dans le bassin du fleuve Pô en Italie; Paramysis kosswigi Bac. dans le Grand Méandre en Turquie; espèces de Schizopera, Nitocra et Ectinosoma dans le lac de Tibériade etc... d'autres ont continué à être formées dans l'espace des mers du Pliocène et du Quaternaire recent (les dizaines d'espèces de mysides, de cumasés, d'amphipodes, de polyphémides, de pyrgulides, de gobiides, etc. de type caspien, des rivières et des limans des bassins pontique et azovien et même méditerranéen)."

"Cependant, un fait est sûr, notamment, bien que déjà endémisées en genres nouveaux ou espèces nouvelles - ou bien ressemblant encore aux taxons caspiens d'aujourd'hui, les formes relictes éparses dans l'ancien aréal sarmatien, ou même au-delà de ce basin, font preuve d'une grande vigueur; elles se trouvent en plein processus de spéciation et sont capables de peupler de nouveaux bassins naturels ou artificiels" (fin citation).

The Pontic Sea had a fauna similar to that of the present Caspian Sea. At the end of the Pliocene, it separated from the Black Sea; the disappearance of the Manytch strait allowed the formation of the Palaeouxin basin; at the same time, a new communication appeared with the Marmara Sea; the salinity grew and an important "Mediterraneanization" began during the inter-glacial Riss-Würm, forming the Karangat Sea. During the post-ice age, a second Mediterraneanization of the Black Sea took place when the ancient fauna withdrew to the estuaries, about 50-30 thousand years ago acquiring more euryhaline characteristics.

Anyway, the Black Sea, daughter of the tertiary Mediterranean - heir of some ancient elements and host of many new elements that came through the Bosphorus strait - is available now to eastern Mediterranean species born meanwhile in the closed Pliocene seas seas which locally preceded it.

The palaeontological research conducted in the Paratethys Sea area revealed the presence of many species related to those of the present Ponto-Caspian domain; recently the statolith analyses identified Ponto-Caspian mysids in various Tortonian layers (Voicu, 1981), a fact that will contribute to the clarification of their origin (Voicu, 1974).

Unfortunately, a part of the infralittoral ecosystems specific to the Black Sea, such as the ecosystems of Phyllophora or the ecosystem of deep-sea Mytilus, are seriously threatened by other factors; Phyllophora, by a weakening of the bottom current and Mytilus for biological reasons, i.e. the penetration of the rapacious gasteropod Rapana venosa into the Black Sea. It has succeeded (during 10-15 years) to completely destroy (though consumption) the ecosystem dominated by Ostrea edulis taurica of Gudansk (north-eastern Black Sea) and is on the point of destroying the ecosystems of the deep-sea mussels (Mytilus galloprovincialis frequens).

The penetration of the bivalve Mya arenaria, 25 years ago, unbalanced and completely replaced the upper infralittoral ecosystem of Corbula (Lentidium) mediterranea (Costa), the most characteristic ecosystem of north-western Black Sea, and this because of the simple competition for the biotope and food.

Several successive freezings in the littoral waters in recent years, associated with massive littoral management which strongly disturbed the waters, contributed to the disappearance of 80% of the ecosystem dominated by Cystoseira barbata and of that of Zostera on the Rumanian coastline.

The accentuation of the Marmara Sea pollution towards the mouth of Bosphorus seriously threatens also the Prebosphoric circalittoral ecosystem, which is such an important base for acclimatization and supply of the Pontic sector, with fauna from the Aegean Sea; this ecosystem had 8.2% benthonic species which had not been found in other zones of the Black Sea.

The pelagic ecosystem was also modified, being deprived of the main commercial fishes (Scomber, Trachurus Trach. mediterraneus, Pelamys etc.) that were still coming regularly from the Marmara Sea 25 years ago.

Synthetical papers concerning the ecosystems of the Mediterranean in general have been written (Pacem in Maribus, 1972), on benthos (Pérès, 1961), on plankton (Furnestin, 1979) or partially: on the Adriatic (Lovrič, 1981; Gamulin-Brida and Span, 1981), on the Levant (Por, 1978), on the Aegean Sea (Pérès and Picard, 1958; Haritonides, et al., 1981) or relative to the Pontic basin (Băcescu et al., see the 5 volumes of marine ecology published from 1967 to 1975 in which 1369 animal species and 700 algae, found in 1000 quantitative stations are recorded).

These synthetical papers will facilitate the knowledge of the present evolution of the ecosystems, providing a good reference.

The ever-more-obvious development of the oceanographic and biological-oceanological research in eastern Mediterranean, in the Aegean Sea and in the Black Sea - on which the ICSEM activity is also focused - will surely offer new starting points. This will probably supply new ways and means to fight against the alteration of the ecosystems which are important for the economy of the sea and for that of man: e.g. the results of the "Ramoge" campaign for the depollution of the Gulf of Genoa are encouraging enough in this respect.

In conclusion, it results that the present situation of the semi-closed basin of the Mediterranean is not but a temporary stage in the evolution of Terra and of the perpetually changing life.

SUMMARY

The geological events causing the separation of the Mediterranean from the Tethys Sea and its links to the Paratethys are briefly summarized. The part played by the eventful Miocene and Pliocene periods (with ice ages) in the achievement of the present composition of some Mediterranean ecosystems is pointed out.

The ecosystems of the Adriatic Sea are analysed and compared to both the western Black Sea and the rest of the Mediterranean.

The faunistic and floristic composition of eastern Mediterranean (Levant Sea in particular) is discussed, including survivors of the Tethysian warm fauna and erythraeic elements.

The third area of discussion is that of the Aegean and Pontic Sea, with the reciprocal influences that determined the composition of the ecosystems; the past and present exchanges between the Black Sea and Aegean Sea basins over the Marmara Sea carried many Sarmatian and palaeo-Euxinic elements into the Aegean Sea, which always filters Mediterranean elements into the Black Sea. Palaeontological research contributes fully to the understanding of the present distribution of the marine plants and animals of the Mediterranean.

REFERENCES

Allain, Ch., Furnestin, M.L., et Maurin, Cl., 1966, "Essai de synthèse sur l'océanographie physique et biologique dans la zone d'influence du courant atlantique en Méditerranée du Sud et du Levant,"C.I.E.S.M.," Paris.

Amoureux, L., Rullier, F., et Fishelson, L., 1978, Systématique et écologie d'annelides polychètes de la presqu'île du Sinai, Israel J. zool., 27:57.

Băcescu, M., 1961, Contribution à la connaissance des Tanaidacés de la Méditerranée orientale. I. Les Apseudidae et Kalliapseudidae des côtes d'Israel, Bull. Res. Coun. Israel, 10:137.

Băcescu, M., 1966, Die Kaspische Reliktfauna im pontoasowschen Becken und in anderen Gesässern, Kieler Meeresf., 22:176.

Băcescu, M., 1977, Contribution à la connaissance des peracarides des eaux du Nord-Est de Libye, Rapp. Comm. int. Mer Médit., 24:2.

Băcescu, M., 1978, Réflexions sur la zoogéographie et l'Ecologie de la Grèce: présent et perspectives, in: "Symp. int. zoogéogr. Ecol. Grèce," Athens.

Barash, A.D., and Danin, Z., 1972, The Indo-Pacific species of Mollusca in the Mediterranean, with notes on a collection from the Suez Canal, Israel J. Zool., 21:301.

Ben-Tuvia, A., 1971, Revised list of the Mediterranean fishes of Israel, Israel J. Zool., 20:1.

Boudouresque, C.F., 1981, Rapport bibliographique sur le Phytobenthos méditerranéen (années 1978-1980), Rapp. Comm. int. Mer Medit., 27:69.

Demir, M., 1952, Bogaz ve Adaler Gahillerinin Omurgasiz Dip Hayvanlari (Bosporus), Hidrob. Mecm. Istambul, A, 2:1.

Feldman, J., 1937, Recherches sur la végétation marine de la Méditerranée. La côte des Albères, Rev. algolog., 10:8.

Forbez, E., 1844, Report on the Mollusca and Radiata of the Aegean Sea and their distribution considered as bearing on geology, Brit. Ass. Adv. Sci., 30:193.

Furnestin, M.L., 1979, Aspects of the zoogeography of the Mediterranean plankton, in: "Zoogeography and diversity in plankton," S. Vander Spoel and P. Bults, eds., Utrecht.

Gamulin-Brida, H., et Span, A., 1981, Situation biogéographique de la Méditerranée et de l'Adriatique en tant que sous-région distincte, Journées Etud. System. Biogéogr. Medit. C.I.E.S.M., Cagliari,:73.

Geldiay, R., and Kocatas, A., 1968a, Report on a collection of Natantia from the Bay of Izmir and its neighbourhood, Sci. Rep. Fac. Sci. Egé Univ., 51:1.

Geldiay, R. and Kocatas, A., 1968b, The species of Peneidae family from the Mediterranean coast of Turkey, Sc. Rep. Fac. Sci. Ege Univ., 58:3.

George, R.Y., and Menzies, R.J., 1968, Additions to the Mediterranean Deep-Sea isopod fauna (Vema-14), Rev. Roum. Biol. Zool., 13:367.

Giaccone, G., 1981, Originalité biogéographique dans la composition des peuplements floristiques en Adriatique et dans le détroit de Messine, Journées Etud. System Biogéogr. Medit. Cagliari, C.I.E.S.M.,:95.

Godreanu, R., et Godreanu, M., 1956, Sur l' Anisarthrus pelseneeri, épicaride, parasite abdominale de la crevette Athanas nitescens, Bul. St. Sect. Biol. Acad. R.P.R., 8:567.

Haritonides, S., Tzavellas, K., and Tsekos, I., 1981, Contribution to the study of marine flora of Pagassitikos Gulf and of N. Sporades Islands, Greece, Rapp. Comm. int. Mer Medit., 27:151.

Holmquist, Ch., 1966, Die sogenannten marin-glazialen Relikte nach neuren Gesichtspunkten, Arch. Hydrobiol., 62:285.

Kokatas, A., 1981, Liste préliminaire et répartition des crustacés décapodes des eaux turques, Rapp. Comm. int. Mer Medit. 27:161.

Kosswig, C., 1948, Uber Tethysrelikte in der türkischen fauna, C.R. Soc. Turq. Sci. Phys. Natur., 10:31.

Koukouras, A., 1979a, Some interesting zoogeographical notes on some species of the benthic fauna of the North Aegean Sea, Biol. Gallo-Hellenica, 8:49.

Koukouras, A., 1979b, Bionomic study of the macrofauna of the mediolittoral soft substratum in Strymonikos and Thermaikos gulfs, Doct. Thès. Univ. Thessaloniki.

Koukouras, A., Voultsiadou, H., Doumas, C. Gogon, A., and Chindiroglou, H., 1979, Preliminary results on the quantitative and qualitative composition of the fauna associated with littoral sponges at Chalkidiki peninsula, Biol. Gallo-Hellenica, 8:40.

Lakkis, S., 1981, Le plancton des eaux libanaises. Caractéristiques biogéographiques, Journées Etud. System. Biogeogr. Medit., Cagliari, C.I.E.S.M.:59.

Lovric, A.Z., 1981, World interpolation of the Adriatic Phytobenthos, Rapp. Comm. int. Mer Medit., 27:209.

Mardukhai-Boltovskoi, D., 1964, Caspian fauna beyond the Caspian Sea, Intern. Rev. ges. Hydrobiol., 49:1.

Menzies, R.J., 1972, Biological history of the Mediterranean Sea with reference to the abyssal benthos, Rapp. Comm. int. Mer Medit., 21:717.

Menzies, R.J., George, R.Y., and Rowe, G.T., 1973, The importance of the Tethys warm-water fauna, in: "Abyssal environment and Economy of the World oceans," J. Wiley and Son, ed., New York.

Motas, C., 1962, O problemă de biogeografie: discontinuitatea ariei de răspîndire a speciilor si unitătolor superioare, Ann. Univ. Iasi (Ser. n.), II, 8:205.

Negoescu, I., 1981, New data about the fauna of Anthuridian isopods from the Libyan coasts and the Suez Canal, Trav. Mus. Hist. nat. "Gr. Antipa," 23:109.

Ninkovitch, D., and Heezen, B.C., 1967, Physical and chemical properties of volcanic glass shards from Pozzuolana ash and from upper and lower ash layers in Eastern Mediterranean deep-sea sediments, Nature, 283:582.

Oliver, P., 1983, Les ressources halieutiques de la Méditerranée occidentale, Etud. Rev. Cons. gen. pêch. Medit., 59:1.

Pacem in Maribus, III, 1972, N. Ginsburg, S. Holt and W. Murdoch, Eds. Split.

Pérès, J.M., 1961, Océanographie biologique et Biologie marine, I. La vie benthique. Univ. Press, Paris.

Pérès, J.M., et Picard, J., 1955, Biotopes et biocoenoses de la Méditerranée occidentale comparés à ceux de la Manche et de l'Atlantique, Arch. Zool. exp. gén., 92:28.

Pérès, J.M., et Picard, J., 1958, Recherches sur les peuplements benthiques de la Méditerranée Nord-Orientale, Res. camp. Sci. Calypso, III, Ann. Inst. Océan., 34:213.

Por, F.D., 1975, An outline of the zoogeography of the Levant, Zoologica Scripta, 4:5.

Por, F.D., 1978, "Lessepsian migration", Springer, Berlin.

Por, F.D., 1981, The Lessepsian biogeographic province of the Eastern Mediterranean, Journées Etud. System. Biogéogr. Medit. Cagliari, C.I.E.S.M.: 81.

Rampal, J., 1981, Biogéographie mediterranéenne d'après d'étude des thécosomes (mollusques pélagiques), Journées Etud. System. Biogéogr. Medit., Cagliari, C.I.E.S.M.:45.

Schäfer, W., 1955, Eine Qualle aus dem Indischen Ozean in der Agäis, Natur u. Volk, 85:241.

Sharf El Din, S.H., and Karam, A.M., 1979, Seasonal variation of temperature and salinity structure of the Mediterranean waters in the Levantine Basin, Rapp. Comm. int. Mer Medit., 25/26:143.

Steinitz, W., 1929, Die Wanderung indopazifischer Arten ins Mittelmeer seit Beginn der Quatour periodi, Itern. Rev. Hydrob., 23:8.

Stock, J.H., 1983, Discovery of a Bogidiellid amphipod crustacean in inland waters of the East Indian Archipelago: Bogidiella (Medigiella) sarawacensis, n. sp., Crustaceana, 44:198.

Tortonese, E., 1951, I caratteri biologici del Mediterraneo orientale et i problemi relativi, Attualità Zool., 7:207.

Tortonese, E., 1959, Osservazioni sui Bentos del mar di Marmara e del Bosporo, Rev. Sci. nat. "Natura", 50:18.

Voicu, G., 1974, Identification des Mysides fossiles dans les dépots du Miocène Supérieur de la Paratethys central et orientale et leur importance paléontologique, stratigraphique et paléogéographique, Geol. zborn. Geol. Carpat., Bratislava, 25:231.

Voicu, Gh., 1981, Contributions stratigraphiques et paléogéographiques pour la connaissance du Miocène moyen des régions carpathiques, Trav. Mus. Hist. nat. "Grigore Antipa", 23:339.

Zibrowius, H. et Bitar, Gh., 1981, Serpulidae (Ann. Polychaeta) indo-pacifiques établis dans la région de Beyrouth, Liban, Rapp. Comm. int. Mer Medit., 27:159.

ENVIRONMENTAL CONTROL OF THE MESOSCALE DISTRIBUTION OF PRIMARY PRODUCERS AND ITS BEARING TO PRIMARY PRODUCTION IN THE WESTERN MEDITERRANEAN

Ramon Margalef

Department of Ecology, Faculty of Biology
University of Barcelona, Spain

INTRODUCTION

The Mediterranean has often been called as a small-scale model of an ocean. Indeed it offers the possibility to study selected phenomena in another dimension. Size is important, as it limits the windows of energy input. A larger ocean allows for a wider spectrum of inputs and much higher local intensities. Tides and anticyclonic gyres are not important in our sea. The Mediterranean has properties of its own: It can be considered as a two-layered, inverse estuary opening into the Atlantic; exchange of water with the Atlantic, in each direction is about 70 times the excessive evaporation that drives it. One consequence is the oligotrophy of the Mediterranean. An oligotrophic system is a good reagent to test consequences of fertilization, providing an ideal place for a detailed analysis of control of primary production. The Mediterranean has a unique feature in the winter mixing in the Gulf of Lions that reaches to extraordinary depths. The annual thermal excursion of surface waters is very large; the thermal cycle controls the density of surface waters and the basic features of the annual biological cycle.

I have been fortunate enough, during the course of many years to study several aspects of primary production, and of primary producers, in different places of the Western Mediterranean. Thanks to Drs M. Estrada and M. Alcaraz, I have been able to study a series of plankton samples, obtained with large bottles (20 l) and fractionated over net fabric of 40 μm mesh. Only large dinoflagellates were expected, but some diatoms and Tintinnida were also found and counted. The final counts are probably too low, because of technical problems in the collection, in washing the pieces of net, in the counts themselves, etc. But they provide a view of the mesoscale distribution of large and relatively scarce planktonic organisms that are poorly-represented, or neglected, in conventional research carried on with sedimentation chambers and inverted microscope. Together with supplementary information provided by M. Estrada, these observations afford an occasion to discuss again old and not so old questions about what organisms can tell us about the functioning of the Mediterranean ecosystem.

ENERGY LEVELS AND PRIMARY PRODUCTION

The dependence of plankton primary production on external energy has been discussed in other papers (Margalef, 1978; Margalef and Estrada, 1981). The laboratory culture bottle is necessary, but is not a good model for study of plankton. At sea, the small volumes of water move up and down, and diffuse, depending on the local climate of turbulent energy. The work done by the available energy sets in function a chain of selective events, operating on the available organisms, and leading to a segregation of the different life-forms. The diffusion of nutrients in relation to the field of light, dispersal and sedimentation of cells, is of paramount importance. The different life-forms of phytoplankton are understood to be selected under different combinations of external conditions. The main sequence spans from fertile and turbulent water (mainly diatoms) to extremely oligotrophic and stratified environments (principally dinoflagellates). The statistical study of the degree of association among different species, with the aim of recognising recurrent communities, apart from pointing out the obvious - the possible projection of one plane of the "niche space" on the "mandala of phytoplankton" - proves to be a rather disappointing exercise. Different sets of data provide non-coincident associations (Margalef, 1966, 1971; Venrick, 1982; Belluan et al., 1982) and the unsatisfactory results cannot be blamed solely on statistical difficulties (zeroes, disregard of the real spatial position of samples, etc.). We do not know how serious competition is in the pelagic environment nor do we know the meaning of the congeneric clusters of species, so frequent in plankton. Nothing is gained in adopting the term "guild" of frequent use by contemporary terrestrial ecologists, because it is more a name in quest of a substance than anything else.

Primary production is dependent on the external energy available to do work locally, and on the covariance of reactants, that in practice amounts to the presence of nutrients and cells in the photic zone. This relation is valid on very large scales. The best predictor of oceanic primary production is a map showing where heat is exchanged across the surface of the sea, and where work is done. Inputs of external energy, usually in the form of mixing, generate peaks of primary production. The usual sequence is a deceleration of the flow of energy, related to a decay of energy and to a segregation of the reactants, in the form

$$dP/dt = \underbrace{(dA/dt)\cdot(cov)}_{\text{decay of energy}} + \underbrace{(d(cov)/dt)\cdot A}_{\text{ecological segregation}}$$

All these expressions are by nature spectral, and the decay of energy and growth of small scale segregation means the compartmentalization of the whole system, leading to mesoscale distributions. Cells in upwelling systems, eddies, rings, layers or sheets, all the medium-grained distributions based on the rich, coastal, hydrographic structures, belong to this segment of the spectrum, on which the main interest of this paper will be focussed. As for size we can state a scale from ½ to 50 km, and as an appropriate time scale, between 10^5 and 10^6 seconds, in the order of the life span of diatoms to dinoflagellates. A certain amount of discontinuity goes with nature and probably the best is to associate the boundaries of the compartments with the organization of turbulent diffusion. How to do so practically, and in the sea, is another question. Turbulent diffusion is not isotropic, and particular values of it can be assigned to each direction around a point. The direction with the minimum coefficient can be selected as normal to the boundary. And the boundary could follow a set of points in continuity for which, consistently, there are large differences between maximum and minimum values of the coefficients of diffusion according to different directions. The frontiers may be closed but not necessarily so, as they can also, theoretically, split in two sheets and gradually vanish.

Inside the mesoscale compartments there is much small-scale heterogeneity, usually studied along transects (Steele, 1978). At a certain level, point properties seem to be at random, but with a strong dependence on the neighboring points, leading to a kind of profile that can be qualified as brownian, fractal or of mountain-range type. In such situations, the variances computed on series of points is proportional to the mean, raised to a power between 1.6 and 2. This small-scale structure is below, and out of the field of inquiry of this paper.

The Western basin of the Mediterranean provides an excellent example of the dependence of primary production on external energy. Primary production estimated with traditional methods gives figures between 60 and 80 g C $m^{-2} year^{-1}$ (Estrada, 1981); consideration of the oxygen budget, heterotrophic activity or nutrient cycles, neither questions seriously such an estimate, nor supports nor clarifies it. The yield of the fisheries is in proportion, or rather on the high side, perhaps as a consequence of an efficient exploitation along a relatively extended shoreline.

At least on the relatively extended shelf south of the Ebro river, primary production can be associated with different events. The first is local mixing in the late fall, after breakdown of the thermocline. Later, and usually in February, there is a very local and modest upwelling, whose effect depends on the level reached by the nutrient rich water dispersed from, and dependent on the winter vertical mixing in the Gulf of Lions. A third and quite regular event is the extension of surface water less nutrient-poor than usual, presumably of Atlantic origin, in April-May. The three events occur with regularity year after year, and one interesting property is that they seem to be independent. In consequence the variance in total yearly production is probably relatively smaller than the variance of each separate contributing phenomenon; this helps to stabilize total yearly production. The local contributions of the separate events depend on the distance, in space and in time, probably following a law of exponential decay. The nutrient inflow that depends on winter mixing in the Gulf of Lions probably plays a key role in large parts of the Spanish coast. Cold winters, with high energy budgets, are very productive (1957,1963,1970), both in the sea and in more or less meromictic freshwater lakes in the Mediterranean basin.

This model has to be modified for other areas. Rivers are less important as direct carriers of nutrients (anyway most of the phosphate is suspended as apatite) than for enhancing coastal mixing. Later in this paper it will be shown that the summer deep chlorophyll maximum is not simple and is related to fertilizing events that proceed even in summer; the contribution to the annual production may be important. As thermal structure of the Mediterranean is reconstructed every year, more or less as in a mictic lake, the Mare Nostrum is perhaps more fertile, and undoubtedly more exploitable, than a truly subtropical sea. More energy works in the process. Upwelling has been identified on the Southern French coast, close to the Spanish border, and the importance of the uncommonly strong vertical mixing in the chimneys of the Gulf of Lions has never been questioned.

On land, the relation between external energy exchanged in transpiration, and internal energy in photosynthesis is around 50:1; the elaborate transport system of the terrestrial plant may produce a favourable ratio, and we should be prepared to find higher quotients in the oceanic producers. Primary production in the Mediterranean is certainly below 1 W m^{-2}; and evaporation alone, an important part of the driving machine in the Mediterranean uses around 100 W m^{-2}.

OBSERVATIONS ABOUT MESOSCALE DIFFERENTIATION

During productive events, in turbulent waters, populations are very different from place to place, and shift rapidly. Summertime provides a better opportunity to study the decay of the movement into medium-sized compartments that usually are considerably stabilized by the thermal stratification. In particular, the thermocline, around 25-30 m "fossilizes" very effectively water lenses, layers, and remnants of previous events. If each compartment is limited by boundaries of lowered diffusivity, some properties of the contents are preserved, and populations evolve rather slowly. The thermocline lasts more than six months, provides a shear surface, and supports internal waves, particularly intense towards the end of August. In summer, currents may be important in the top layers, but circulation slackens below, to oscillations following the pendulum day.

The examination of the samples collected in July 1982, during a cruise of the Research Vessel "Garcia del Cid", and following a transect from Barcelona to the channel between Majorca and Minorca provides a beautiful picture of the degree of heterogeneity - really large - to be expected at this time. Surely, a more close analysis would provide much more detail, but probably not a very different picture. The distributions shown in the figures 1 to 9, without a more formal statistical approach, can be related to the former experience of the writer in the area of Castelló and elsewhere. Some other species, whose distributions are not represented, follow the same patterns. The transect can be easily divided into compartments, providing material for comment and containing several suggestions for future work, even in the field of physical oceanography.

The different life-forms are distributed as usual, diatoms are more abundant in mixed and nutrient-rich water, and dinoflagellates become dominant in the better-stratified and warmer surface water. This water is very poor in nutrients, and one key to success is to store phosphate in the cells. This contrast is similar to the features observed, also in summer, at a site close to Castelló (Margalef, 1969) and that provided an opportunity for a detailed comparative study of the spectra of diversity. My personal interpretation is that any input of energy is followed by a peak in primary production, the diatoms leading the way. When diatoms decrease, other organisms, mainly the dinoflagellates, become relatively more important, although not necessarily in absolute biomass, and differentiate gradually into different assemblages in separate compartments. Applying such general points of view to the interpretation of our distributions, we could perhaps find the following connections, in a very tentative way and for the sole purpose of description.

dinoflagellates
- top layer dinoflagellates
  - Ceratia in surface (fig. 1)
  - *Gonyaulax* and *Heteraulacus* around the thermocline
- "shadow", dusk, dinoflagellates (*Ceratium platycorne*)

diatoms
- *Rhizosolenia alata*, *Hemiaulus*
- *Chaetoceros*, *Bacteriastrum*, *Thalassionema*, *Nitzschia*

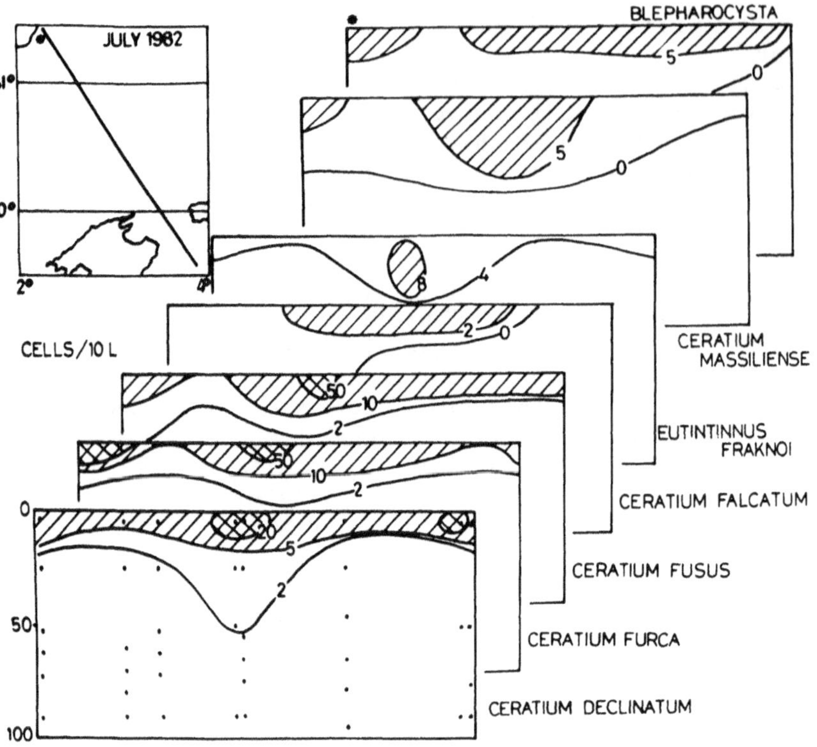

Fig. 1 Distribution of several species of dinoflagellates in the top layers, along the transect of the map. In one of the sections the sampled points have been marked.

The distribution of congeneric species can provide clues about ecological differentiation. Ceratium furca, C. fusus, C. declinatum, C. concilians, C. karsteni, persist under very strong light in extremely poor water. C. horridum, C. platycorne, and other species show a completely different distribution. This agrees well with previous experience, although not completely. The analysis of 1086 small samples (100 ml) of which 467 contained Ceratia, in the same area (Margalef, 1974) only partially supports the associations found in the present occasion. The proximity of the coast probably introduces more complications, when comparing, for instance, the studies made in the Pacific Ocean (Venrick, 1982). Comparison with other attempts at a grouping of species (Margalef, 1966, 1971) leads to frequent inconsistencies in the limited recurrent groups, but the overall picture is much more convincing. In the last cited papers, some large compound groups stand out, and these can be recognized clearly in this study: groups of coastal species, C_1, C_2; species coming with nutrient rich water (F, mostly diatoms), and a number of epipelagic forms (P) here clearly represented by the dinoflagellates of the surface layers. In the present study, of course, small

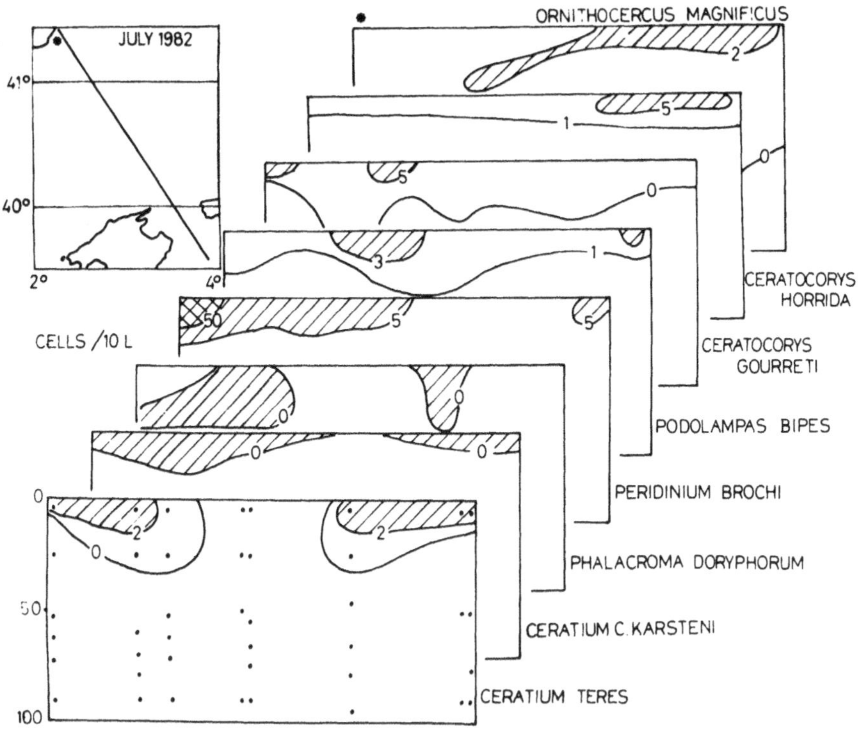

Fig. 2 Distribution of several species of dinoflagellates in the top layers, along the transect of the map. The sampled points are marked in one of the sections.

organisms, such as cryptomonads, coccolithophorida, and cyanophycea, were not available for observation. In fact, this paper is only a complement, concerning the large items, to the study of phytoplankton, to be published later by Estrada. Ceratocorys gourreti was very scarce 20 or 30 years ago, and now seems increasingly common; for the rest, notwithstanding the turmoil about the "dying Mediterranean", the plankton looks as healthy and wonderful as 30 years ago.

The particular cruise to which I owe these samples had as an objective the study of the deep chlorophyll maximum. In fact, not a simple structure, but a stack of layers, with much horizontal heterogeneity was found. A certain contradiction with existing assumptions could be found. Diatoms are preserved in compartments, if the boundaries are not very diffusive; and dinoflagellates thrive in the top layers where a considerable amount of mixing is going on. Such situations emphasize the need for avoiding classificatory dogmatisms and for looking directly for the responsible processes. In this case the essential thing is the coincidence of the factors of production, as measured by the distribution of

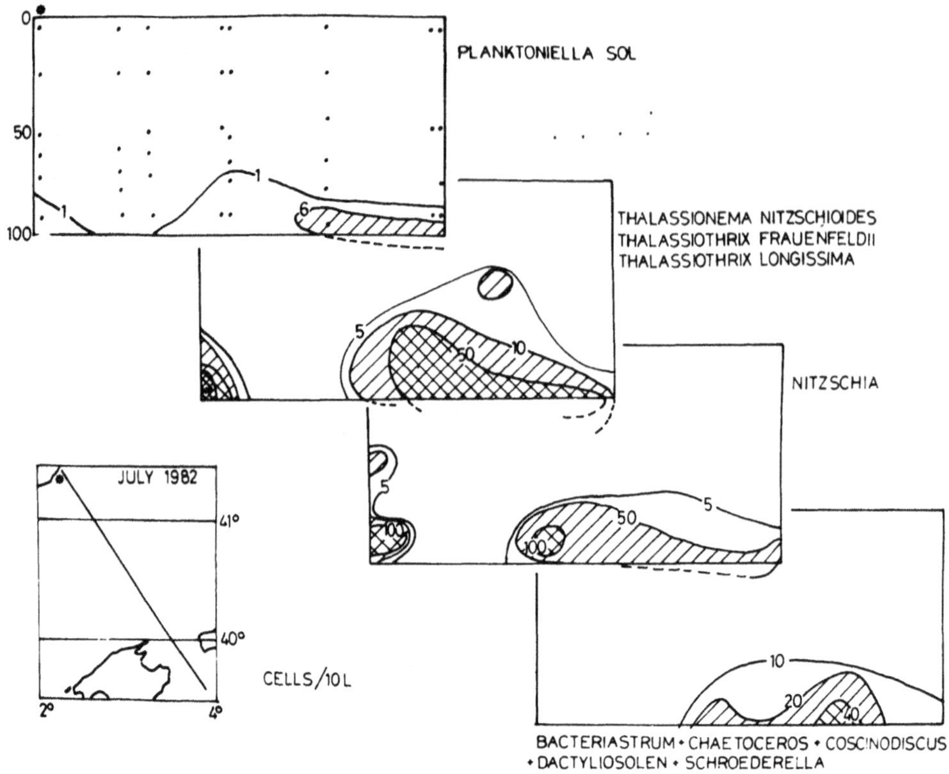

Fig. 3 Distribution of species and groups of species pooled together, representative of the diatoms, and present in the deep layers of the transect of the map. Nitzschia includes colonial species of the groups delicatissima and seriata. The sampled points are marked in one of the sections.

nutrients against light. The product (intensity of light) × (concentration of N) may be a good predictor of the development of diatoms, or else of the bare survival of dinoflagellates.

Tintinnida have been recorded; they are most common below a core of high production but their appropriate food has not been collected or studied on this occasion. Zooplankton able to migrate can combine functionally-different groups, but this is also another story.

RELATIONS WITH HYDROGRAPHY

The different patterns that were observed have been combined in fig. 10, that may provide a useful summary. Groups 1 and 2 are found above the thermocline, in surface; group 1 in the center of the

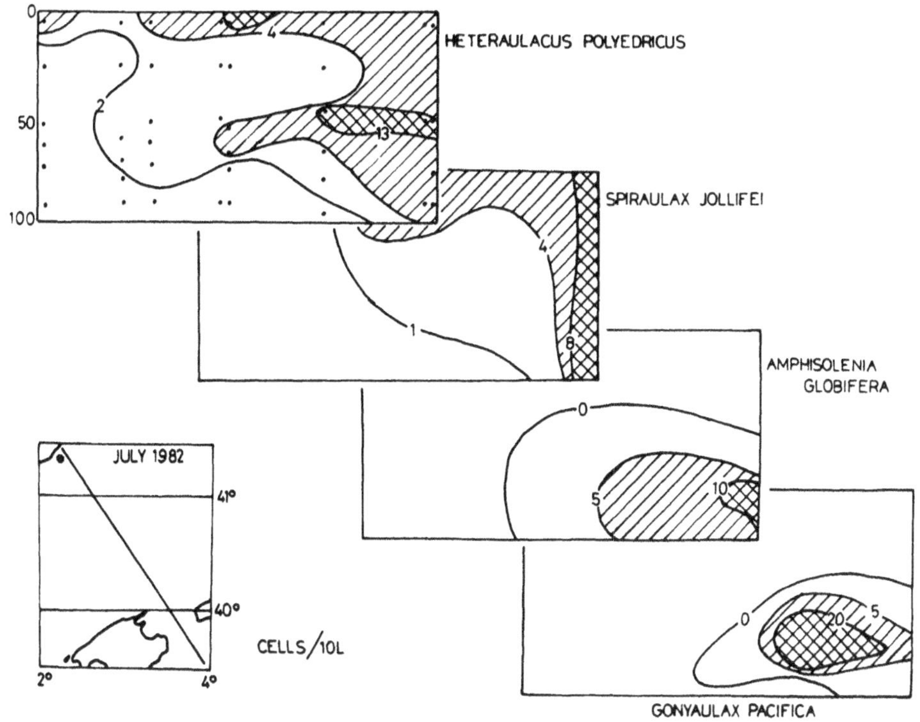

Fig. 4 Distribution of several species of dinoflagellates that, at the time of the survey, were more common in the Balearic area, along the transect of the map. In one of the sections the sampled points have been marked.

section; group 2 at both sides, closer to the coast of Barcelona, and to the sill between the Balearic islands. An important group of species (with several subgroups) is plentiful along the coast, and extends sloping to outer reaches, with a characteristic inflexion midway. Important contributors to the deep chlorophyll maximum are populations of diatoms developing around 100 m depth (group 4); other diatoms, specifically Rhizosolenia alata and one species of Hemiaulus constitute a core of high concentration around the thermocline (group 5). Some species come apparently from farther reaches (group 6) and the denser populations of Tintinnida (7) are placed just below the diatoms of group 5.

The hydrographic sections provide a clue to the whole distribution. There is a sharp discontinuity, a front, running approximately two thirds right of the studied section. The T-S diagrams point to the essential difference between an offshore gyre (apparently anticyclonic, from the section), and more coastal water (flowing, as usual, towards south). The limit of the said gyre is formed by a

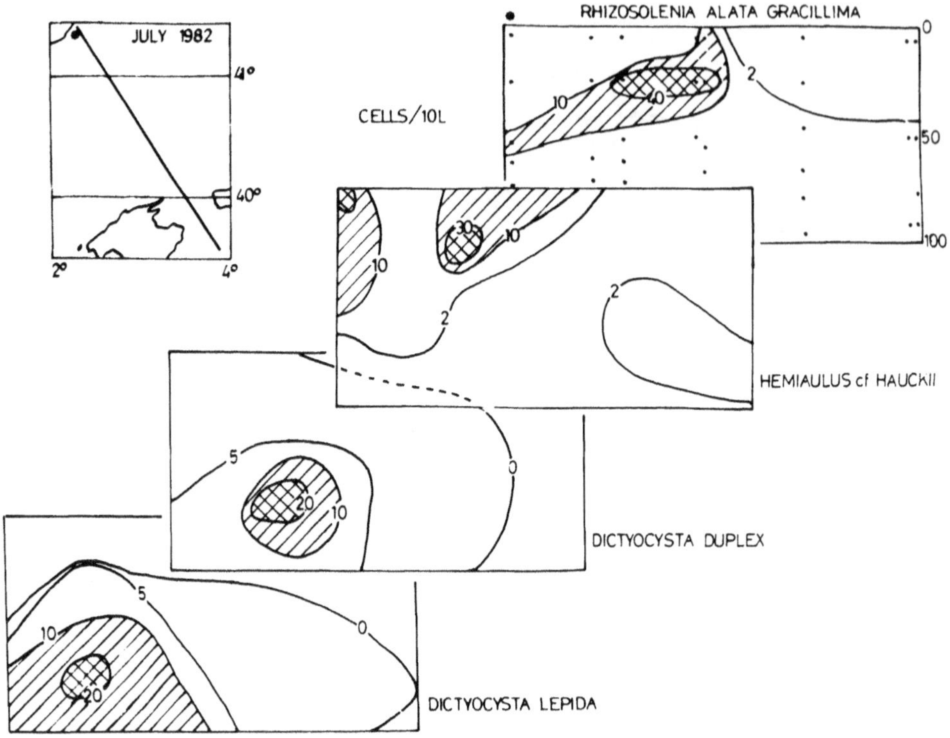

Fig. 5 Distribution of two species of diatoms (Rhizosolenia, Hemiaulus) and two species of ciliata tintinnoinea of the genus Dictyocysta, along the transect of the map. The sampled points are marked in one of the sections.

protrusion of deep water, approximately in the center of the section. The distribution of silica and nitrate shows clearly the introduction of nutrients, through this form of upwelling, in the photic zone, and explains the bloom of Rhizosolenia, Hemiaulus, and, perhaps, smaller phytoplankton, close to the level of the thermocline. The deeper layers of diatoms are associated, as well, with local enrichments. The patterns of the species of group 3 show the deformation introduced by the presumed ascending movement of water at this site. It is expected that the publication of the hydrographic sections will provide a sounder basis for this interpretation.

In my opinion, the observed distributions of planktonic organisms, together with the preliminary data on hydrography pose a number of interesting questions, some of them inexplicably neglected until now in the study of the Mediterranean environment.

A very tentative interpretation has the usual shore current directed south, and an offshore current, with relatively low salinity water, directed north. A strong front develops between both

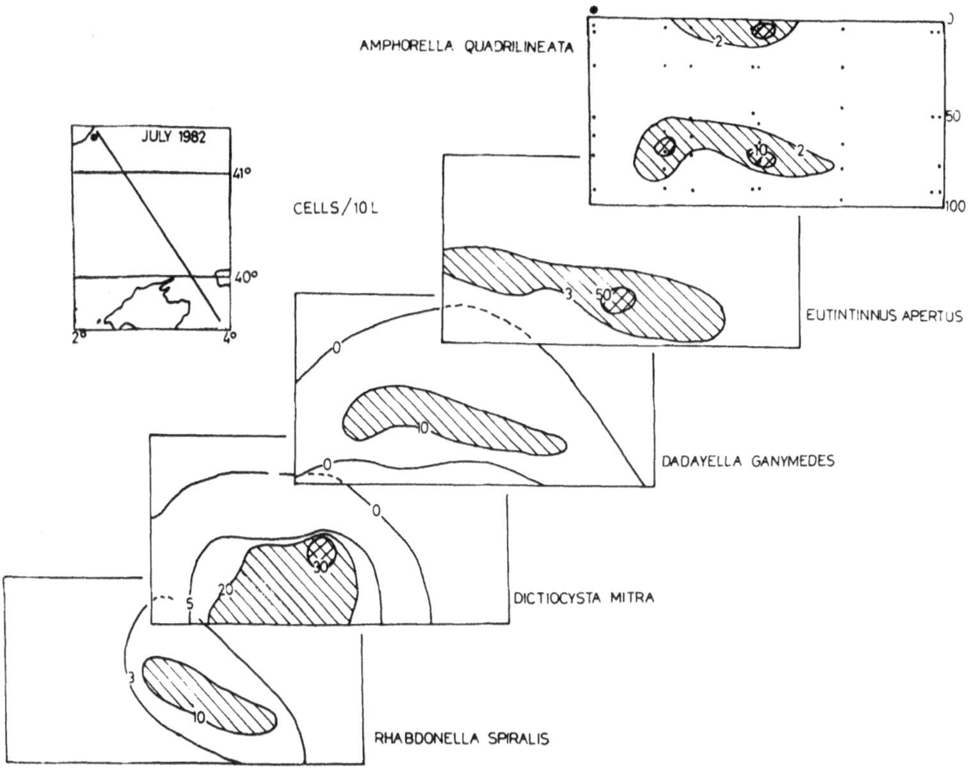

Fig. 6 Distribution of several species of ciliata tintinnoinea along the transect of the map. The occurrence of these species may be indicative of fair densities of phytoplankton of the appropriate size as food. The sampled points have been marked in one of the sections.

currents, and the Ekman effect has to create a divergence with upwelling, West of the front. This is the explanation for the relatively large populations of diatoms there. I would speculate that part of the upwelled water is redirected East, beneath the warm surface layers of the offshore current. Probably the studied section is but one temporary aspect of highly dynamic processes, and successive events, associated with gusts of wind, or Kelvin waves, or simply wobbling of the offshore gyre, have to produce layers of water that escape offshore, and move vertically according to their density, and are a reason for the extremely rich layer-structure observed in these waters. What I mean, also on the basis of other anectodal evidence, is that the current directed North, at a certain distance from the coast, is not very strong, and is changing continuously, also its distance from land. When the current moves closer to the shore, it should suck water below and push it offshore; when separating itself from the shore, upwelling has to be enhanced in the

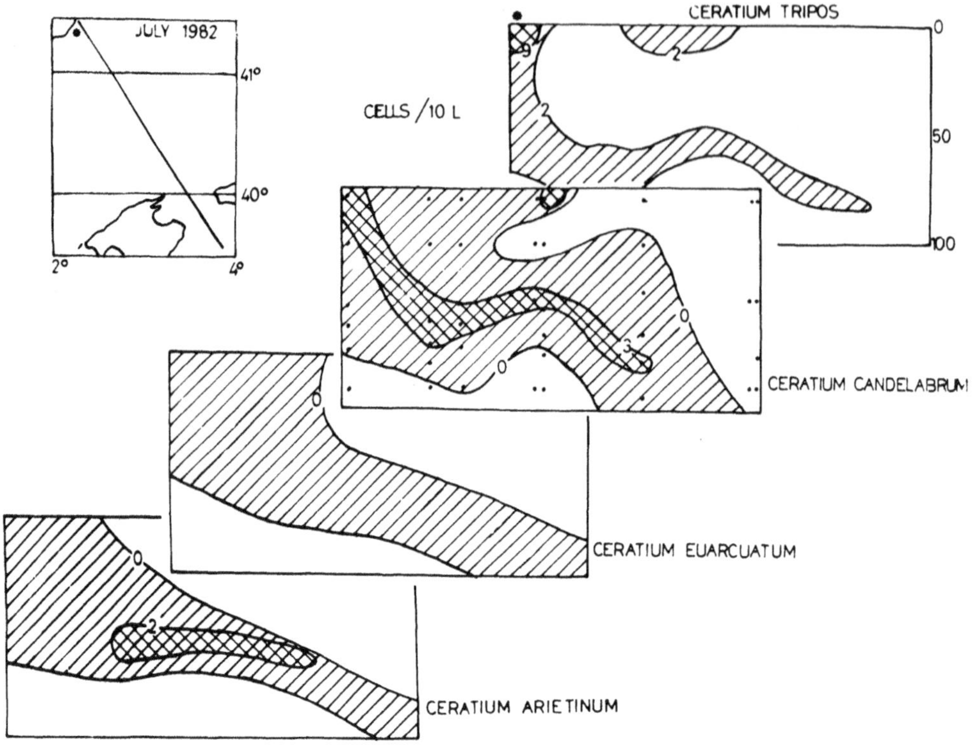

Fig. 7 Distribution of several species of the dinoflagellates genus Ceratium along the transect of the map. The sampled points have been marked in one of the sections.

shoreward directed side. This complex pump effect should be very effective in creating boluses of water of varied characteristics and containing divergent populations.

The existence of an anticyclonic gyre in the Balearic area, or, at least, of a current in the Northern direction, and carrying water of more or less remote Atlantic origin cannot be doubted, although it has been inexplicably neglected or forgotten. We do not have precise information on the currents, at the moment of the survey, nor has the dynamic topography been studied in detail or properly interpreted. There is good evidence that, at some times, the anticyclonic gyre overlays the coastal water completely and reaches the shore line. Fishermen know that usually at the end of March, and the beginning of April, blue water, with tuna, reaches the coast. This was important, when tuna were caught with trapnets that were set until a few decades ago in Hospitalet; it was also the time, and still is, to change the fishing gear for sardine, this fish then beginning to shoal. Strandings of Velella were a good indicator of the inversion of the usual direction of the coastal

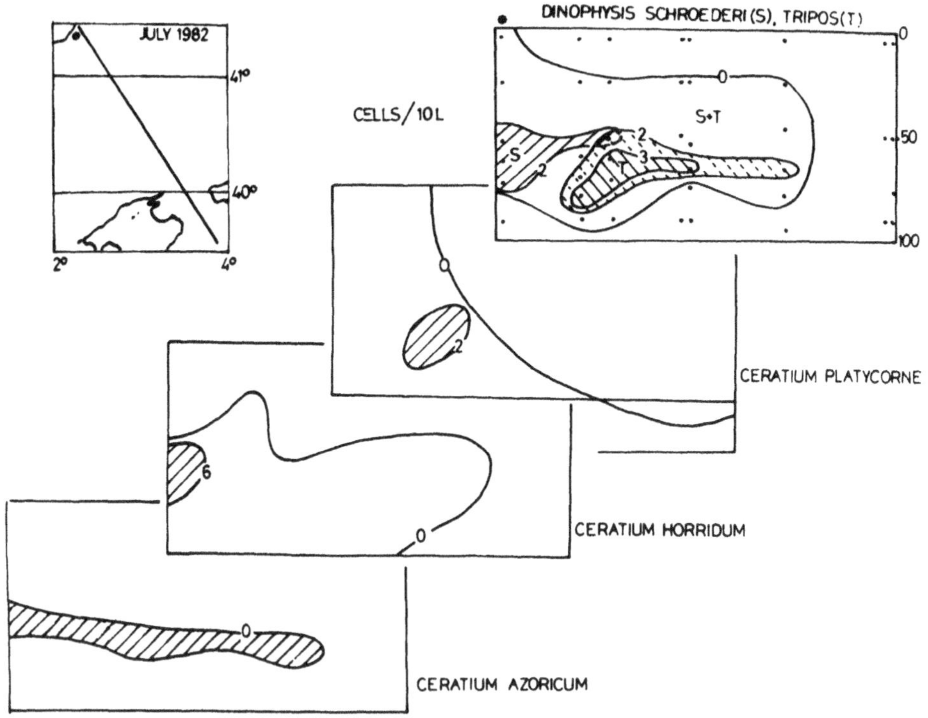

Fig. 8 Distribution of several species of dinoflagellates along the transect of the map. The sampled points have been marked in one of the sections.

current, and there is further evidence provided by drift bottles and also by direct measures of currents in front of Castelló. Such events were discontinuous, but more or less regular, at the end of March, in July, the Northern-directed current coming regularly to the shore.

During a cruise in the Tyrrhenian (Margalef et al., 1966) essentially similar situations were observed, North of Sicily, and at the West of the Italic Peninsula. An anticyclonic gyre was found on the top of a probably cyclonic deep layer, with a lateral wedge of coastal water. In this occasion, attention was given to the meaning of the NO_2 layer, as reflecting with great fidelity events in the pelagic realm. The same can be said in this occasion, and the distribution of NO_2^- can be related to the general patterns of distribution that have been discussed. The question will be considered in another paper.

In consequence, during the summer months, the Mediterranean is not as poor as usually assumed, and mechanisms of fertilization

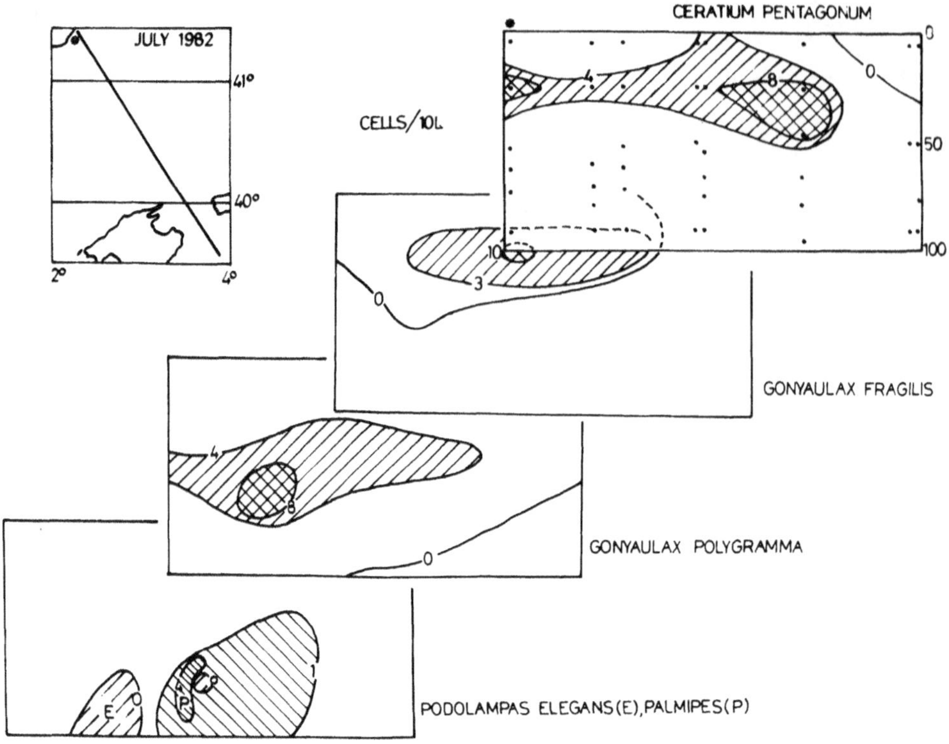

Fig. 9 Distribution of several species of dinoflagellates along the transect of the map. Depth in meters. The sampled points have been marked in one of the sections.

are always operative. Consideration of the patterns observed in July 1982 remind us of the basis for the long and inconclusive discussions, about one decade ago, about, 1-, 2- or 3-celled upwelling systems. We have here a nice example of cell formation in boundary currents,but on a small scale. The Mediterranean basin, after all, cannot contain a large anticyclonic gyre, it is too small and encumbered by islands. We need badly observations along the coast, on the effect of wind, on eventual Kelvin waves, and on the effect of rivers.

It is barely necessary to recall the analogies between the coastal structure in the Mediterranean, and the major oceans, with coastal water, slope water, and main currents. The approximation or the separation of the anticyclonic gyres can influence the population of fishes - like El Niño. A more accurate analysis of extended data may help to explain the alternative abundances of sardine and anchovy in our area. I suggest that sardine populations are more linked to coastal water and anchovy could be favoured in

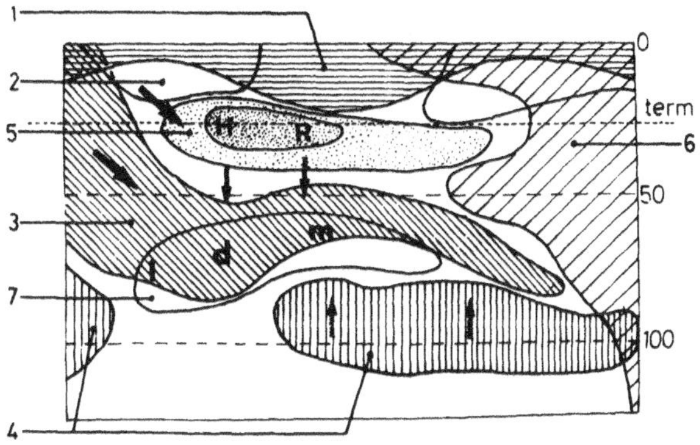

Fig. 10 Summary of the distributions shown in figures 1 to 9. The numbers 1 to 7 in the graph refer to the main patterns or types of distribution, as referred to in the beginning of the last section of the text. Further, H = Hemiaulus, R = Rhizosolenia alata, term = thermocline, 1 = Dictyocysta lepida, d = D. duplex, m = D. mitra (fig. 5,6).

the years in which the anticyclonic gyre becomes more important. The study of zooplankton populations could contribute also to the understanding of the generation of structures, and their diversity from year to year, along the Mediterranean coasts.

The coastal current directed right (in the northern hemisphere) is related to the coastal supply of freshwater; in arid countries, the anticyclonic gyre touches the coast and generates upwelling. Upwelling is believed to create descending atmospheric circulation and contribute to local aridity. We are at a loss to decide if wind generates upwelling, on a large scale, or if upwelling generates coastal breezes. Of course, the problem has to be placed in a more comprehensive framework.

Internal waves, travelling towards the coast, have to experience much perturbation and refraction at the level of the structure found along the coastal fronts. This has to result in generalized and chaotic turbulence close to the shore. In fact, it has been puzzling for me to find that sometimes these internal waves disappear. This may be another mechanism for mixing parcels of water of similar density, but different combinations of temperature and salinity, that slope out (cabbeling), finding their density level and extending as very thin sheets, carrying their own populations offshore, and perhaps a substantial amount of nutrients. These layers have to be tracked along the transects, and they contribute very much to breaking otherwise regular gradients.

Pollution is fed into this complex mechanism and its dispersal and dilution can be understood only if the real complexity of the physical environment is properly appreciated, and the activity of the organisms in the different compartments is better known.

SUMMARY

Patterns in the distribution of phytoplankton extend over all the spatial range, in close association with the changing input of external energy, from advection to turbulence, as seem from a human scale. The size of the Mediterranean does not allow very large structures nor local inputs of high intensity, neither tides nor upwellings comparable in intensity and effectiveness with those associated with the main oceanic Eastern boundary currents. Small scale distributions present the same characteristics and problems as everywhere. In between, the mesoscale distributions, in the range of 0.5 to 50 km and time commensurate with the individual life of dinoflagellates and ciliates, may be of particular interest, as associated with fronts, domes or eddies of importance in the supply of nutrients and control of primary productivity. The low level of primary production in the Mediterranean may emphasize the importance of such structures, as well as make their study easier. In the usual sequence of events in Western Mediterranean, along the coasts of Spain, the most important fertilizing events are the local mixing after breakdown of the thermocline, in late fall; later on, normally in February, local upsurges of water resulting from the winter mixing in the Gulf of Lions; and in spring a variable influence of moderately nutrient rich water from Atlantic origin. In summer, surface layers are almost empty of nutrients, but a characteristic deep chlorophyll maximum develops below the thermocline, made of discrete layers or "pancakes" in an environment of low vertical diffusivity. The variance of the total production is lower than the separate interannual variance of the contributing fertilizing events, and probably such a damping of the variability of the primary production is characteristic of most oligotrophic environments.

The study of a number of samples of 20 l water, filtered through a net of 40 μm wide mesh has provided some illustration of the distribution, in July 1982, of a number of dinoflagellates and ciliates, along a transect from Barcelona to the Balearic islands. Diatoms were less-quantitatively represented in the samples. The distribution of the organisms reflects quite well the hydrographic structures which will be reported on in another paper (Estrada, this book). The statistical association among species was not always exactly coincident with previous observations in the area, but the same high level communities could be recognized. In this particular situation, between the coastal current, towards SW, and an offshore anticyclonic gyre, a wedge or dome of more fertile water provided an important potential contribution to primary production. The existence of anticyclonic gyres than can come close to the coast has been too much neglected by hydrographers working in Western Mediterranean, but is well known to the fishermen and to naturalists. Phytoplankton composition has not changed noticeably in the last forty years.

REFERENCES

Belluan, M., Blanc, F., and Coste, C., 1982, Structures hydrologiques à petite échelle dans la zone de divergence de la mer Ligure. Etude par l'analyse en composantes principales des séries spatio-temporelles multiples Marine Biol., 70:382.

Estrada, M., 1981, Biomasa fitoplanctónica y producción primaria en el Mediterráneo occidental, a principios de otonõ, Inv.Pesq., 45:211.

Estrada, M., This book, Deep phytoplankton and chlorophyll maxima in the Mediterranean.

Margalef, R., 1966, Análisis y valor indicador de las comunidades de plancton mediterráneo, Inv. Pesq., 30:429.

Margalef, R., 1969, Estudios sobre la distributión a pequenã escala del fitoplancton marino, Mem. Acad. Cienc. Artes Barcelona, 40:1.

Margalef, R., 1971, Composition et analyse par groupes du phytoplancton au large des côtes méditerranéennes espagnoles, en 1965-1967, Rapp. Comm. int. Mer Médit., 20:307.

Margalef, R., 1974, Asociación o exclusión en la distribución de especies del mismo género en algas unicelulares, Mém. real Acad. Cienc. Artes Barcelona, 42:353.

Margalef, R., 1978, Life-forms of phytoplankton as survival alternatives in an unstable environment, Oceanologica Acta, 1:493.

Margalef, R., Herrera, J., Steyaert, M., and Steyaert, J., 1966, Distribution et caracteristiques des communautés phytoplanctoniques dans le bassin Tyrrhenien de la Méditerranée en fonction des facteurs ambients et à la fin de la stratification estivale de l'année 1963, Bull. Inst. r. Sci. nat. Belg., 42:1.

Margalef, R., and Estrada, M., 1981, On upwelling, eutrophic lakes, the primitive biosphere and biological membranes, in: Coastal Upwelling," F.A. Richards, ed., Amer. Geophys. Union, Washington.

Steele, J.H. (ed.), 1978, Spatial pattern in plankton communities, Plenum Press, New York & London.

Venrick, E., 1982, Phytoplankton in an oligotrophic ocean: observations and questions, Ecol. Monographs, 52:129.

PHYTOPLANKTON PRODUCTION IN OLIGOTROPHIC MARINE ECOSYSTEMS: THE MEDITERRANEAN SEA

Trevor Platt

Marine Ecology Laboratory
Bedford Institute of Oceanography
Dartmouth, Nova Scotia B2Y 4A2, Canada

INTRODUCTION

One of the major unresolved questions in biological oceanography today concerns the magnitude of phytoplankton production in the sea (Eppley, 1980). The problem can be stated in the following way. The accepted figures for the rates of carbon fixation in the pelagic zone of the ocean are much too low when compared with the estimates of the fluxes of other, biologically-important elements, such as oxygen and nitrogen: the discrepancy is most acute in the so-called oligotrophic areas of the ocean. It is axiomatic that the fluxes of carbon, nitrogen and oxygen be interrelated in a quantitative way. Although ambiguity surrounds the estimation of all three fluxes, the conventional explanation has been that the carbon numbers are at fault, perhaps low by a factor of ten (Williams et al., 1983). The ^{14}C method is thought by many to be subject to systematic errors that become more acute in oligotrophic waters. Some authors (e.g. Platt et al., 1983a; Smith et al., in press) have argued the case that the available data are not strong enough to reject selectively the ^{14}C results.

An unfortunate complication in this controversy concerns the use of the word "oligotrophic". What does it mean in an organic context? It was introduced as a descriptor for a particular type of annual oxygen cycle in the hypolimnion of lakes. Granted that oligotrophic lakes owe their characteristic in part to the primary production cycle, the fact that the boundary conditions for lake circulation are quite different from those in the ocean makes a direct transfer of the term from lake to ocean a matter of some difficulty. How do we recognise oligotrophy in the ocean? Does it mean low autotrophic biomass? low nutrient concentrations? low absolute growth rates of autotrophic biomass? or low specific growth rates ? Once one or more of these criteria have been fixed on, how much deviation are we willing to accept as temporal fluctuation before we discount a given station as being oligotrophic ? In what follows we use the word "oligotrophy" freely, in its conventional sense, but recognise at the same time that a precise definition is lacking.

THE NATURE OF PRIMARY PRODUCTION

If the idea of oligotrophy requires clarification, the idea of primary production is no less in need. It is particularly important to be sure of the meaning of primary production when comparing the fluxes of different elements, because different methods estimate different aspects of primary production.

The rate of carbon assimilation by autotrophs, uncorrected for any losses, is called the gross primary production, P_g. Of course various kinds of carbon losses can occur simultaneously with carbon fixation (respiration, excretion). The difference between P_g and the sum of these loss rates is the net production, P_n. In isotopic tracer kinetic experiments (Smith and Platt, 1984) it is found that the substrate for respiration equilibrates very quickly, so that normal or prolonged ^{14}C productivity experiments will give a number which is certainly smaller than P_g. A further complication is that within the duration of incubation, heterotrophic organisms can grow at the expense of (labelled) metabolites excreted by the autotrophs (Smith et al., in press). For this reason, P_n is sometimes referred to as "net production rates of small particles". At longer time-scales we approach the concept of net community production, P_c. This quantity is corrected for the respiration of all heterotrophs as well as autotrophs. For a system in perfect dynamic equilibrium, P_c will be identically zero when averaged over periods longer than one day. The concept of time-scale is important. Eppley and Sharp (1975) considered that a 24h ^{14}C experiment in the oligotrophic north Pacific was long enough to be contaminated by the effects of respiration by heterotrophs which had previously nourished themselves on labelled phytoplankton or their labelled metabolites. In this case, therefore, the ^{14}C method would return a rate smaller than P_n. The result would be the closer to P_c the larger the production of the heterotrophic biomass that were small enough to find its way into the incubation vessel.

Various authors (Pomeroy, 1974; Williams, 1981; Platt et al., 1983a,b) are coming to the conclusion that a very large proportion of pelagic metabolism is carried by the few smallest size classes of organisms. To the extent that this is true, prolonged ^{14}C experiments will give results that are certainly smaller than P_n, and possibly rather close to P_c. Inconsistent with this view, however, is the recent paper by Williams et al. (1983) in which the conclusion is reached that, for an oligotrophic station in the north Pacific off Hawaii, 12h ^{14}C incubations gave results close to P_g.

When a particular system is not in dynamic equilibrium, net accumulations (positive or negative) of certain elements should be observable over sufficiently long time-scales. For example, P_c is equal to the rate of accumulation of organic carbon within the photic zone plus the rate of export of organic carbon from the photic zone. Again, because P_c is the excess of gross production P_g (oxygen evolution) over community respiration, it is observable as the accumulation rate of oxygen in the photic zone. Finally, where nutrient supply is a limiting factor for production, P_c can be measured indirectly as the supply rate of nitrogen from below the photic zone.

COMPARISON OF OXYGEN AND CARBON FLUXES IN NORTH PACIFIC

Indirect measurements of P_c of the type just described have the potential advantage that they do not involve any <u>in vitro</u> manipulations of metabolising organisms. They are observations of the effect of metabolism on the environment. A further potential advantage is that because they are typically applied over time-scales of weeks or months, they are relatively free from the effects of random sampling errors rooted in the spatial heterogeneity of the populations and from those of temporal fluctuations in metabolic rate. In applications made so far, this remains only a <u>potential</u> advantage.

A recent attempt to evaluate primary production from oxygen accumulation for the oligotrophic north central Pacific is that by Shulenberger and Reid (1981). Because dissolved oxygen in the surface layer can exchange with the atmosphere, accumulation is often apparent only below a density discontinuity strong enough to prevent mixing with of the water below that above. Therefore Shulenberger and Reid chose to estimate primary production from the oxygen developed below the seasonal thermocline. This gives an estimate of production averaged over the time period since the water column was well mixed. They concluded that the oxygen accumulation rate implied a rate of primary production many times higher than that measured by a set of ^{14}C measurements at the end of the accumulation period, and that therefore the ^{14}C method was giving serious underestimates of "true" production.

Some features of this analysis were challenged by Platt et al. (1983a) and a detailed criticism was prepared by Platt (in press). The salient points are

(1) The time-scale of accumulation is not known because the onset of stratification was not observed.

(2) The initial oxygen concentration was not known, therefore the calculation was based on relative saturation with the initial value equal to 100% saturated. However, it is known (Broecker and Peng, 1982) that on the average the surface layer of the ocean is 103% saturated with respect to oxygen.

(3) The initial temperature of the water is not known, but the relative saturation of oxygen depends on temperature such that summer warming could increase the apparent oversaturation.

(4) The analytical error is significant compared to the amount of oxygen accumulated.

(5) The internal variance of the ^{14}C numbers used for comparison with the oxygen flux is high.

(6) The internal variance of the final oxygen data is high.

(7) The extrapolation of results of 6h ^{14}C incubation to daily rates is uncertain. This is related to the problem of deciding whether the ^{14}C method gives P_g, P_n or close to P_c.

(8) The constancy of carbon assimilation during the oxygen accumulation period was not demonstrated, but was implicit in the analysis.

A reexamination of the available data by Platt (in press) lead to the conclusion that a discrepancy could not be demonstrated between the oxygen and carbon fluxes, given the various sources of possible error.

PRELIMINARY CALCULATION OF P_c FROM OXYGEN DATA: MEDITERRANEAN SEA

The National Oceanographic Data Center in Washington, D.C., USA maintains archived oceanographic stations for the Mediterranean Sea. We have made a preliminary analysis of the available data within the area bounded by 10° and 20°E longitude and 35° and 40°N latitude. The stations run from 1910 to present. Not all the stations have oxygen determinations, and the coverage is extremely uneven both in space and time. Nevertheless, from the 790 stations that we examined with oxygen, we have identified certain blocs that can be used for interpretation. The calculations are based on the absolute concentration of oxygen, not on the relative saturation.

The procedure used was to integrate the oxygen concentration from surface to 100 m depth, average these integrals in each month for each of three regions (Strait of Sicily, Tyrrhenian Basin and Ionian Basin, Fig. 1) and compare the monthly averages within regions. The station integrals and averages are given in the tables, which

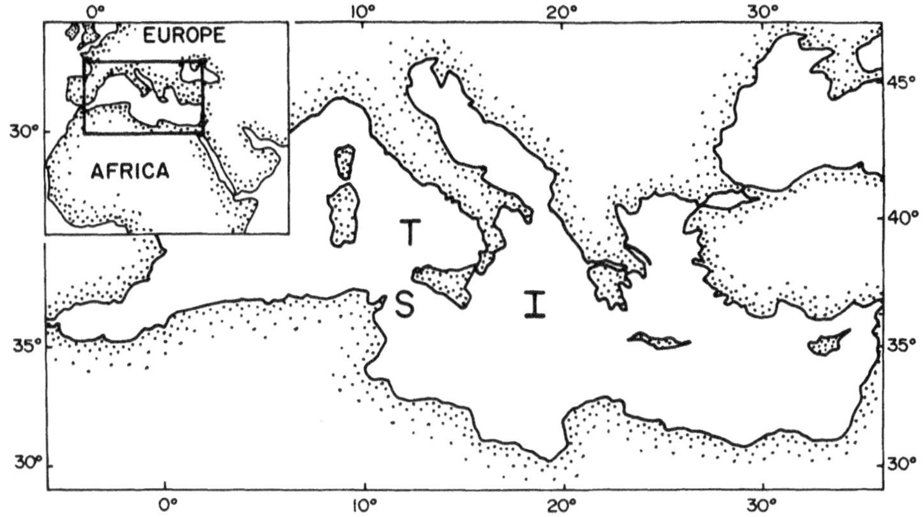

Fig. 1. Showing the three regions referred to in the text. T is the Tyrrhenian Basin; S is the Strait of Sicily; and I is the Ionian Basin.

also indicate whether the water column was stratified or not in a particular month. A possible problem in integrating from the surface is that the integral could be biassed to a low value because of surface losses to the atmosphere. We checked this by making the integrals also from 10 m to 40 m. This did not lead to any substantive change in the conclusions to that stated below probably because particularly when the water column is unstratified, we found it to be undersaturated with respect to oxygen.

A different case holds when the water column is stratified. Here there is invariably found a subsurface peak in the oxygen profile below the thermocline (Fig. 2). The decrease in oxygen concentration above the thermocline may be caused at least partially by loss to the atmosphere (Shulenberger and Reid, 1981). In this case we also list the peak oxygen concentration and the depth at which it was found. A related problem is that the highest oxygen concentration in the water column can be easily missed through poor sampling resolution. This means that the variance associated with the integrals of the stratified water column can be expected to be greater than in the unstratified case, and that comparison of observed peaks between months is of limited use. The usage of the peak values here is slightly different from that of Shulenberger and Reid (1981) who compared the peak oxygen on a particular profile with a ^{14}C productivity result obtained at the same nominal depth. In many cases the peak values are found below 50 m and this is the main reason we choose to quote the integrals to 100 m. In the text the averages are quoted as the mean ± two standard errors.

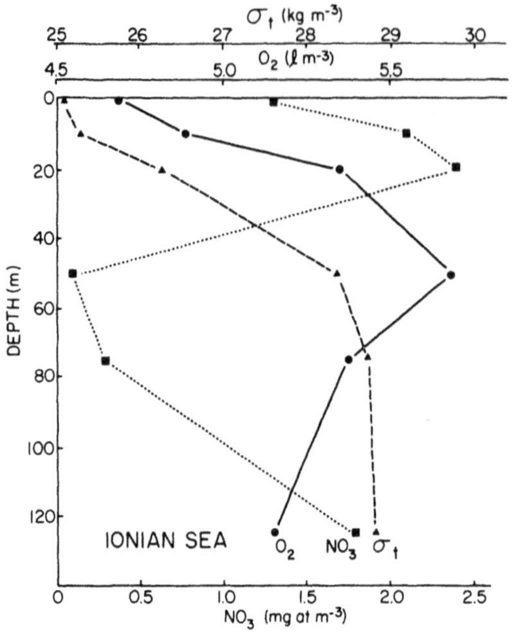

Fig. 2. Vertical profiles of dissolved oxygen, nitrate and σ_t at a station in the Ionian Sea, August 1972.

(i) Strait of Sicily

For the Strait of Sicily we find useful blocs of data only in 1970 (Table 1). Between January and April, the unstratified water column increased in oxygen content from 544±4 to 573±4 $\ell\ m^{-2}$. The corresponding figures for 10 m to 40 m integrals (data not shown) went 5.50±8 and 5.80±9 $\ell\ m^{-2}$. By July the column had stratified and the oxygen content had falled to 535±4 $\ell\ m^{-2}$. On the other hand, the average value at the oxygen peak was 5.79 ml ℓ^{-1}, equivalent to 579 $\ell\ m^{-2}$ if extrapolated over 100 m. The depth of the peak value is variable, and there is no obvious connection with the magnitude of the peak and the depth at which it is found.

(ii) Tyrrhenian Basin

Data from three years (1970, 1959 and 1958) are useful in the Tyrrhenian Basin. In 1970 (Table 2), the unstratified water column increased in oxygen content from 543±6 to 573±2 $\ell\ m^{-2}$. These figures are remarkably similar to those for the Strait of Sicily in the same year. In July the stratified water column had an oxygen content of 540±4 $\ell\ m^{-2}$. The average value at the peak was 5.76 ml ℓ^{-1}, and in this case the peak value tended to be higher if it lay closer to the surface.

Table 1. Oxygen content of upper 100 m (ℓ m^{-2}) in Strait of Sicily 1970.

January (Unstratified)					N	Mean	Standard Deviation
553	532	539	538	551			
548	549	551	552	549			
551	536	533	534		14	544	8.1
April (Unstratified)							
562	566	570	574	585			
573	568	570	572	580			
586	562	582			13	573	8.1
July (Stratified)							
545	539	547	546	531			
525	529	537	530	515			
528	538	547	552	530			
520	533	513	536	536			
547	544	546	541	528			
536					26	535	10.2
Peak values only (ml $ℓ^{-1}$) with depth of occurrence (m) in brackets							
July (Stratified)							
5.96(38)	5.73(38)	5.82(51)	5.77(50)	5.83(72)			
5.89(56)	5.67(95)	5.73(66)	5.71(56)	5.74(66)			
5.75(28)	5.75(70)	5.75(75)	5.82(90)	5.74(120)			
5.84(64)	5.90(30)	5.85(57)	5.82(53)	5.81(51)			
5.61(45)	5.89(31)	5.79(75)	5.80(46)	5.69(28)			
5.98(30)						5.79	0.0868

In 1959 (Table 3) we have data only from months in which the water column is stratified. Oxygen content was 525±7 in July and 529±8 ℓ m^{-2} in August; that is the months were not significantly different. This case is most interesting in that the subsurface peak oxygen concentrations were found at the same nominal depth in each sample for both months. We can therefore demonstrate an increase (barely significant) in the stratum around 50 m depth from 5.65±0.080 ml $ℓ^{-1}$ in July to 5.80±0.11 ml $ℓ^{-1}$ in August.

There are useful data from several months in 1958 (Table 4). The unstratified column progressed from 565±8 ℓ m^{-2} in January to 538±11 in February to 559±9 in April to 574±8 in May. The column was stratified in August and contained 536±8 ℓ m^{-2}. In September

Table 2. Oxygen content of upper m ($\ell\ m^{-2}$) in Tyrrhenian Basin 1970.

					N	Mean	Standard Deviation
January (Unstratified							
550	545	539	538		4	543	5.6
April (Unstratified)							
573	572	570	575	576	5	573	2.4
July (Stratified)							
540	533	547	539	542			
536					6	540	4.8
Peak values only (ml ℓ^{-1}) with depth of occurrence (m) in brackets							
July (Stratified)							
5.76(50)	5.83(28)	5.89(29)	5.70(47)	5.72(48)			
5.65(50)						5.76	0.0884

it was not significantly different at 532±7 $\ell\ m^{-2}$. Peak oxygen values were variable in August and September, the magnitudes being apparently unrelated to depth of occurrence.

(iii) Ionian Basin

For the Ionian Basin we have useful data in the years 1970, 1962, 1960 and 1959. The 1970 (Table 5), the water column was unstratified in January and contained 553±7 $\ell\ m^{-2}$. The stratified column of July had 552±6 $\ell\ m^{-2}$; it was essentially the same in August at 556±7. On the other hand from the peak values we can deduce an increase from 5.86±0.008 ml ℓ^{-1} in July to 5.96±0.007 in August with a tendency for the depth of the peak to increase between the months.

In 1962 (Table 6) the unstratified January water column had 523±7 $\ell\ m^{-2}$ oxygen. This had increased to 542±9 in February and 553±12 in March.

In 1960 (Table 7) the unstratified water column in March had 568±6 $\ell\ m^{-2}$ oxygen. In May and June it was stratified and contained 532±6 and 534±5 $\ell\ m^{-2}$ oxygen respectively. The subsurface oxygen peaks were mostly close to 50 m and contained 5.60±0.078 ml ℓ^{-1} in July and 5.56±0.076 ml ℓ^{-1} in August.

Table 3. Oxygen content of upper 100 m (ℓ m^{-2}) in Tyrrhenian Basin 1959.

July (Stratified)					N	Mean	Standard Deviation
520	519	527	538	521	5	525	7.9
August (Stratified)							
527	528	513	527	536			
543					6	529	10.1
Peaks only (ℓ m^{-2}) and depth of occurrence (m) in brackets							
July (Stratified)							
5.53(50)	5.62(48)	5.71(50)	5.76(53)	5.62(55)		5.65	0.0893
August (Stratified)							
5.86(50)	5.54(50)	5.76(48)	5.88(48)	5.84(52)			
5.91(52)						5.80	0.1363

In 1959 (Table 8) the water column was unstratified in April and May. There was a barely significant decrease from 542±6 to 527±11 ℓ m^{-2}. In July and August it was stratified and contained respectively 526±8 and 535±11 ℓ m^{-2}. The difference between the average peak values was equally insignificant (5.60±0.144 and 5.74±0.127 ml ℓ$^{-1}$).

(iv) Implications

The changes in oxygen content of the water column are interpreted as net community productivity, P_c, in Table 9.

DISCUSSION

The most striking feature of Table 9 is that positive community productivity was only detectable in the early months of the year. In only one case (April-May 1958) was positive community production observed for a period beginning later than February. For some of the other periods the sense of the change may have been positive, but the between station variance was high enough that the changes could not be called significant. Note that the confidence intervals associated with each estimate represent only this sampling error.

Table 4. Oxygen content of upper 100 m (ℓ m^{-1}) in Tyrrhenian Basin 1958.

January (Unstratified)					N	Mean	Standard Deviation
584	597	556	578	541			
557	550	577	544	556			
566	544	567	569	583			
578					16	565	16.6
February (Unstratified)							
546	524	536	547	547	4	538	10.7
April (Unstratified)							
552	548	543	524	546			
552	558	561	561	579			
594	600	561	559	547			
553					16	559	18.9
May (Unstratified)							
583	573	574	559	564			
591	576				7	574	10.8
August (Stratified)							
575	529	523	537	526			
549	516	532	567	514			
535	520	537	562	537			
544	534	502	545		19	536	18.4
September (Stratified)							
534	527	549	520	546			
526	523	528			8	532	10.6
Peaks only (ml ℓ^{-1}) and depth of occurrence (m) in brackets							
August (Stratified)							
6.52(50)	6.00(30)	6.03(29)	5.95(50)	5.72(50)			
6.11(47)	5.81(53)	5.91(29)	5.80(49)	5.94(50)			
6.14(30)	5.82(50)					5.98	0.2130
September (Stratified)							
5.92(51)	5.65(33)	5.75(35)	6.01(50)	6.00(34)			
5.58(49)	5.64(35)	6.01(35)				5.82	0.1845

Table 5. Oxygen content of upper 100 m (ℓ m^{-2}) in Ionian Basin 1970.

January (Unstratified)					N	Mean	Standard Deviation
544	555	565	550	549	5	553	8.0
July (Stratified)							
548	561	558	544	551	5	552	7.0
August (Stratified)							
566	552	557	549		4	556	7.4
Peak values only (ml ℓ^{-1}) at depth of occurrence (m) in brackets							
July (Stratified)							
5.73(28)	5.89(50)	5.84(30)	5.84(24.40)	5.98(28)		5.86	0.0907
August (Stratified)							
6.02(50)	5.88(75)	6.01(75)	5.94(40)			5.96	0.0655

Table 6. Oxygen content of upper 100 m (ℓ m^{-2}) in Ionian Basin 1962.

January (Unstratified)							
519	526				2	523	4.9
February (Unstratified)							
531	531	531	514	528			
542	553	565	553	535			
552	565				12	542	15.9
March (Unstratified)							
592	591	548	567	556			
553	535	530	542	535			
543	541				12	553	20.8

Table 7. Oxygen content of upper 100 m (ℓ m^{-2}) in Ionian Basin Basin 1960.

March (Unstratified)					N	Mean	Standard Deviation
584	582	591	592	569			
562	585	551	545	555			
566	587	553	563	572			
566	574	585	585	565			
562	572	552	564	554			
543					26	568	14.5
May (Stratified)							
539	525	537	520	538			
532					6	532	7.8
June (Stratified)							
519	529	532	522	532			
540	543	529	532	546			
544	533	544			13	534	8.6
Peak values only (ml ℓ^{-1}) and depth of occurrence (m) in brackets							
May (Stratified)							
5.75(20)	5.56(50)	5.60(20)	5.46(50)	5.60(50)			
5.64(50)						5.60	0.0952
June (Stratified)							
5.35(30)	5.54(50)	5.63(50)	5.70(50)	5.58(50)			
5.58(50)	5.46(50)	5.60(50)				5.56	0.1077

The only useful data for calculating production at a particular depth is that for the Tyrrhenian Basin, July and August 1959, at the 50 m horizon. The estimate rate here is +0.0050±0.0046 m ℓ^{-1} d^{-1} (0.0027 g C m^{-3}d^{-1}), a barely significant result.

The positive net community productivity values calculated from the column integrals fall in the range 0.17 to 0.34 g C m^{-2} d^{-1}. These numbers should be taken as minimum estimates for gross primary production, P_g, for the periods concerned.

With respect to the potential systematic errors in these estimates, as discussed earlier for the north Pacific calculations of

Table 8. Oxygen content of upper 100 m (ℓ m^{-2}) in Ionian Basin 1959.

April (Unstratified)					N	Mean	Standard Deviation
553	528	523	552	550			
541	536	558	557	546			
541	531	539	529		14	542	11.3
May (Unstratified)							
515	534	553	538	535			
540	522	501	524	537			
557	553	527	521	486			
487					16	527	21.5
July (Stratified)							
513	513	519	545	512			
524	503	513	525	546			
538	542	546	525		14	526	14.8
August (Stratified)							
505	529	564	536	511			
517	553	527	565	537			
552	527				12	535	19.8
Peaks only (ml ℓ$^{-1}$) and depth of occurrence (m) in brackets							
July (Stratified)							
5.98(25)	5.36(50)	5.58(25)	5.56(25.49)	5.61(50)			
5.49(50)	5.65(48)					5.60	0.1909
August (Stratified)							
5.62(77)	5.55(52)	5.73(55)	5.55(26)	6.03(50)			
5.95(50)	5.69(50)	5.83(50)				5.74	0.1794

Shulenberger and Reid (1981), the time scale of the oxygen accumulation is still somewhat obscure as we are dealing with monthly averages and time is not resolved to within ±15 days. The calculations are not biassed by any assumptions related to calculation of relative oxygen saturation. On the other hand, they are biassed by possible losses of oxygen to the atmosphere during summer and possible advective losses. They are still end-point calculations and can

Table 9. Summary of estimates of net community production, P_c, calculated from integrals 0 to 100 of water column oxygen, and carbon equivalent, assuming photosynthetic quotient equal to unity.

Region	Period	P_c	
		$\ell\ O_2\ m^{-2}d^{-1}$	$g\ C\ m^{-2}d^{-1}$
Strait of Sicily	January-April (1970)	+0.32±0.069	0.17
	April-July (1970)	-0.42±0.067	
Tyrrhenian Basin	January-April (1970)	+0.33±0.067	0.18
	April-July (1970)	-0.37±0.050	
	July-August (1959)	n.s.	
	January-February (1958)	-0.90±0.451	
	February-April (1958)	+0.35±0.238	0.16
	April-May (1958)	+0.50±0.416	0.23
	February-May (1958)	+0.40±0.130	0.21
	May-August (1958)	-0.42±0.178	
	August-September (1958)	n.s.	
Ionian Basin	January-July (1970)	n.s.	
	July-August (1970)	n.s.	
	January-February (1962)	+0.63±0.583	0.34
	February-March (1962)	n.s.	
	January-March (1962)	+0.50±0.231	0.27
	March-May (1960)	-0.60±0.142	
	May-June (1960)	n.s.	
	April-May (1959)	-0.50±0.411	
	May-July (1959)	n.s.	
	July-August (1959)	n.s.	

n.s. means that no net productivity was detectable for the period within the error of the calculation, standard errors being added in quadrature.

estimate only average rates at best.

Concerning random errors, it is clear from even the simple error analysis given here that variation between stations is high enough to obscure many potential net changes. The error is particularly acute for selection of subsurface oxygen peaks in stratified water columns. Only one significant change could be demonstrated for any period using peak values, once the spatial variation had been taken into account. Even for this station, the expected error was ∿95% of the mean. This underlines the possibility for self deception in drawing conclusions from single end-point determinations on peaks. Both the magnitude and depth of occurrence of the peaks were variable even for stations with ten samples in the upper 100 m.

FUTURE WORK

The analysis discussed above suggests that something could be learned about the community productivity of the Mediterranean Sea from a carefully designed time-series of observations of dissolved oxygen within particular regions. A series of at least 12 months should be sought. Special emphasis should be given to the early months of the year, when positive net community production could be detected in our calculations. The data we have used were entirely fortuitous: they are entirely lacking in *a priori* statistical design. Considerably better resolution of changes in time could be expected if data were collected with this application in mind.

The data would be considerably more interesting if they were supplemented with a simultaneous time series of ^{14}C productivity. This would help in the resolution of the problem of relative size of P_n and P_c in the impoverished Mediterranean waters. The net microbial production and the net community production will be close if small organisms dominate the metabolism. Thus there is need to analyze production and metabolism by size-fractionation, giving particular attention to the so-called picoplankton (Platt et al., 1983a, 1983b; Li et al., 1983).

SUMMARY

The current controversy about the absolute level of primary production in oligotrophic waters is discussed. The importance of a clear definition of primary production is emphasized. Difficulties with a recent comparison of molar fluxes of oxygen and carbon in the central north Pacific are summarized.

Using historical data from the Strait of Sicily, the Ionian Basin and the Tyrrhenian Basin, average net community production rates are estimated from vertical profiles of dissolved oxygen for various periods of the year. Positive net community production was found only in the winter and early spring months. For the rest of the year these basins appear to be a sink for oxygen.

Suggestions for future work are made.

REFERENCES

Broecker, W.S., and Peng, C.D., 1982, "Tracers in the Sea," Eldigio Press, New York.

Eppley, R.W., 1980, Estimating phytoplankton growth rates in the central oligotrophic oceans, in: "Primary productivity in the sea", P.G. Falkowski, ed., Plenum Press, New York.

Eppley, R.W., and Sharp, J.H., 1975, Photosynthetic measurements in the central north Pacific: The dark loss of carbon in 24-h incubations, Limnol. Oceanogr., 20:981.

Li, W.K.W., Suba Rao, D.V., Harrison, W.G., Smith, J.C., Cullen, J.J., Irwin, B., and Platt, T., 1983, Autotrophic picoplankton in the tropical ocean, Science, 219:292.

Platt, T., (in press), Primary productivity in the Central North Pacific: Comparison of oxygen and carbon fluxes, Deep-Sea Res.

Platt, T., Lewis, M., and Geider, R., 1983a, Thermodynamics of the pelagic ecosystem: Elementary closure conditions for biological production in the open ocean, in: "Flows of energy and materials in marine ecosystems: Theory and practice," M.J.R. Fasham, ed., Plenum Press, London.

Platt, T., Subba Rao, S.V., and Irwin, B., 1983b, Photosynthesis of picoplankton in the Oligotrophic ocean, Nature, 301:702.

Pomeroy, L., 1974, The ocean's food web, a changing paradigm, Bioscience, 24:499.

Shulenberger, E., and Reid, J.L., 1981, The Pacific shallow oxygen maximum, deep chlorophyll maximum, and primary productivity, reconsidered, Deep-Sea Res., 28a:901.

Smith, R.E., and Platt, T., 1984, Carbon exchange and ^{14}C trace methods in nitrogen-limited diatom Thalassiosira pseudonana, Mar. Ecol. Progr. Ser., 16:75.

Smith, R.E.H., Geider, R.J. and Platt, T., in press, Microplankton productivity in the Oligotrophic Ocean, Nature.

Williams, P.J. LeB., 1981, Incorporation of microheterotrophic processes into the classical paradigm of the planktonic food web, Kieler Meeresforsch, 5:1.

Williams, P.J. LeB, Heinemann, K.R., Marra, J., and Purdie, D.A., 1983, Phytoplankton production in oligotrophic waters: Measurements by the ^{14}C and oxygen techniques, Nature, 305: 49.

DEEP PHYTOPLANKTON AND CHLOROPHYLL MAXIMA IN THE WESTERN MEDITERRANEAN

Marta Estrada

Instituto de Investigaciones Pesquera, Paseo Nacional s/n, Barcelona-3, Spain

INTRODUCTION

A stratification period and a mixing period alternate during the seasonal cycle in the Mediterranean. Typically, in the Western basin, the development of the thermocline starts around April; the stratification of the water column progresses throughout the summer and begins to disappear in October. Deep chlorophyll maxima (DCM), accompanied by nitrite and oxygen maxima, are typical features of the hydrographical situation in the Mediterranean during the summer months. Although such structures have been studied in a number of tropical and temperate marine areas (Anderson, 1969; Dandonneau, 1979; Goering et al., 1970; Hobson and Lorenzen, 1972; Margalef et al., 1966; Revelante and Gilmartin, 1973; Riley et al., 1949; Saijo et al., 1969; Sorokin, 1970; Steele and Yentsch, 1960; Venrick et al., 1973; among others) there is not yet available a clear picture of the mechanism or mechanisms leading to their origin and maintenance. Another open question concerns the importance of the relative contribution of the deep chlorophyll maximum (DCM) layers to the total primary production in a particular region.

Since 1982, the Instituto de Investigaciones Pesqueras of Barcelona has sponsored two cruises devoted to the study of the hydrographical structure and the dynamics of the plankton community during the stratification period in an area of the western Mediterranean extending from the Catalan coast to beyond the channel between Mallorca and Menorca. The first of these cruises ("Produccio estival profunda",PEP-82) was carried out from 12 to 28 July 1982 and the second (PEP-83), from 30 June to 17 July 1983. Information based on a previous mesoscale survey covering a larger region in the western Mediterranean basin has been published in Estrada (1981, 1982),

Font and Miralles (1978), Salat and Cruzado (1981) and Vives (1978, 1979). The present report is based on data from the July 1982 cruise and deals with the distribution of chlorophyll and the composition of the phytoplankton (as examined with the inverted microscope technique) in relationship to the vertical structure of the water column. Some aspects of the primary production measurements will also be discussed, but the overall results of these determinations during the PEP cruises will be treated in detail elsewhere. R. Margalef (this book) will provide a study of the larger and less common phytoplankton organisms, obtained by filtering 20 l of water through a net of 40 µm mesh size.

MATERIAL AND METHODS

The area visited during the PEP 82 cruise is shown in fig. 1. The transect AB was covered four times. Two of the sections (the "hydrographic" ones) were completed in less than three days. In the other two sections, some of the stations lasted 12 or 24 hours ("biological" stations). In the 24 hour stations, bottle casts were repeated at three hour intervals. The station locations marked in fig. 1 are those of the first hydrographical section, stations 1-9 and 12; 12-7-82/15-7-82; and the first biological section, stations 36,38,22 (24 hours), 21, 19 (12 hours), 17,15 (12 hours); 15-7-82/20-7-82.

At each station, and at an additional number of positions, the temperature and salinity profiles were monitored by means of a CTD recording system. Underwater irradiance was measured at the biological stations with a LI-COR underwater quantum sensor. Water samples for chemical and biological analyses were taken with 4 l Niskin bottles at intervals of 10 m between the surface and 110 m depth and at larger intervals from this depth down to 400 m. Discrete temperature readings were made from reversible thermometers mounted on the bottles. Oxygen was analyzed by Winkler's method. Nitrate, nitrite, phosphate and silicate were determined automatically as described in Strickland and Parsons (1972), and ammonia was analyzed manually by Koroleff's method (Koroleff, 1970). Chlorophyll a was measured fluorometrically (Yentsch and Menzel, 1963; Strickland and Parsons, 1972); concentrations were evaluated with and without correction for phaeopigments; the differences were relatively small, but the phaeopigment subtraction diminished the precision of the measurements and was applied only to those chlorophyll values used in the estimation of cellular chlorophyll content. At the 12 hour stations, a second cast was made with 20 l metacrilate Van Dorn bottles or with 30 l Niskin bottles. Six selected depths (three of them covering the DCM) were sampled and the water was transfered to polyethylene containers for use in several determinations, including chlorophyll a by spectrophotometry, particulate organic C and N, and ATP. Water for ^{14}C uptake experiments was taken from the Niskin

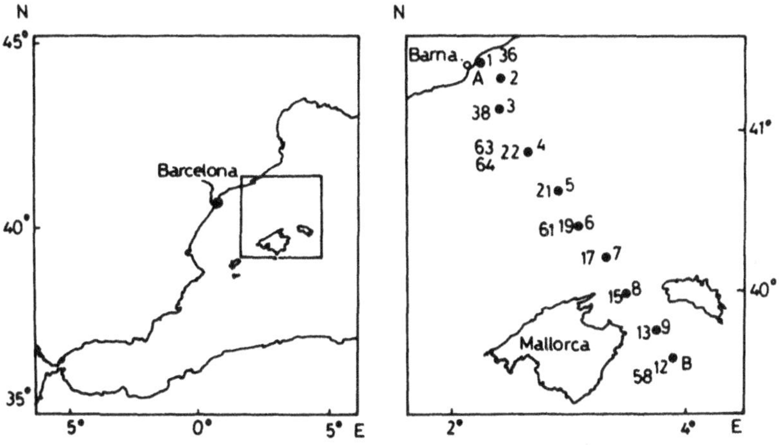

Fig. 1. Cruise PEP 82 (Instituto de Investigaciones Pesqueras of Barcelona, July 1982). Position of the stations referred to in the text.

bottles at the 24 hour stations and from the polyethylene containers (which were protected from light) at the other biological stations. The uptake of ^{14}C was measured as described in Estrada (1981), in clear and opaque 100 ml Pyrex bottles which were inoculated with 4 µCi of Na $H^{14}CO_3$. The bottles were incubated for 2 to 3 hours on the deck of the ship in incubators which were cooled by circulating surface water and had been provided with neutral density stainless steel screens to simulate the light intensity at the depths from which the samples were drawn. Dark bottle corrections were not applied, in view of the uncertainty of their meaning. However, in practically all cases the dark uptake was very low and showed little variability.

Samples of 120 ml of water for phytoplankton counts taken from the different bottle casts were put in Pyrex bottles and fixed with Lugol's solution (without acetic acid). 100 ml of water were sedimented in a chamber and counted with an inverted microscope. The entire chamber bottom was scanned at low magnification (125X) to count the larger organisms. One transect of the chamber (approximately one ml of sample) was made under high magnification (500X) to enumerate the smaller forms. An additional transect (about 5 ml of sample) was examined at 125X to count cells of intermediate size. The organisms were classified to the level of species when this was possible, but many of them had to be lumped in groups such as "flagellates" or "small dinoflagellates" (mostly non-thecate forms). Another limitation of the sedimentation method is that it is not adequate for the smaller (<1.5 µm) picoplankton organisms which have been shown (Johnson and Sieburth, 1979; Li et al., 1983; Platt et al., 1983) to contribute significantly to phytoplankton biomass and production in some marine areas. The numerous problems associated with phytoplankton enumerations (identification, deterioration of naked cells, statistical and counting errors, etc.) have been discussed in several publications (e.g. Lund et al., 1958; Margalef, 1978a; Travers, 1972; Venrick, 1978). However, as the same taxonomic and counting criteria were used for all the samples, the data can be assumed to be useful for comparison. It should also be pointed out (Margalef, 1978a) that species which cannot be distinguished in a routine count belong generally to the same life-forms and tend to have similar ecological responses.

The phytoplankton data set used in this report consisted of 140 samples from the first hydrographical section and from the biological stations of the second station (fig. 1). Two types of standard multivariate statistical methods were used to condense the information contained in the phytoplankton inventories, principal component analysis (PCA) and correspondence analysis (CA). Accounts of the mathematical techniques involved can be found in Cooley and Lohnes (1971), Benzécri (1973) and Legendre and Legendre (1979). From the 160 species or groups recorded, the 50 most frequent ones were selected for the statistical analyses (excluding a few taxa whose identification was inconsistent). The data were logarithm transformed for the PCA ($X \rightarrow \log(X+1)$, X=number of cells/100 ml), which was performed on the correlation matrix. The computer programs were adapted from the IBM Scientific Subroutine Package (IBM, 1969) for PCA and from Lebart and Fénelon (1971) for CA (Lleonart, 1979) In addition, a hierarchical classification method, unweighted arithmetic average clustering (UPGMA), was applied to the presence-absence data of the 50 selected taxa (Sneath and Sokal, 1973; Legendre and Legendre, 1979) using the point correlation coefficient as index of similarity. The programs employed are described in Lleonart (1980).

Similarities between adjacent samples were evaluated by means of Spearman's rank correlation coefficient (Siegel, 1956). The calculations were based on the abundance data of the 50 selected species.

RESULTS

The Physical Environment

The temperature distribution in the hydrographic sections presented a narrow mixed layer of approximately 10 m on top of an extended thermocline (figs. 2-3). Salinity, as is typical in the zone, increased with depth from the surface to the deep waters (fig. 4); a sharp salinity drop around $3^{o}20'$E reflected the presence, in the east, of low salinity waters of Atlantic origin. A prominent feature of the hydrographic profiles was the presence of a doming of the isotherms centered about midway from Barcelona to the channel between the islands. This feature could be explained

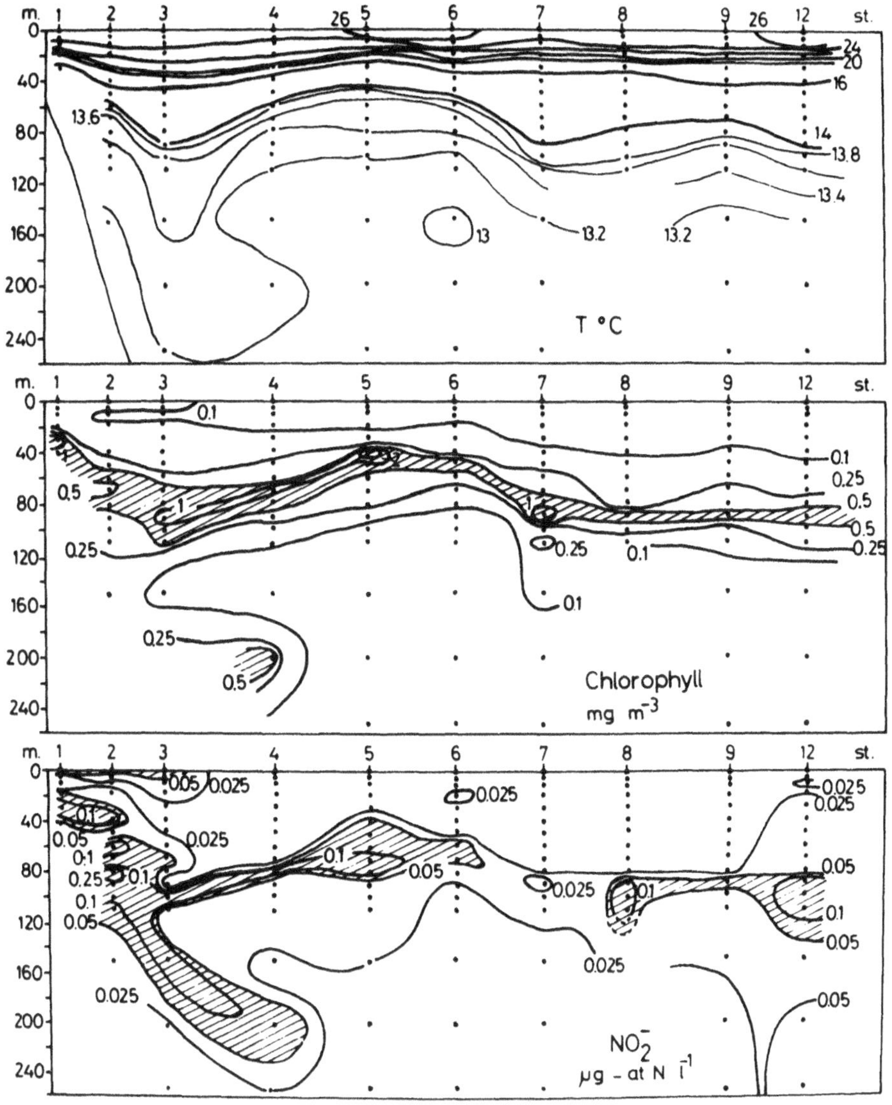

Fig. 2. Distribution of temperature (T^oC), chlorophyll a ($mg \cdot m^{-3}$) and nitrite (NO_2^-, μg-at $N \cdot l^{-1}$) along the first AB transect (fig. 1).

as a result of the divergence created in the contact between the coastal current, directed to the south, and a northward flow which could be part of an offshore anticyclonic gyre (Margalef, in preparation). Although no direct current measurements were available, previous results from geostrophic calculations (Font and Miralles, 1978) and other kinds of data were consistent with such a circulation pattern. The CTD profiles revealed considerable microstructure within the thermocline. The biological significance of this physical

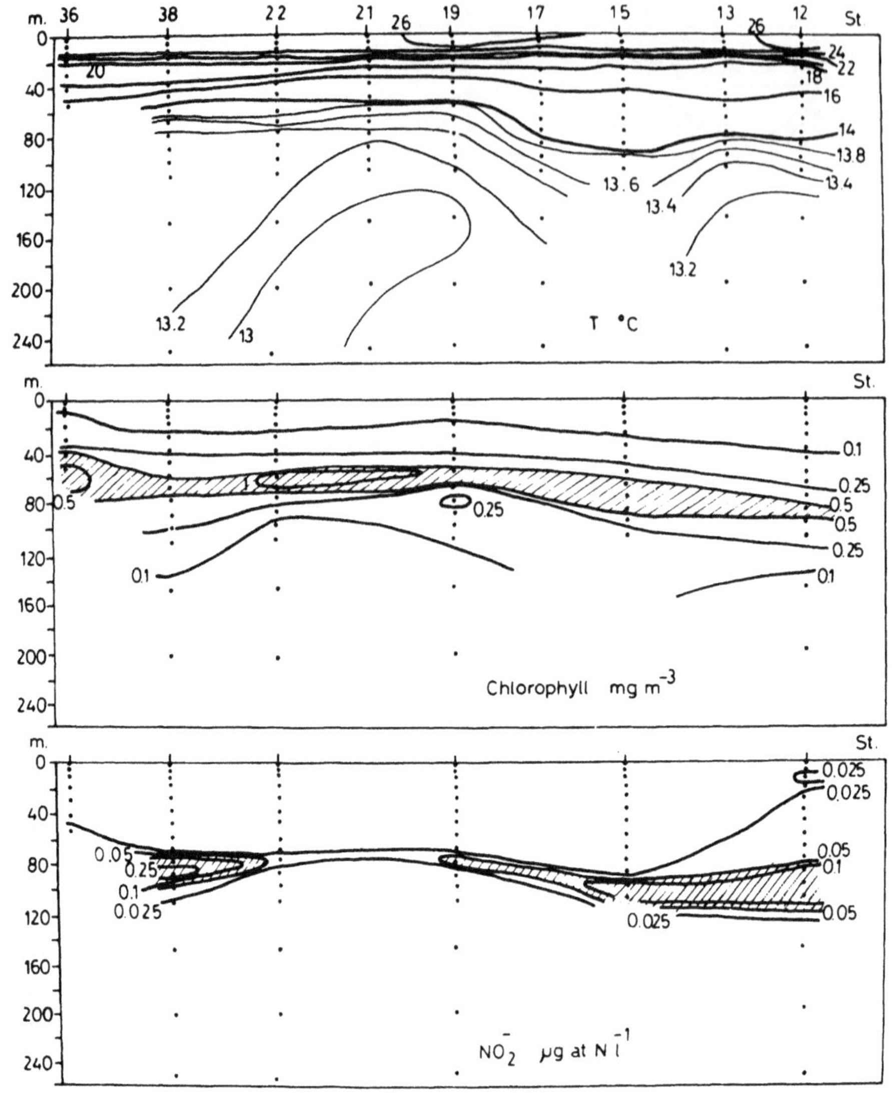

Fig. 3. Distribution of temperature ($T^{o}C$), chlorophyll a ($mg \cdot m^{-3}$) and nitrite (NO_2^-, μg-at $N \cdot l^{-1}$) along the second AB transect (fig. 1).

heterogeneity has been discussed by Margalef (this book).

Vertical Distribution of Phytoplankton Biomass and Primary Production. Relationships with Physicochemical Parameters

The chlorophyll a profiles (figs 2-3) showed the existence of a pronounced maximum at depths ranging from 40 to 90 m, well below

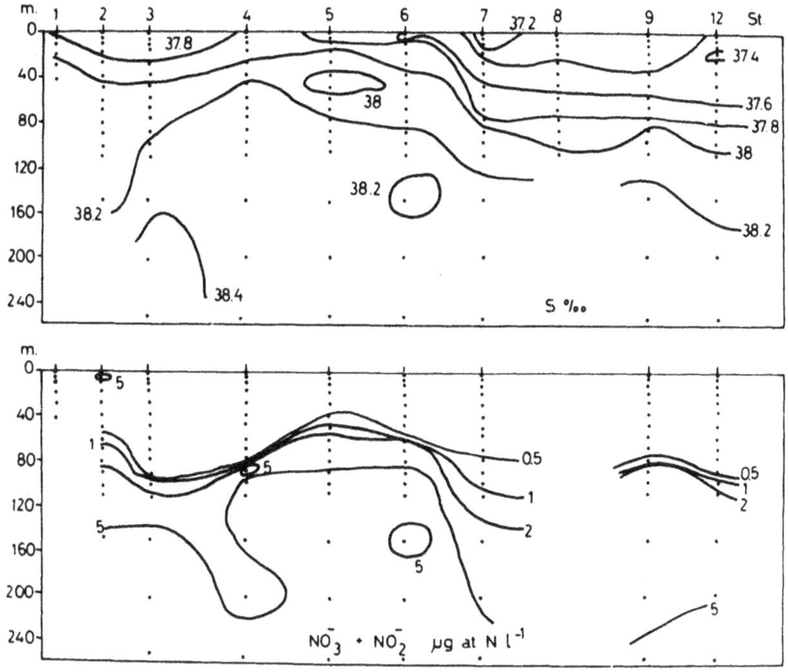

Fig. 4. Distribution of salinity (S o/oo) and nitrate along the first transect.

the zone of maximum vertical gradient of the thermocline or the pycnocline. As pointed out in the methods section, phaeopigment concentrations were generally low relative to total pigments; their distribution paralleled that of chlorophyll a (Fig. 5).

The position of the chlorophyll maximum in the water column followed the patterns of the temperature profiles and shoaled with the cold water doming centered around stations 4-5 or 19-22. The highest chlorophyll a concentrations were found in this area, where the relatively high nutrient waters reached shallower levels. At some points in the deep chlorophyll maximum (DCM), chlorophyll a exceeded 2 $mg \cdot m^{-3}$, a remarkably high value, taking into account the usually very low chlorophyll concentrations reported in the offshore zones of the Mediterranean. To the east of the sill between Mallorca and Menorca, overall chlorophyll concentrations were much lower, reflecting probably the lower nutrient concentrations in the underlying waters (fig. 4).

As can be seen in figs 2-3, chlorophyll concentration within the DCM showed considerable heterogeneity. Repeated stations at the same nominal position, taken at three hour intervals (fig. 6), also showed large, short-term fluctuations, but did not present any consistent chlorophyll distribution patterns which could be related

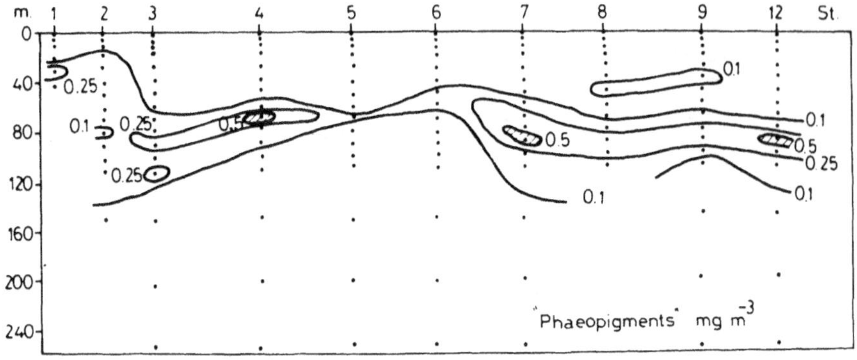

Fig. 5. Distribution of phaeopigments along the first transect.

to a diel cycle. Of course, all this variability included the possible effects of factors such as ship's drift and internal wave activity, but nonetheless, it appeared to reflect the existence of considerable spatial patchiness. The spatio-temporal variability on the time scale of several days revealed by the comparison of the repeated sections (figs 2 and 3) was also high; both advection of water masses and temporal changes probably contributed to it.

The nitracline (depth of maximum vertical nitrate gradient) started at approximately the level of the DCM (± the sampling interval of 10 m). The nitrite maximum was located within 10 m of the chlorophyll maximum and the nitracline (figs 7-8). In one case (station 4) secondary nitrite and chlorophyll maxima were found at a depth of 200 m (fig. 7); examination of the profiles of temperature and other parameters suggested that this phenomenon was due to a transport to lower layers of water from the shallower chlorophyll maximum. Silicate concentrations showed the same general patterns as nitrate, although the vertical gradients at the DCM level were smoother in the case of silicate. Phosphate determinations were available only at a few stations on the two first transects; in these and in the remaining stations, phosphate was practically undetectable throughout the euphotic zone; appreciable levels ($\geq$0.1 µg-at P l^{-1}) appeared generally about 30 m below the DCM.

An oxygen maximum, typically located 30 m above the DCM, was well developed in all the AB sections except the first one (stations 1-9 and 12), in which the oxygen distribution appeared to be very patchy, a fact that could perhaps be attributed to analytical problems. It is interesting to note that in the stations with highest chlorophyll concentrations ($\geq$2 mg·m^{-3}) at the maximum, (nos. 5,22,63) the depth of the oxygen maximum coincided with the DCM or was found only 10 m above it.

Light levels at the DCM depth ranged from 0.4 to 2.3% of the

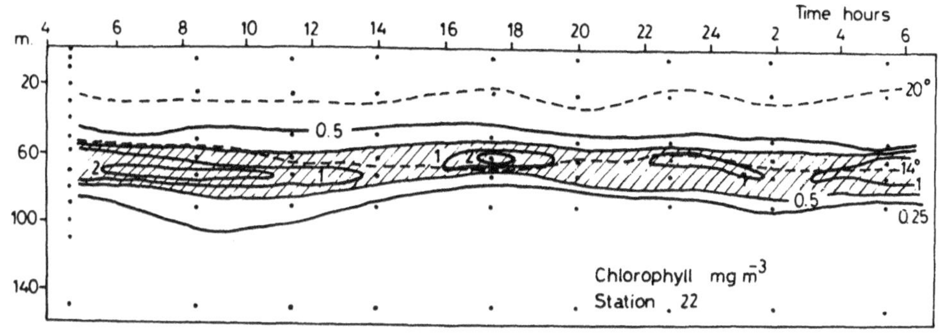

Fig. 6. Time series (h, GMT) of the chlorophyll a concentration at biological station no. 22. The dashed lines indicate the position of the 20°C and 14°C isotherms.

irradiance at surface; the geometric means of the absolute and relative noon light intensities measured at six stations (nos. 12, 15,19,44 and 65) were respectively 18 μE $m^{-2}s^{-1}$ and 0.9% of surface irradiance.

The profiles of primary production showed in general a peak (which sometimes represented the highest value for the water column) at the level of the chlorophyll maximum (fig. 9). The average productivity index (mg C fixed mg Chlor. $a^{-1}h^{-1}$) of the DCM samples was significantly lower than that of surface samples (Table 1), although there was considerable scatter. Low light levels were likely to limit carbon assimilation in the DCM and deeper samples; mid-day 10-15 GMT) incubation of DCM samples at higher light levels (typically 8 to 13% of surface irradiance) yielded only slightly higher productivity indices; however, these results could be biased by light inhibition effects or (especially in the richer samples) nutrient exhaustion in the incubation bottles.

Standing Stock of Phytoplankton

As can be seen in figs 7,8,10 and 11, phytoplankton cell numbers showed a deep maximum located at the same depth or above the DCM. Comparison of figs 2-3 with 10-11 shows that an increased phytoplankton abundance contributed significantly to the DCM, especially in the zones where phytoplankton and cell concentrations were important. In addition, average chlorophyll a content per cell appeared to be higher in the deeper parts of the euphotic zone (Table 2), as was found in a previous cruise and has been reported of other DCM (Steele, 1964; Anderson, 1969; Saijo et al., 1969; Kiefer et al., 1976; Venrick, 1982).

The biomass represented (in terms of carbon for example) by phytoplankton cell numbers varies with the composition of the

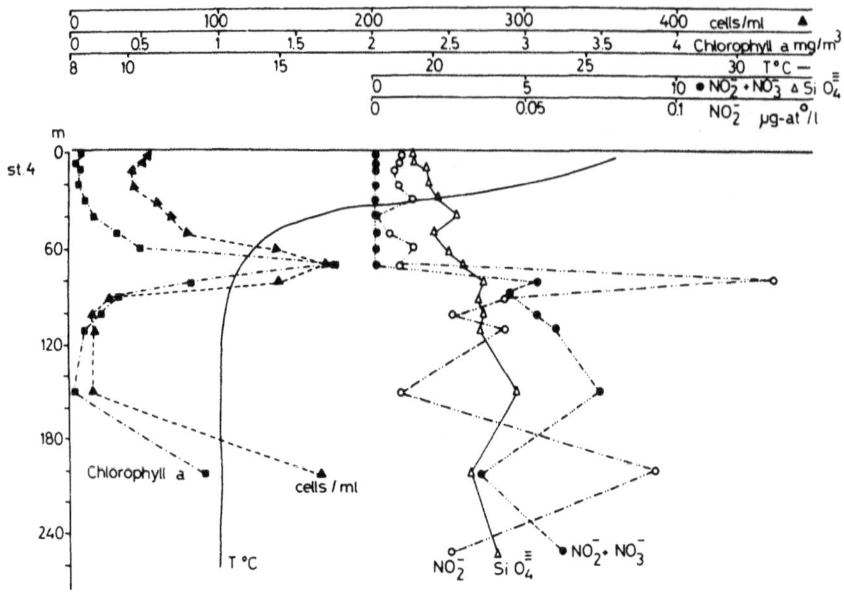

Fig. 7. Vertical profiles of temperature (T^oC), nitrite (NO_2^-, µg-at $N \cdot l^{-1}$), nitrate+nitrite ($NO_2^- + NO_3^-$, µg-at $N \cdot l^{-1}$), silicate (SiO_4^{-3}, µg-at $Si \cdot l^{-1}$), chlorophyll a (mg m^{-3}) and number of cells per ml at station 4.

community; however, especially when large density gradients are involved, both parameters, cell number and phytoplankton C may be expected to vary in the same sense. Additional measurements related to phytoplankton abundance (besides other seston components, of

Table 1. Mean values of carbon fixed per unit chlorophyll a per unit time (productivity index, in µg C µg Chl $a^{-1}h^{-1}$ ± standard error of the mean) for samples from 8 biological stations (15,19,22-1,22-2,45,58,61 and 64) in which incubations were carried out between 10 and 15 h GMT. The number of observations in each case is given in parentheses.

Sampling depth	Productivity index µg C µg Chl $a^{-1}h^{-1}$
5m	2.07±0.64(7)
25 m	2.00±0.64(7)
between 50 m and the DCM	0.57±0.13(9)
DCM:	
incubation 1% surface light	0.23±0.03(8)
8-13.5% surface light	0.28±0.09(6)

Table 2. Mean phytoplankton abundance (cells/ml ± standard error of the mean). Mean chlorophyll *a* concentration (mg m^{-3} ± standard error of the mean; fluorometric determinations corrected for "phaeopigments") and mean chlorophyll *a* content per 10^6 cells (µg Chlor. *a*/10^6 cells ± standard error of the mean) in samples from several depth ranges of hydrographical stations 1 to 9. The DCM sample of station 9 (90 m) gave a very high ratio (64.88 µg chl *a*/10^6 cells) and was excluded in one set of calculations. See fig. 10 for the location of the sampling depths. The number of observations in each case is given in parentheses.

Depth range	Chlorophyll *a* mg m^{-3}	Total phytoplankton cells/ml	g chlor. *a*/ 10^6 cells
Above the DCM	0.12±0.01 (64)	46.67±2.62 (66)	3.13±0.44 (64)
DCM	0.84±0.59 (9)	189.78±93.15 (9)	13.31±66.65 (9)
DCM excluding st.9,90m	0.90±0.21 (8)	212.76±102.35 (8)	9.11±2.74 (8)
Below the DCM	0.18±0.03(42)	27.22±5.12 (44)	8.21±0.88 (42)

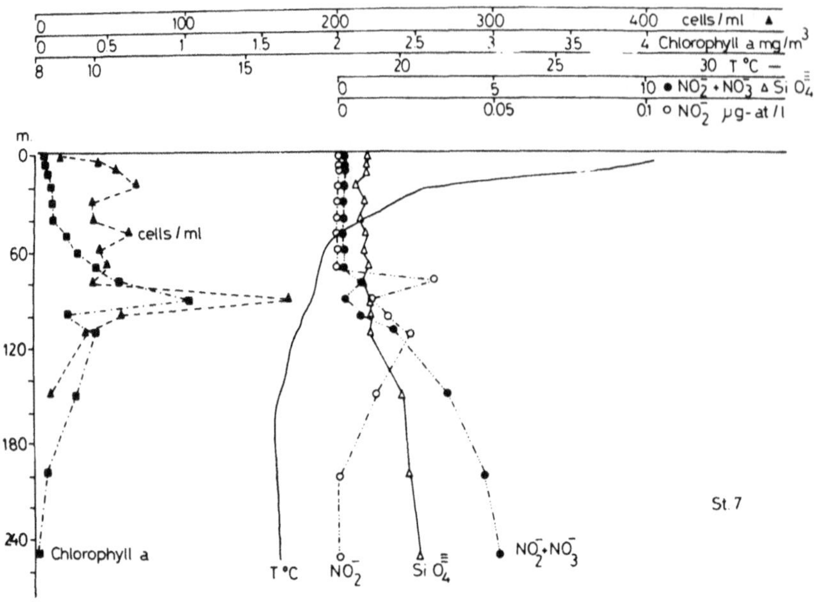

Fig. 8. Vertical profiles of temperature ($T^{o}C$), nitrite (NO_2^-, μg-at $N \cdot l^{-1}$), nitrate + nitrite ($NO_2^- + NO_3^-$, μg-at $N \cdot l^{-1}$), silicate (SiO_4^{-3}, μg-at $Si \cdot l^{-1}$), chlorophyll a ($mg \cdot m^{-3}$) and number of cells per ml at station 7.

course) such as particulate ATP (Vallespinós, in preparation), particulate organic C and N and particle volume from Coulter counts (Flos, unpublished data), which were available for the biological stations, showed maxima which coincided in general with the chlorophyll maxima.

Composition of the Phytoplankton Community

Numerically, the phytoplankton community (see figs 10-12 and table 3) was dominated in the upper euphotic layers by small flagellates and non-thecate dinoflagellates. Coccolithophorids were present in appreciable numbers only at the inshore stations. The two silicoflagellates recorded, *Dictyocha fibula* and *D. speculum*, were found at the lower layers of the euphotic zone, a fact typical of their summer distribution in the Mediterranean (Nival, 1965; Travers and Travers, 1968); however they were not confined to the DCM. Diatoms were scarce in the second transect; in the first one (fig. 10) they showed the highest concentrations at the coastal station and in a narrow band in the richest parts of the DCM; a weakly silicified *Chaetoceros* was the dominant form in these DCM zones. Although the population density of dinoflagellates was rather uniform throughout the euphotic layers, diatoms, and to a lower degree the flagellates, showed wide concentration fluctuations. It is interesting to note

Table 3. Statistical data for the most frequent and/or abundant taxa in the set of 141 samples; m = mean in cells per ml, s = standard deviation, computed for the number of occurrences (n). A: Taxa included in the statistical analyses. B: Taxa not included in the statistical analyses.

		n	m	s
A:				
1	Amphidoma caudata Halldal	17	0.47	0.64
2	Amphisolenia globifera Stein	6	0.01	0.005
3	Blepharocysta spp.	7	0.02	0.01
4	Ceratium furca (Ehrenb) Clap. @ Lachm.	6	0.01	0.004
5	C. fusus (Ehrenb.) Dujardin	8	0.01	0.004
6	Cochlodinium spp.	66	0.21	0.46
7	Dinophysis ovum Schütt	13	0.06	0.08
8	D. parvula (Schütt) Balech	13	0.02	0.01
9	Heteraulacus sphaericus (Murr.@ Whit.) Leblich III	7	0.01	0.003
10	Gonyaulax cf. fragilis	12	0.03	0.02
11	Gyrodinium sp. 1	9	0.01	0.01
12	G. spp.	73	0.33	0.53
13	Mesoporos sp.	7	0.40	0.84
14	Oxytoxum scolopax Stein	12	0.01	0.002
15	O. variabile Schiller	42	1.72	1.22
16	O. spp. (small)	46	0.85	0.84
17	Protoperidinium brochii (Kofoid @ Swezy) Balech	6	0.03	0.03
18	P. steinii (Jorg) Balech	8	0.03	0.07
19	P. spp. (small)	31	0.16	0.42
20	Prorocentrum compressum (Bailey)	17	0.07	0.09
21	P. spp. (small)	33	0.71	0.52
22	Torodinium robustum Kofoid @ Swezy	29	0.09	0.22
23	Palaeophala chroma ?	14	0.03	0.05
24	Cachonina spp.	88	5.85	4.28
25	Dinoflagellate "sp. 1"	15	0.60	0.66
26	Small dinoflagellates	139	16.81	12.70
27	Emiliania huxleyi (Lohmann) (Hay @ Mohler)	9	2.60	2.87
28	Helicosphaera carteri (Wallich) Kamptner	8	1.02	1.42
29	Unidentified coccolithophorids	15	1.45	1.10
30	Chaetoceros spp.	12	3.19	3.55
31	Chaetoceros spp.	19	46.57	132.95
32	Coscinodiscus spp.	7	0.04	0.07
33	Cylindrotheca closterium (Ehrenb.) (Reimann @ Lewin)	6	1.47	2.10
34	Hemiaulus sp.	11	0.27	0.43
35	Nitzschia sp. 1	22	1.50	2.23

(continued)

Table 3 (Cont.)

		n	m	s
36	Nitzschia sp. 2	24	2.02	3.45
37	Rhizosolenia alata Brightwell	19	0.58	0.96
38	R. setigera Brightwell	5	2.69	4.28
39	R. fragilissima Bergan	6	0.04	0.01
40	R. spp.	10	0.23	0.33
41	Pennate sp. 1	10	0.06	0.06
42	Thalassiosira spp (small)	15	1.72	1.12
43	Th. spp. (large)	10	0.43	0.49
44	Pennate diatoms	37	0.20	0.47
45	Dictyocha fibula Ehrenberg	39	0.83	0.80
46	D. speculum Ehrenberg	5	0.25	0.49
47	Flagellates	131	18.85	30.21
48	Flagellate spp. 1	20	2.28	1.36
49	Kephyrion like forms	39	3.14	4.49
50	Pterosperma spp.	17	0.74	0.95
B:				
	Dinophysis spp. 1	17	0.05	0.08
	D. spp. 2	17	0.11	0.28
	Gonyaulax spp.	17	0.04	0.07
	Scrippsiella spp. (small)	27	1.43	1.39
	Torodinium spp. (small)	16	0.44	0.53
	Unidentified dinoflagellates (large)	56	0.08	0.15
	Unidentified centric diatoms	15	1.24	2.07
	Pennate diatoms (very small)	15	0.97	0.83
	Cyanobacteria	4	1001.11	1728.77

that the chlorophyll peak of station 22 (50m) was due to the presence of 4000 cells/ml of small (1.5-2 μ in diameter) spherical and crescent-shaped forms which appeared to be Cyanobacteria.

A common feature (figs 7,8,10,12) of the phytoplankton profiles was the narrowness of the layers of high phytoplankton concentrations and the sharpness of the vertical gradients around the abundance peaks. This made the comparison with some of the previous surveys in the area difficult; in these, samples below 25 m were taken routinely at depth intervals of 25 m or more.

The ordination of the phytoplankton taxa resulting from the principal components analysis (PCA) is illustrated in fig. 13. The variances explained by the first three axes are listed in Table 4. Examination of fig. 12 shows that there is a segregation of the major phytoplankton groups. This finding, which has been reported from other similar analyses (Margalef and Estrada, 1981; Estrada, 1982) points to the fact that taxonomic groups such as diatoms,

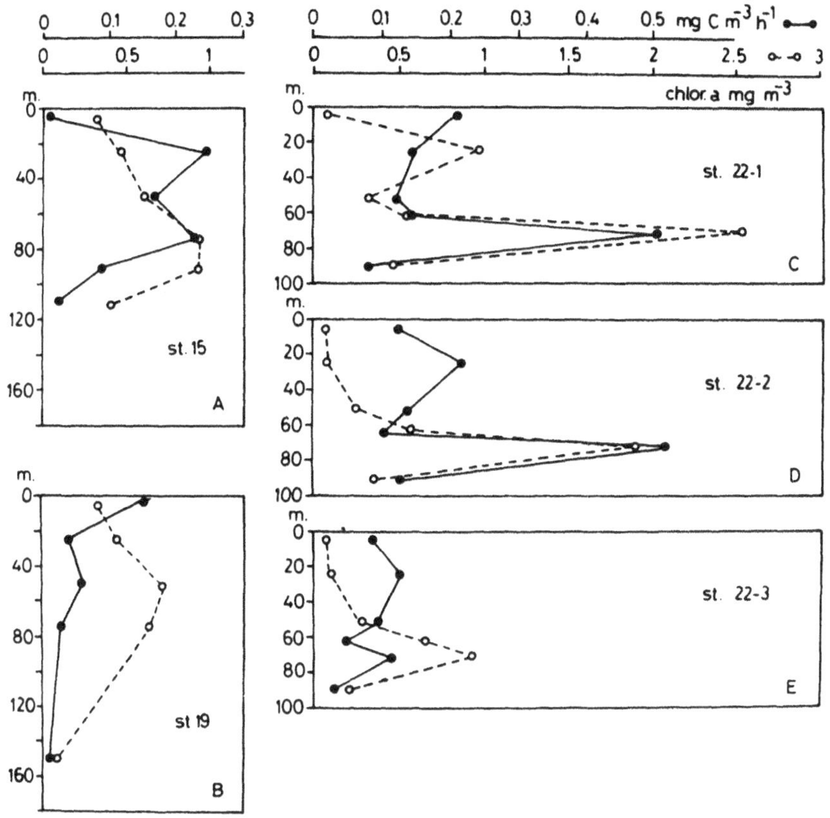

Fig. 9. A,B,C: Vertical profiles of primary production (mg $C \cdot m^{-3} \cdot h^{-1}$) for the biological stations of the second transect at which the incubations were carried out between 10 and 15 h GMT. C,D and E show the production profiles for samples taken at three different times of day at station 22. Incubation starting times (GMT): station 15, 11:43; station 19, 11:59 GMT; station 22-1, 9:30 GMT; station 22-2, 12:05 GMT; station 22-3, 14:35 GMT.

dinoflagellates or coccolithophorids represent life-forms (Margalef, 1978b) that show a degree of common response to environmental conditions.

The general features of the distribution of the first components (figs 14-15) were similar in the two sections studied. Because of the better sampling coverage, comments will be referred only to the first section (fig. 14).

The distribution of the first principal component (fig. 13)

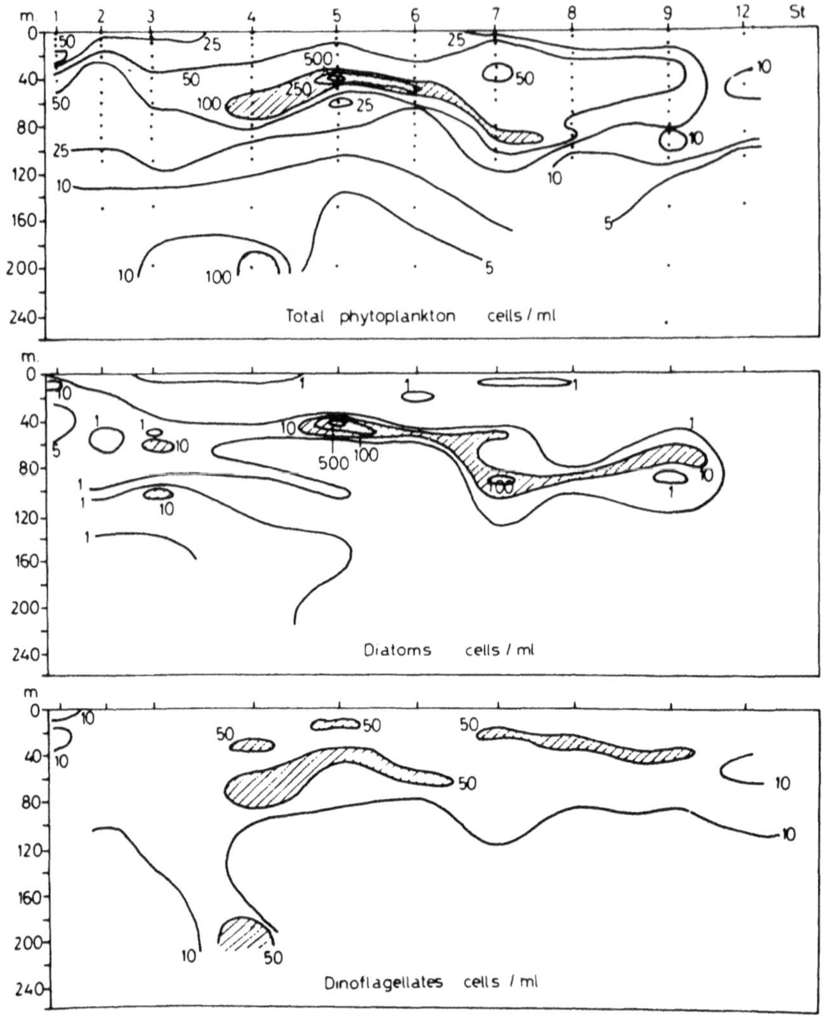

Fig. 10. Distribution of the population density (cells ml^{-1}) of total phytoplankton, diatoms and dinoflagellates along the first transect.

showed marked similarities with the distributions of chlorophyll and total phytoplankton; the highest positive values were found in the DCM layers. Most of the species were positively associated with this component (fig. 13), but the highest positive correlation coefficients corresponded to one dinoflagellate, Oxytoxum scolopax, and several diatoms, Chaetoceros sp. 1, and spp., Thalassiosira sp. 1 and species of Nitzschia and Rhizosolenia, which were especially abundant at the DCM.

The second principal component appeared to express the contrast

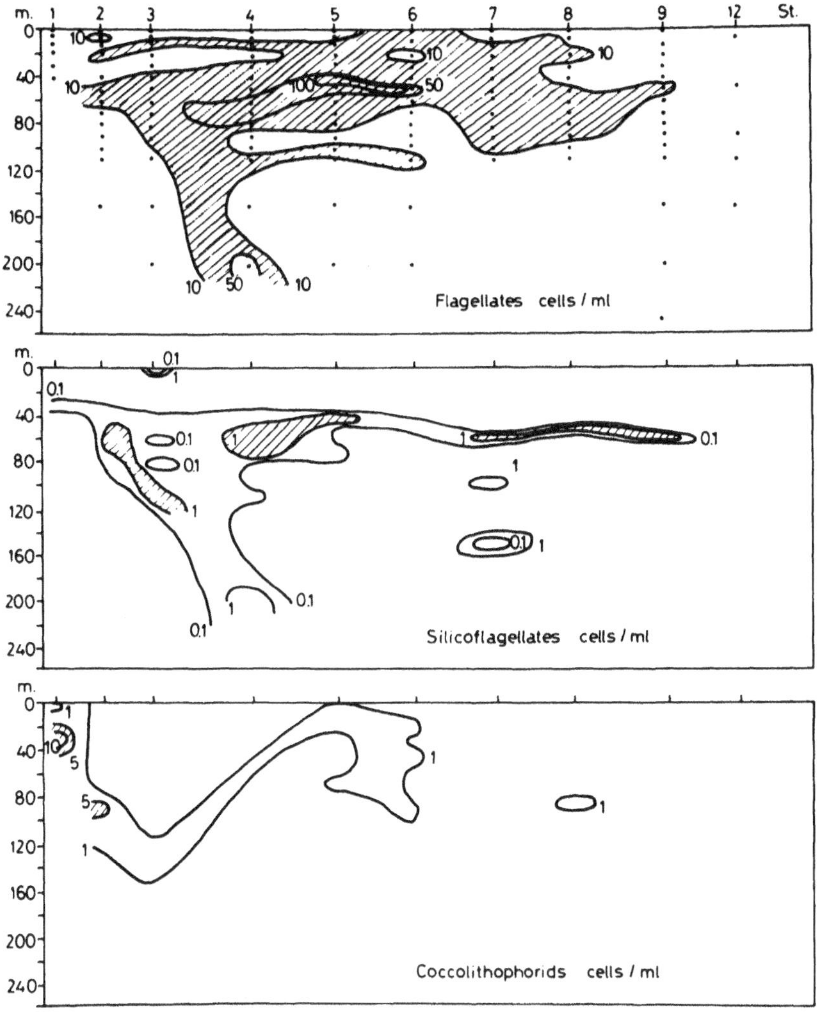

Fig. 11. Distribution of the population density (cells ml^{-1}) of flagellates, silicoflagellates and coccolithophorids along the first transect.

between some species of the neritic zone showing negative correlations with it (Rhizosolenia setigera, Emiliania huxleyi, Helicosphaera carteri, besides some diatoms like Nitzschia sp. 1 and 2, and Chaetoceros spp. also present in the DCM) and a group of dinoflagellates and flagellates distributed throughout the euphotic layers of the water column in the offshore stations.

The third component had the strongest positive associations with several large dinoflagellates (Ceratium fusus, Heteraulacus sphaericus, Protoperidinium brochii). The distribution of this

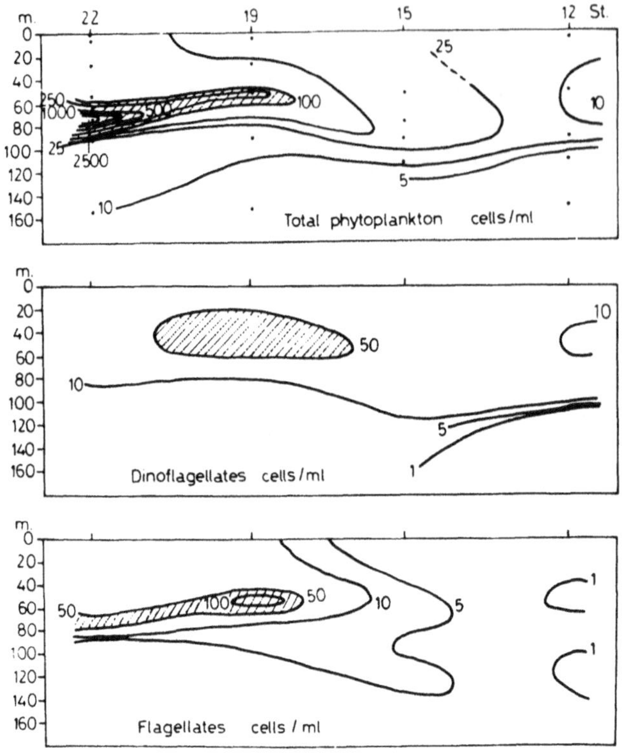

Fig. 12. Distribution of the population density (cells ml^{-1}) of total phytoplankton, dinoflagellates and flagellates along the second transect.

component was very patchy, but its higher positive scores were found in the surface layers of the water column, suggesting that a segregation of phytoplankton occurred also in the poor upper part of the euphotic zone.

With reversed signs of the coordinates, the two first axes of the correspondence analysis (CA) gave (Tables 4 and 5) an ordination similar to that of the PCA, although in the CA, some species with peculiar distributions (like Rhizosolenia setigera) explained a very large part of the variability of the axes. The third factor of the CA had no parallel in the PCA and expressed differences in the distribution of the littoral forms, Rhizosolenia setigera, on one side, and the coccolithophorid taxa on the other.

Examination of the PCA distributions showed that, in addition to a vertical structuring of the phytoplankton assemblages, there was horizontal inshore-offshore variability. The distinctive features of the phytoplankton composition of the coastal station can be explained by the intervention of different fertilization processes

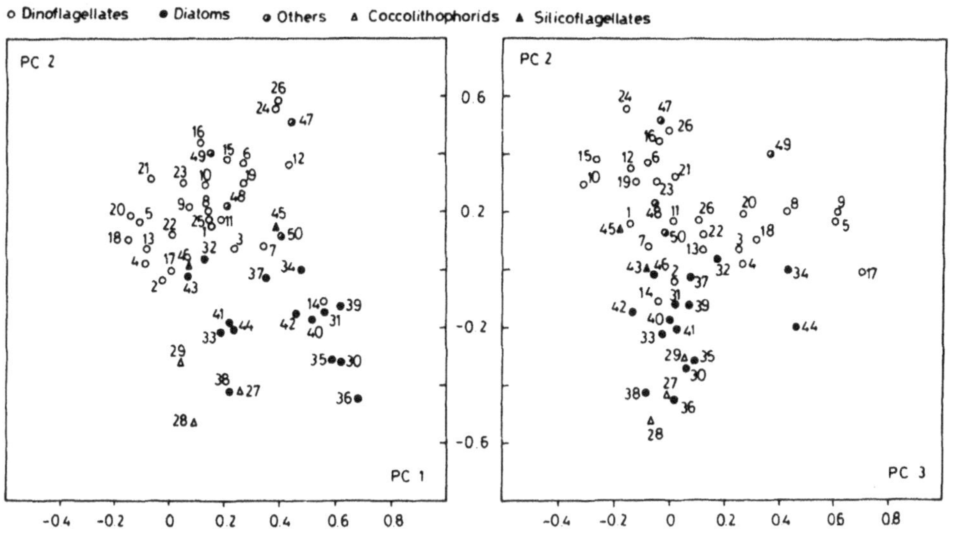

Fig. 13. Position of the extremes of the 50 species vectors in the spaces determined by the first and second principal components and by the second and third principal components:
Number key: (1) *Amphidoma caudata*, (2) *Amphisolenia globifera*, (3) *Blepharocysta* sp., (4) *Ceratium furca*, (5) *Ceratium fusus*, (6) *Cochlodinium* spp. (7) *Dinophysis ovum*, (8) *Dinophysis parvula*, (9) *Heteraulacus sphaericus*, (10) *Gonyaulax* cf. *fragilis*, (11) *Gyrodinium* sp. 1, (12) *Gyrodinium* spp., (13) *Mesoporos* sp., (14) *Oxytoxum scolapax*, (15) *Oxytoxum variabile*, (16) *Oxytoxum* spp. (17) *Protoperidinium brochii*, (18) *Protoperidinium steinii*, (19) *Protoperidinium* spp., (20) *Prorocentrum compressum*, (21) *Prorocentrum* spp. (small), (22) *Torodinium robustum*, (23) *Palaeophalachroma* cf. *unicinctum*, (24) *Cachonina*(?) spp., (25) Dinoflagellate sp. 1, (26) Small dinoflagellates, (27) *Emiliania huxleyi*, (28) *Helicosphaera carteri*, (29) Unidentified coccolithophorids, (30) *Chaetoceros* spp., (31) *Chaetoceros* sp. (32) *Coscinodiscus* spp., (33) *Cylindrotheca closterium*, (34) *Hemiaulus* sp., (35) *Nitzschia* sp. 1 (36) *Nitzschia* sp. 2, (37) *Rhizosolenia alata*, (38) *Rhizosolenia setigera*, (39) *Rhizosolenia fragilissima*, (40) *Rhizosolenia* spp., (41) *Pennate* sp. 1, (42) *Thalassiosira* spp. (small), (43) *Thalassiosira* spp. (large), (44) *Pennate diatoms* (small), (45) *Dictyocha fibula*, (46) *Dyctiocha speculum*, (47) Flagellates, (48) Flagellate sp. 1, (49) Kephyrion-like forms (50) *Pterosperma* spp.

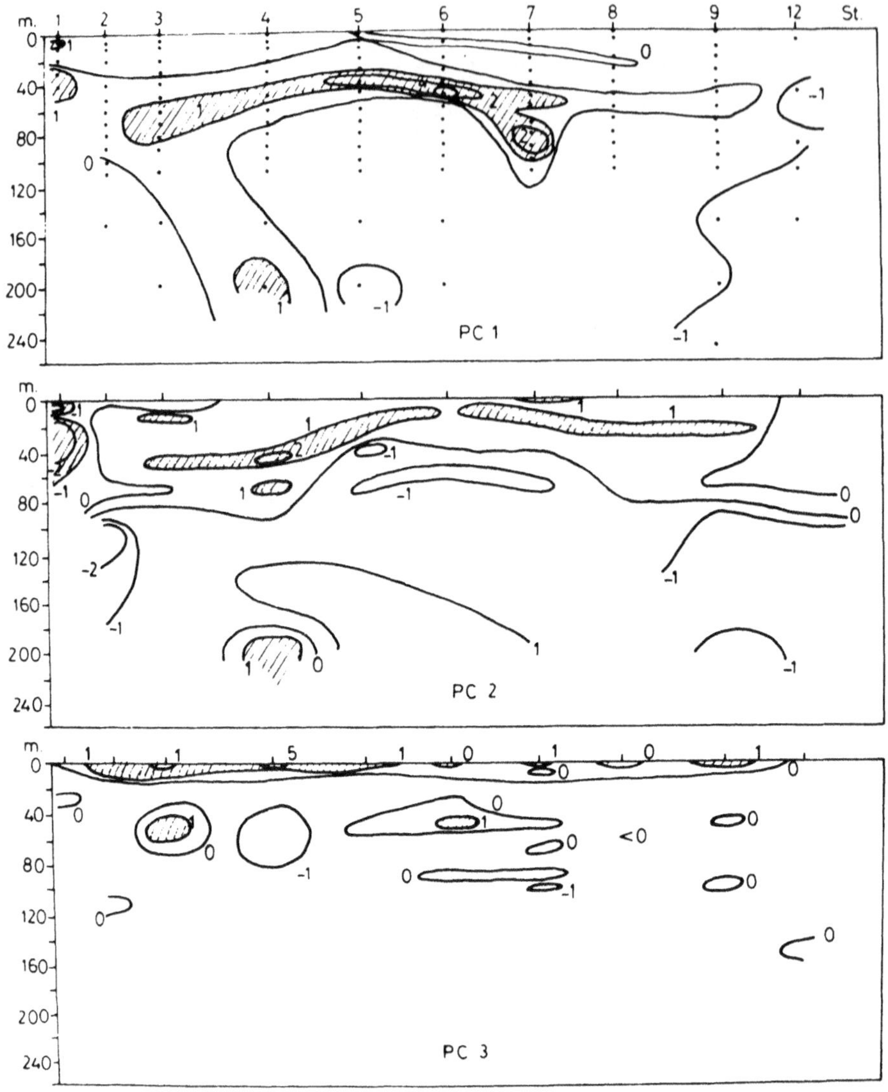

Fig. 14. Distribution of the scores of the first three principal components along the first transect.

Table 4. Percentage of variance explained by the first three axes of the principal components (PCA) and correspondence (CA) analyses

	1	2	3
PCA	10.0	7.7	5.3
CA	22.0	15.7	8.7

Table 5. Coordinates of the 50 taxa for the first three axes of the correspondence analysis. Species ordained according to their coordinate for the first factor

		1	2	3
31	Chaetoceros spp.	-1.89	-0.12	0.07
33	Cylindrotheca closterium	-1.37	0.09	0.76
30	Chaetoceros spp.	-0.95	0.61	0.65
34	Hemiaulus spp.	-0.87	0.01	-0.01
39	Rhizosolenia fragilissima	-0.83	-0.07	-0.05
36	Nitzschia spp. 2	-0.55	3.45	1.22
40	Rhizosolenia spp.	-0.31	0.44	-0.49
38	Rhizosolenia setigera	-0.09	10.98	-5.18
14	Oxytoxum scolopax	-0.07	0.34	0.51
35	Nitzschia spp. 1	-0.06	1.98	-0.77
50	Pterosperma spp.	-0.03	-0.10	-0.06
41	Pennate spp. 1	0.09	-0.12	0.04
47	Flagellates	0.11	-0.11	-0.14
12	Gyrodinium spp.	0.17	0.58	-0.40
29	Coccolithophorids	0.20	2.01	1.67
42	Thalassiosira spp.	0.21	1.31	0.95
27	Emiliania huxleyi	0.22	2.44	5.03
2	Amphisolenia globifera	0.25	-0.17	-0.11
44	Pennate diatoms	0.26	0.07	0.27
7	Dinophysis ovum	0.28	0.08	-0.23
46	Dictyocha speculum	0.31	0.15	0.22
45	Dictyocha fibula	0.31	-0.09	-0.12
28	Helicosphaera carteri	0.33	3.26	6.37
3	Blepharocysta spp.	0.33	-0.14	-0.02
37	Rhizosolenia alata	0.36	1.32	0.54
11	Gyrodinium spp. 1	0.40	-0.14	0.05
43	Thalassiosira spp. (large)	0.41	-0.11	0.08
26	Small dinoflagellates	0.43	-0.06	0.04
22	Torodinium robustum	0.44	-0.11	-0.25
19	Protoperidinium spp.	0.45	-0.11	-0.05
15	Oxytoxum variabile	0.45	-0.18	-0.11
23	Palaeophalachroma?	0.49	-0.19	-0.11
24	Cachonina?	0.49	-0.03	0.03
6	Cochlodinium spp.	0.50	-0.17	-0.08
48	Flagellate sp. 1	0.51	0.09	-0.14
10	Gonyaulax cf. fragilis	0.51	0.19	-0.34
13	Mesoporos sp.	0.51	-0.23	-0.20
4	Ceratium furca	0.52	-0.01	0.13
1	Amphidoma caudata	0.53	0.01	0.12
16	Oxytoxum spp.	0.53	-0.17	-0.15
25	Dinofagellate sp.	0.54	-0.15	-0.12
32	Coscinodiscus spp.	0.57	-0.09	-0.09
18	Protoperidinium steinii	0.58	-0.16	-0.05
21	Prorocentrum spp. (small)	0.59	-0.11	0.13

(continued)

Table 5 (Cont.)

		1	2	3
8	Dinophysis parvula	0.59	-0.08	-0.03
49	Kephyrion like forms	0.61	-0.07	2.01
17	Protoperidinium brochii	0.62	0.05	0.37
9	Heteraulacus sphaericus	0.63	-0.03	-0.06
5	Ceratium fusus	0.64	-0.19	-0.03
20	Prorocentrum compressum	0.65	-0.18	0.00

than those in the offshore stations. At the other extreme of the transect, station 12, beyond the Mallorca-Menorca channel, was characterized, in a rather negative way, by the low abundances of most taxa.

The results obtained with the UPGMA clustering (fig. 16) agreed with the main patterns of the ordination methods. Distinctive groups were formed by the DCM taxa (Oxytoxum scolopax, Chaetoceros, sp. 1 and spp., Nitzschia sp. 1 and 2, Rh. fragilissima and Rh. spp.), the neritic taxa (Rhizosolenia setigera and the coccolithophorids) and the surface dinoflagellates (Ceratium furca, C. fusus, Protoperidinium brochii, P. steinii, Prorocentrum compressum).

Another way of looking at the vertical structure of the phytoplankton community was provided by the examination of the similarity between adjacent samples within the same profile, which was evaluated using Spearman's rank correlation coefficient. Neighboring samples with significant ($p < 0.05$) rank correlation were joined by continuous lines in fig. 17. While neighboring surface or DCM samples showed often significant associations, rank correlation coefficients were not significant between DCM and upper euphotic zone or DCM and deeper samples, indicating a differentiation between the phytoplankton assemblages of the DCM and the neighboring upper and lower zones of the water column.

DISCUSSION

The results of the PEP 82 survey, showed that, in the period of study, the DCM was in general associated with an elevated concentration of phytoplankton. Average chlorophyll per cell was also increased with respect to the upper parts of the euphotic zone. An elevated chlorophyll content at the DCM may be due to adaptation of the phytoplankton to suboptimal light intensities or to exposure to increased nutrient concentrations (Steemann Nielsen and Hansen, 1959; Hobson and Pariser, 1971; Jensen and Sakshaugh, 1973). However, other possibilities such as an accumulation of chlorophyll-

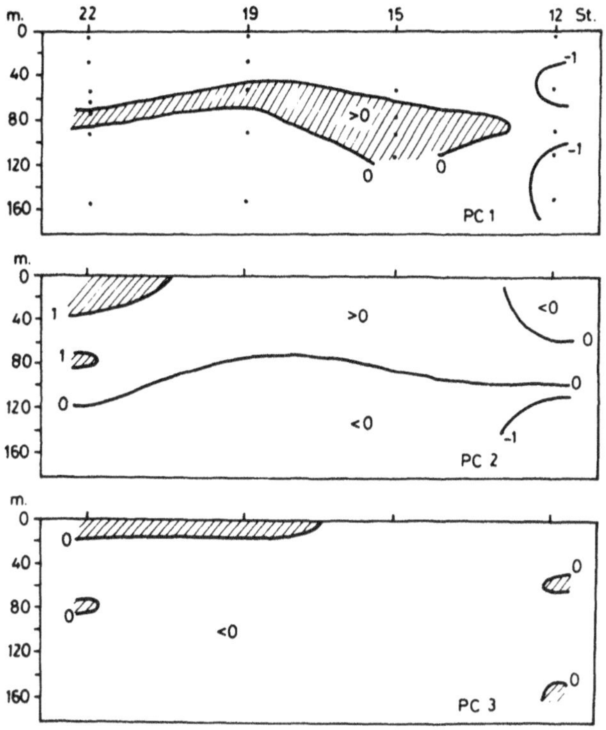

Fig. 15. Distribution of the scores of the first three principal components along the second transect.

containing detritus (Gieskes et al., 1978) cannot be excluded on the basis of the available data.

An increased cellular chlorophyll rather than higher cell abundance has been considered as the origin of the DCM in areas like the Central North Pacific (Beers et al., 1975; Kiefel et al., 1976; Venrick, 1982). Comparison of the results from a previous cruise (Estrada, 1982) and those presented here indicated that the relative contribution of elevated chlorophyll per cell versus increased phytoplankton abundance was important when the phytoplankton biomass was low, a conclusion also reached by Cullen and Eppley (1981).

Many of the most frequent phytoplankton forms cited in the present study were also abundant in other DCM areas (Beers et al., 1975; Cullen et al., 1982). A direct comparison with the organisms recorded by Margalef (this book) is not possible because of the different sampling methods; both data sets should be considered as complementary, although the distribution of the larger species may be expected to be more persistent in time and thus point to longer term environmental changes. Keeping this in mind the general patterns of distribution of the smaller phytoplankton forms were consistent with those reported by Margalef.

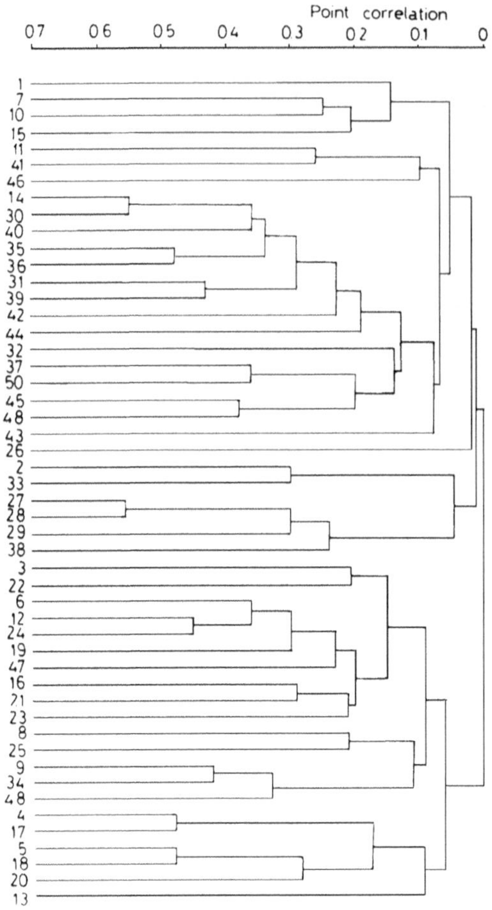

Fig. 16. Clustering of the 50 species using the UPGMA algorithm. The point correlation coefficient was used as a similarity index. Number key as in fig. 13.

The overall picture of the distribution of the phytoplankton responded to well-documented ecological differences among the major life-forms. Diatoms dominated in the coastal station and in a series of DCM samples, suggesting that there was (or had been recently) a pulse of high primary production in these areas. The dinoflagellates were ubiquitous throughout the nutrient-poor euphotic zone and several large forms were typically found only in surface layers. The taxonomic affinities of the organisms with abundance peaks at the DCM were very varied (diatoms, flagellates, dinoflagellates, Cyanobacteria) indicating that adaptation to the DCM conditions was not confined to particular algal groups. However, a detailed ecophysiological evaluation of the composition of the phytoplankton assemblages is hindered by the lack of physiological data for many organisms and by the fact that many abundant

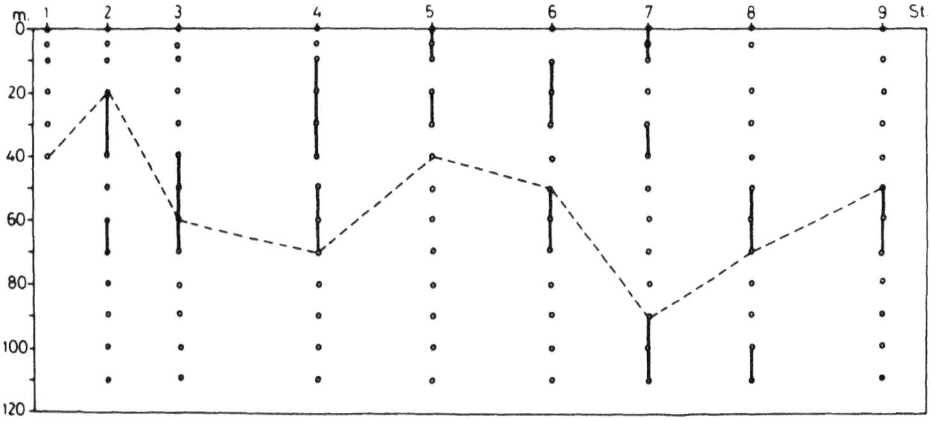

Fig. 17. Vertical variation of the Spearman rank correlation coefficient between neighboring samples. Adjacent samples within the same profile with significant positive correlations have been joined by a solid line. The dashed line indicates the depth of the maximum chlorophyll concentration.

forms (flagellates, small dinoflagellates, small centric diatoms) could not be identified to the level of species.

The differences between the phytoplankton distributions found in July 1982 and in some of the stations of the October 1976 survey that covered approximately the same positions (Estrada, 1982) can be attributed in large part to seasonal variation. The mixed layer was wider and cooler in October and the chlorophyll concentrations at the DCM appeared to be lower, although it is possible that sharp chlorophyll peaks could have been missed because of the larger sampling depth intervals in the October cruise. The most prominent difference in the overall composition of the phytoplankton was the relative abundance of coccolithophorids in the offshore stations, in October; curiously, two of the dominant coccolithophorid species in these stations (Emiliania huxleyi and Helicosphaera carteri) were the same as those that characterized the coastal assemblages in the present study.

In the area of study, the phytoplankton composition of the DCM differed qualitatively and quantitatively from that of the upper euphotic zone. Similar results have been reported for the Central Pacific (Venrick et al., 1973), the Western North Pacific (Furuya and Marumo, 1983) and the sea off Southern California (Cullen et al., 1982). In contrast, in the Central North Pacific, Beers et al. (1975) found that none of the major groups of phytoplankton showed an increase in the maximum as compared to the upper waters. Venrick (1982), also in the Central North Pacific, distinguished a shallow and a deep phytoplankton association, but the chlorophyll maximum did not fall within either of them. Taniguchi and Kawamura (1972) in the Oyashio waters, found no significant differences in species composition between the DCM and shallower layers; they concluded that the DCM in their area was largely a consequence of accumulation of sinking cells from the upper waters. The reasons for the discrepancies in the evaluation of the differences in phytoplankton composition of the DCM with respect to other layers have been related to different mechanisms of origin and maintenance of the chlorophyll maxima (Cullen and Eppley, 1981; Furuya and Marumo, 1983) but other phenomena such as seasonal (Beers et al., 1975) and shorter term temporal variability in the stage of development of the DCM community could also play an important role.

The presence of diatoms and the localized high cell densities found in some of the DCM samples of the present study suggested that they were the result of both in situ growth and low diffusion losses. The phytoplankton maxima were located below the steepest part of the pycnocline, but still within a zone of density stratification where a reduced turbulent mixing could reduce the export of cells to less favorable zones. It has been shown (Minas, 1970; Shulenberger and Reid, 1981) that the subsurface oxygen maxima which occur above the DCM arise from the photosynthetic production of oxygen below layers of density stratification which reduce the losses to the atmosphere. In support of a significant biological activity at the DCM is the finding that in the biological stations of the PEP 82 cruise, the abundance of bacteria (estimated by plate counts, Vallespinós, in preparation) and the mid-day concentration of zooplankton C and N (the organic C and N content of the zooplankton retained by filtering 20 l of water through a 200 μm mesh plankton net; Alcaraz, unpubl. data) showed maxima that coincided with the phytoplankton and chlorophyll peaks. Significant in situ growth is difficult to reconcile with the low light levels and the low productivity indices generally recorded at the maxima. It has been suggested, on the basis of oxygen distribution data, that primary production in oligotrophic marine areas may be much greater than the estimates derived from the ^{14}C method (Shulenberger and Reid, 1981; Jenkins, 1982). The cause of the discrepancy is not clear, but suspicion is usually directed at the ^{14}C techniques (Jenkins, 1982). However the analysis of Shulenberger and Reid has recently been criticised by Platt, who concluded from a set of oxygen data for the Tyrrhenian and Ionian basins of the Mediterranean, that no positive community production was detectable during the summer months (Platt, in press; Platt, this book).

The suggestion that the major growth period of the cells could occur when the winter mixing returns them to the euphotic zone has been raised by Venrick (1982). At present, there are not enough seasonal data to clarify this point, although, in addition, some concentration mechanism would be necessary to account for the high biomasses found in many of the cases. Another possibility is that the growth could have occurred at shallower and more illuminated levels than those where the maximum was found later; this is likely to be the case for the deeper second maximum of station 4. Increased production at the DCM could occur in pulses, when hydrographic events such as internal waves or other types of water motion, provide exposure to higher irradiances; in the PEP 82 cruise, several CTD time series suggested the presence of internal wave activity (Salat, unpubl. data). The synergistic importance of increased nutrient supply combined with better illumination is stressed by the fact that the highest biomass concentrations at the DCM were found where the dense, nutrient rich waters were shallower. Margalef (this book) has suggested that mixing processes, resulting, for instance, from the interaction between coastal topography and water circulation patterns, could induce the transport of layers of relatively rich water which would extend offshore into the appropriate density levels. The stratification of the populations of some of the larger species could be interpreted in these terms. All together, the available data indicate that the deep phytoplankton and chlorophyll maxima may be much more dynamic entities than was previously thought.

SUMMARY

A series of stations located along a transect from Barcelona to beyond the channel between Mallorca and Menorca were sampled several times during July, 1982. All of the vertical profiles revealed the existence of a pronounced chlorophyll maximum, approximately coinciding with the nitracline and the nitrite maximum, at depths ranging from 40 to 90 m. The deep chlorophyll maximum was, in general, associated with an elevated concentration of phytoplankton corresponded to well-documented differences among the major life-forms. Diatoms dominated at the coastal station and in a series of samples from the deep chlorophyll maximum. The dinoflagellates and small flagellates were ubiquitous throughout the nutrient-poor euphotic zone and several large dinoflagellate forms were typically found in the surface layers. The presence of diatoms and the localized, high cell-densities found in some of the deep chlorophyll maximum samples suggested that they were the result of _in situ_ growth combined with low diffusion losses.

Although the photosynthetic activities recorded in the deep chlorophyll maximum were relatively low and appeared to be light-limited, pulses of increased growth could occur when internal waves or other types of water motion exposed nutrient-rich water layers to higher light intensities. The synergistic importance of increased nutrient supply combined with higher illumination is emphasized by the fact that the highest biomass concentrations in the deep chlorophyll maximum were found where the dense, nutrient-rich waters were shallower. Several statistical procedures showed that the phytoplankton assemblages in the deep chlorophyll maximum differed from those in the upper euphotic zone.

ACKNOWLEDGEMENTS

This research was supported by the Consejo Superior de Investigaciones Cientificas and a grant from the Comision Asesora de Investigación Cientifica y Tecnica.

I thank Drs. R. Margalef and T. Platt for discussion of the manuscript. All the participants in the cruise cooperated in the collection of the samples and on-board analyses. C. Bas and J. Biosca provided technical assistance in the preparation of the figures.

REFERENCES

Anderson, G.C., 1969, Subsurface chlorophyll maximum in the Northeast Pacific Ocean, Limnol. Oceanogr., 14:386.

Beers, J.R., Reid, F.M.H., and Stewart, G.L., 1975, Microplankton of the North Pacific Central Gyre. Population Structure and Abundance, June 1973, Int. Revue ges. Hydrobiol., 60:607.

Benzécri, J.P., 1973, "L'analyse des données. I. La taxonomie. II. L'analyse des correspondances." (2nd edition, 1976). Dunod, Paris.

Cooley, W.W., and Lohnes, P.R., 1971, "Multivariate Data Analysis". Wiley, New York.

Cullen, J.J., and Eppley, R.W., 1981, Chlorophyll maximum layers of the Southern California Bight and possible mechanisms of their formation and maintenance, Oceanol. Acta, 4:23.

Cullen, J.J., Reid, F.M.H., and Stewart, E., 1982, Phytoplankton in the surface and chlorophyll maximum off southern California in August, 1978, Journal of Plankton Research, 4:665.

Estrada, M., 1981, Biomasa fitoplanctónica y producción primaria en el Mediterráneo occidental, a principios de otoño. Inv. Pesq., 45:211.

Estrada, M., 1982, Phytoplankton of the Western Mediterranean at the Beginning of Autumn, Int. Revue ges. Hydrobiol., 67:517.

Font, J., and Miralles, L., 1978, Circulación geostrófica en el mar Catalán, Res. Exp. Cient. B/O Cornide, 7:155.

Furuya, K., and Marumo, R., 1983, The structure of the phytoplankton community in the subsurface chlorophyll maxima in the western North Pacific Ocean, Journal of Plankton Research, 5:393.

Gieskes, W.W.C., Kraay, G.W., and Tijssen, S.B., 1978, Chlorophylls and their degradation products in the deep pigment maximum layer of the tropical North Atlantic, Neth. J. Sea Res., 12:195.

Goering, J.J., Wallen, D.D., and Nauman, R.M., 1970, Nitrogen uptake by phytoplankton in the discontinuity layer of the eastern subtropical Pacific Ocean, Limnol. Oceanogr., 15:789.

Dandonneau, Y., 1979, Concentrations en chlorophylle dans le Pacifique tropical sud-ouest: comparaisons avec d'autres aires oceaniques tropicales, Oceanol. Acta, 2:133.

Hobson, L.A., and Pariser, R.J., 1971, The effect of inorganic nitrogen on macromolecular syntheses by Thalassiosira fluviatilis HUSTEDT and Cyclotella nana HUSTEDT grown in batch culture, J. exp. mar. Biol. Ecol., 6:71.

Hobson, L.A., and Lorenzen, C.J., 1972, Relationships of chlorophyll maxima to density structure in the Atlantic Ocean and Gulf of Mexico, Deep-Sea Res., 19:297.

IBM, 1969, "1130 Scientific Subroutine Package" (1130-CM-02X). Progammers Manual, IBM Corporation.

Jenkins, W.J., 1982, Oxygen utilization rates in North Atlantic subtropical gyre and primary production in oligotrophic systems, Nature, 300:246.

Jensen, A., and Sakshaug, E., 1973, Studies on the phytoplankton ecology of the Trondsheimfjord. II. Chloroplast pigments in relation to abundance and physiological state of the phytoplankton, J. exp. mar. Biol. Ecol., 11:137.

Johnson, P.W., and Sieburth, J. McN., 1979, Chroococcoid Cyanobacteria in the sea: a ubiquitous and diverse phototrophic biomass, Limnol. Oceanogr., 24:928.

Kiefer, D.A., Olson, R.J., and Holm-Hansen, O., 1976, Another look at the nitrate and chlorophyll maxima in the Central North Pacific, Deep-Sea Res., 23:1199.

Koroleff, F., 1970, Revised version of "Direct determination of ammonia in natural waters as indophenol blue," Int. Cons. Explor. Sea, CM 1969/C:9, ICES. Information on Techniques and Methods for Sea Water Analysis, Interlab Rep., 3:19.

Lebart, L., and Fénelon, P.J., 1971, "Statistique et Informatique Appliquée," Dunod, Paris.

Legendre, L., and Legendre, P., 1979, "Ecologie numérique," Masson, Paris.

Li, W.K.W., Subba Rao, D.V., Harrison, W.G., Smith, J.C., Cullen, J.J., Irwin, B., and Platt, T., 1983, Autotrophic Picoplankton in the Tropical Ocean, Science, 219:292.

Lleonart, J., 1979, "La communitat epibentònica del banc Canario-saharià, tipificació i cartografia mitjancant l'analisi de dades i comentaris crítics sobre metodologia," Ph.D. Thesis, University of Barcelona.

Lleonart, J., 1980, Anàlisi de grups per classificatió jeràrquica (unpublished manuscript).

Lund, J.W.G., Kipling, C., and Le Cren, E.D., 1958, The inverted microscope method of estimating algal numbers and the statistical basis of estimation by counting. Hydrobiologia, 3:390.

Margalef, R., 1978a, Phytoplankton communities in upwelling areas. The example of NW Africa, Oecologia Aquatica, 3:97.
Margalef, R., 1978b, Life-forms of phytoplankton as survival alternatives in an unstable environment, Oceanologica Acta, 1:493.
Margalef, R., (this book), Environmental control of the mesoscale distribution of primary producers and its bearing to primary production in the western Mediterranean.
Margalef, R., Herrera, J., Steyaert, M., et Steyaert, J., 1966, Distribution et caractéristiques des communautés phytoplanctoniques dans le bassin Tyrrhenien de la Méditerranée et à la fin de la stratification stivale de l'année 1963, Bull. Inst. Royal Scien. Natur. Bel., 42:2.
Margalef, R., and Estrada, M., 1981, On upwelling, eutrophic lakes, the primitive biosphere and biological membranes, in: "Coastal Upwelling," F.A. Richards, ed., AGU, Washington.
Minas, H.J., 1970, La distribution de l'oxygène en relation avec la production primaire en Méditerranée Nord-Occidentale, Mar. Biol., 7:181.
Nival, P., 1965, Sur le cycle de Dictyocha fibula Ehrenberg dans les eaux du surface de la rade de Villefranche-sur-Mer. Cah. Biol. Mar., 6:67.
Platt, T., (this book), Phytoplankton Production in Oligotrophic Marine Ecosystems: the Mediterranean Sea.
Platt, T., (in press), Primary productivity in the Central North Pacific: comparison of oxygen and carbon fluxes, Deep-Sea Res.
Platt, T., Subba Rao, S.V., and Irwin, B., 1983, Photosynthesis of Picoplancton in the Oligotrophic ocean, Nature, 301:702.
Revelante, N., and Gilmartin, M., 1973, Some observations on the chlorophyll maximum and primary production in the eastern north Pacific, Int. Rev. ges. Hydrobiol., 58:819.
Riley, G.A., Stommel, H., and Bumpus, D.F., 1949, Quantitative ecology of the plankton of the western North Atlantic, Bull. Bing. Oceanogr. Coll., 12/1/69.
Saijo, Y., Iizuka, S., and Asaoka, O., 1969, Chlorophyll maxima in Kuroshio and adjacent area, Mar. Biol., 4:190.
Salat, J., and Cruzado, A., 1981, Masses d'eau dans la Méditerranée Occidentale, Rapp. Comm. int. Mer Medit., 27:201.
Shulenberger, E., and Reid, J.L., 1981, The Pacific shallow oxygen maximum, deep chlorophyll maximum, and primary productivity reconsidered, Deep-Sea Res., 28A:901.
Siegel, S., 1956, "Nonparametric statistics," McGraw-Hill Kogakusha Ltd., Tokyo.
Sneath, P.H., and Sokal, R.R., 1973, "Numerical Taxonomy," W.H. Freeman and Co., San Francisco.
Sorokin, Y.M., 1970, Some data on primary production in the central Pacific, Oceanol., 10:538.
Steele, J.H., 1964, A study of production in the Gulf of Mexico, J. Mar. Res., 22:211.

Steele, J.H., and Yentsch, C.S., 1960, The vertical distribution of chlorophyll, J. Mar. Biol. Ass. U.K., 39:217.
Steemann Nielsen, E., and Hansen, V.K., 1959, Light adaptation in marine phytoplankton populations and its inter-relation with temperature, Physiologia Pl. 12:353.
Strickland, J.D.H. and Parsons, T.R., 1972, A Practical Handbook of Seawater Analysis, J. Fish. Res. Bd. Canada, 167:1.
Taniguchi, A., and Kawamura, T., 1972, Primary production in the Oyashio region with special reference to the subsurface chlorophyll maximum layer and phytoplankton-zooplankton relationships, in: A.Y. Takenouti, et al., eds., "Biological Oceanography of the Northern North Pacific Ocearn," Idemitsu Shoten, Tokyo.
Travers, M., 1972, Le microplankton du golfe de Marseille. Méthodes d'étude microscopique des organismes, Téthys, 4:689.
Travers, A., and Travers, M., 1968, Les Silicoflagellés du Golfe de Marseille, Mar. Biol., 1:185.
Vallespinós, F., (in preparation), Distribución verticale de la concentration d'ATP dans les eaux méditerranéennes pendant l'été.
Venrick, E.L., 1978, The implications of subsampling, in: "Phytoplankton Manual," A. Sournia, ed., UNESCO, Paris.
Venrick, E.L., 1982, Phytoplankton in an Oligotrophic Ocean: Observations and Questions, Ecological Monographs, 52:129.
Venrick, E.L., McGowan, J.A., and Mantyla, A.W., 1973, Deep Maxima of Photosynthetic Chlorophyll in the Pacific Ocean, Fish. Bull., 71:41.
Vives, F., 1978, Distribución de la población de copépodos en el Mediterráneo occidental, Res. Exp. Cient. B/O Cornide, 7:263.
Vives, F., 1979, Campaña "Mediterráneo I" (octubre-noviembre 1976), Datos Informativos Inst. Inv. Pesq., 7:1.
Yentsch, C.S., and Menzel, D.W., 1963, A method for the determination of phytoplankton chlorophyll and phaeophytin by fluorescence, Deep-Sea Res., 10:221.

FEATURES AND PECULIARITIES OF ZOOPLANKTON COMMUNITIES FROM THE WESTERN MEDITERRANEAN

R. Gaudy

Centre d'Océanologie de Marseille; Faculté des Sciences de Luminy; 13288 Marseille Cedex 9, France

INTRODUCTION

The Mediterranean Sea displays several peculiarities in comparison with the typical oceanic environment: it is a semi-enclosed basin where exchanges of nutrients, fauna and flora with the outside waters can occur only through the Gibraltar strait, and to a minor extent, via the Suez Canal. Its waters present relatively high salinity and high temperature values, and, below 200 m depth, unique homothermic conditions. Mediterranean hydrodynamics are characterized by the absence of tides and by the occurrence of more or less temporary upwellings in the coastal zone, which are favoured by the narrowness of the continental shelf over most of the Mediterranean coast. Moreover, the geological history of the Mediterranean is marked by several strong changes of its climatic and hydrological characteristics, and of its possibility of connection with the other seas.

Although the Mediterranean zooplankton has been the subject of many studies most of them are devoted mainly to the systematic study of separate groups, and research on zooplankton assemblages were still scarce until the recent past. Since the review of research on the zooplankton from the occidental basin (Furnestin, 1968), many new studies have been carried out in coastal and offshore environments, contributing to our understanding of population structures. Despite the incompleteness of knowledge for many parts of the western Mediterranean, it is now possible to attempt a general description of the main zooplanktonic communities. This paper does not pretend to give a complete analysis of this subject. Its main goal is, starting from a description of the main features of zooplankton communities, to assess their uniqueness by reference to other seas, especially the Atlantic Ocean. This analysis is restricted to the western basin; the zooplankton from the eastern Mediterranean are reviewed by Moraitou-Apostolopoulou (this book).

QUANTITATIVE ASPECT

The Mediterranean pelagos is traditionally considered to be a poor fauna, since Jespersen's (1923) volumetric measurements of the macroplankton collected during the Thor and Dana cruises across the Mediterranean. The same trend was observed in the western basin by comparing the copepod abundance from the Alboran Sea to the Central area (Furnestin, 1968). The poverty of the pelagos was also expected from the low concentration of nutrients, particularly phosphates, which characterize the Mediterranean waters (Sournia, 1973). Thus, this concept of low zooplankton standing stock is based on only a few offshore samples. Despite the increasing number of plankton studies, especially in the coastal area, few of them contribute useful quantitative information to a general synthesis: taking into account the seasonal variations of abundance and composition, it is very difficult to compare samples obtained at different seasons. For the neritic areas where these variations are strong, we have analysed only the data based on annual cycles. In the offshore area, annual studies are absent, and it has been necessary to take into account all the available data. Further, we must be aware of the high degree of approximation in the conclusions that can be drawn from a comparison of different authors quantitative data, because the diversity of plankton nets used (several opening diameters and mesh sizes), of collecting techniques (horizontal, oblique or vertical hauls) and of the units chosen for standing stock estimates (dry weight, displacement or settling volumes, total number of individuals). Concerning the sample collections, large progress has been made since the adoption by most of the Mediterranean laboratories of standard nets, the most-used now being the WP2 (Unesco, 1968), and the choice of standardized vertical hauls (50-0 or 200-0, according to the coastal or offshore character of the stations). Concerning the choice of a quantitative unit for estimating the standing stocks, no satisfactory equivalence index between weight, volume and number of individuals can be adopted without restriction (Unesco, 1976). Fig. 1 shows quantitative data from different parts of the Mediterranean Sea expressed in different units. A rather good agreement exists between the three methods of estimation, taking into account the diversity of the stations and of the collecting methods. Thus, for comparative purposes, as each index yields similar information, we used all the results available in any of these three units, without trying to transform the data into one common unit.

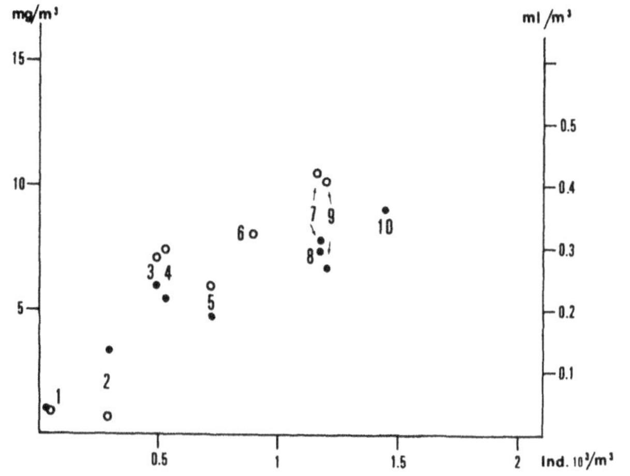

Fig. 1 Relationship between dry weight ($mg \cdot m^{-3}$), volume ($ml \cdot m^{-3}$) and abundance (individuals m^{-3}) in different Mediterranean mesoplankton communities. Open circles=volume; black circle=weight; (1): off Marseille, 500-200m; (Arellano Lenox, 1970). (2): south part of the Eastern Mediterranean, 400-0 m (Porumb et al., 1981). (3): Adriatic Sea (Ganoturska), 50-0 m (Vucetic, 1957). (4): Malaga 0m (Rodriguez, 1979). (5): Marseille, 200-0m (Arellano Lenox, 1970). (6): Lebanon coasts, 0m (Lakkis, 1983). (7): Marseille, 50-0m (Arellano Lenox, 1970). (8): Monaco Cap Martin, 0m (Gilat et al., 1965). (9): Castellon, 50-0m (Vives, 1965, 1966). (10) Ligurian coasts, 50-0m (Carli and Sertorio, 1964).

Neritic area

The zooplankton from the north coast of the occidental basin has been more intensively studied than that from the coast of North Africa or of the Tyrrhenian Sea (Fig. 2). Table 1 and Fig. 3 give available quantitative data for the western basin, and for some neighboring marine areas: the eastern basin and the Atlantic Ocean. It can be observed that mean displacement volumes values for Malaga, Castellón, Marseille and Alger fall in a narrow range: 0.3 to 0.4 $ml \cdot m^{-3}$. A lower value is found near Monaco-Cap Martin, but not when dry weights are considered: they are similar to the values found in Castellón, Marseille and Genova. In the bay of Villefranche, the relatively high value results probably from the more coastal character of this station. Another high value, in Banyuls, seems to be due to the use of a net equipped with a fine mesh (160 μm) instead of the usual size (200 to 250 μm). Two results differ sensibly from the others. Caminas (1981) gives a map for the coastal area near Malaga, where the annual mean value (20 to 60 $mg \cdot m^{-3}$) indicates a standing stock considerably higher than that which is suggested by

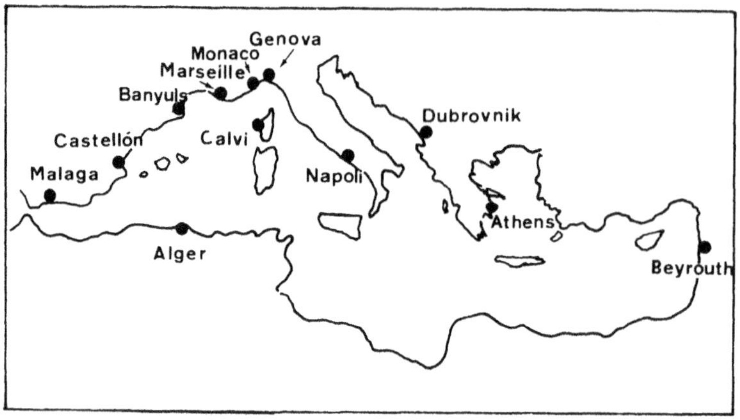

Fig. 2 Map of the stations where annual quantitative and qualitative studies on mesozooplankton were performed.

the volumetric measurements of Rodriguez (1979) for the same area. On the other hand, Carrada et al. (1980) found a value of 3.2 $mg \cdot m^{-3}$ for the Tyrrhenian Sea near Naples, unusually low in comparison with the other parts of the Mediterranean coast. Judging by the numbers of individuals, the abundances of the plankton of Castellōn, Marseille, Monaco and the Ligurian coasts are very similar. Low numbers are found at Calvi, this latter station displaying a less marked neritic character due to the narrowness of the Corsican shelf.

Most of the average standing stock values of the western basin are similar to those recorded in the Oriental basin (Table 1, Fig. 3). It does not seem that strong differences exist between most of the various neritic areas investigated throughout an annual cycle. However, in the eutrophic littoral or brackish areas high biomass values can be observed (El Maghraby and Halim, 1965).

A similar result has been found for the primary production of the Mediterranean neritic regions (Sournia, 1973). This author remarked: "Au sein des eaux neritiques.... il est difficile de retenir des différences géographiques... les valeurs numériques en g $C \cdot m^{-2} \cdot an^{-1}$ forment un ensemble assez homogène... elles gravitent, sauf exceptions, autour d'un même ordre de grandeur, soit 80 ou 90 g $C \cdot m^{-2}$". As primary production and zooplankton standing stock are narrowly linked (Taniguchi, 1973), it is not surprising to find such homogeneity in zooplankton abundance among the different Mediterranean areas investigated until now. However, generalization throughout the whole coastal environment is still hazardous, because many areas remain not, or only poorly, investigated.

The problem of the relative poverty of the Mediterranean pelagos can be reconsidered by referring to some results obtained recently

Table 1. Zooplankton standing stocks for different Mediterranean areas (pluriseasonal or annual means only)

Locality	Sampling characteristics	$ml \cdot m^{-3}$	$mg \cdot m^{-3}$	No $ind.m^{-3}$	Authors
Malaga	JB 250 um H 0-20 m	0.29		530(1)	Rodriguez, 1982
"	JB 250 um V 50-0		20-60		Caminas, 1981
Castellõn	JB 250 um V F-0	0.41	6.6 (2)	1207	Vives 1965,1966
Banyuls	Hens. 160 V F-0		15	2887-1550(3)	Razouls, 1972
Marseille	WP2 220 um V 50-0	0.42(4)	7.8 (4)	1140	Arellano Lenox 1970
Villefranche	WP2 ? H 0		11.2	8043	Fenaux & Quelart, 1977; Seguin, 1981
Monaco Cap M.	GM 160 um H 0	0.10(5)	7.4(5)	1151	Gilat et al., 1965
Genova region	N ? V 50-0			1445(6)	Carli & Sertorio, 1964
" "	WP2 ? V 50-0		9.2(6)		Della Croce et al.,1979
Naples	WP2 250 um V 50-0		3.2		Carrada et al., 1980
Calvi	WP2 180 um H 0		1.1(7)	682	Dauby, 1980
Alger	V 50-0	0.31			Bernard, 1950
Adriatic Sea					
Ganoturska	? V 50-0	0.29	4.9		Vucetic, 1957
Roviny	V 50-0	0.43			Gamulin, 1954
Split	V 50-0	0.26			" "
Pellegrin	V 50-0	0.22			"
Komizi	V 50-0	0.19			"
Eastern Basin					
Athens	WP2 240 um V F-0		2.9-9.1(8)		Yannopoulos, 1976
	WP2 180 um H 0			1287	Moraitou-Apostolopoulou, 1981
Lebanon	WP2 200 um H 0	0.32(9)		905	Lakkis, 1983

JB=Juday Bogorov net; H or V=Horizontal or Vertical hauls; B=Bottom; Hens.=Hensen net; GM=G.M. Manufacturing Company net; (1): from his fig. 36. (2): from his fig. 3; (3): means for station J and C; (4): means of two neritic stations; (5): from his Table 7. (6): values calculated for a year period from the authors' data. (7): from his fig. 2. (8): values for the outer and central part of the Saronikos gulf; (9): average value for neritic stations.

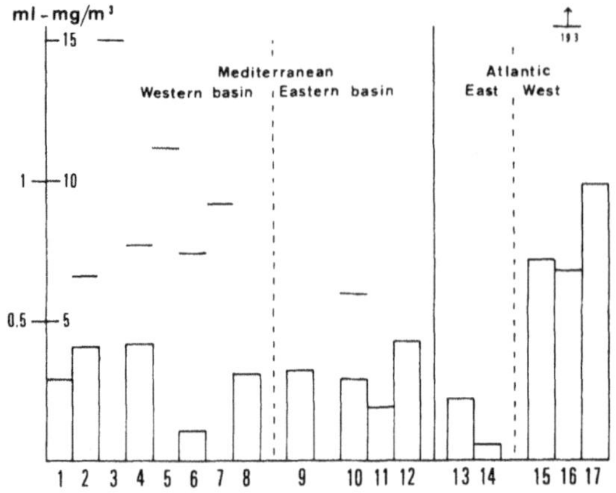

Fig. 3 Comparison of standing stock values of neritic mesozooplankton in some Mediterranean and Atlantic areas. Vertical bars = volume (ml m^{-3}); horizontal bars = weight (mg m^{-3}). (1): Malaga (Rodriguez, 1979). (2): Castellōn (Vives, 1966). (3): Banyuls (Razouls, 1972). (4): Marseille (Arellano Lenox, 1970). (5): Villefranche (Fenaux & Quelart, 1977). (6): Monaco Cap-Martin (Gilat et al., 1965). (7): Ligurian coast (Della Croce et al., 1979). (8): Alger (Bernard, 1955). (9): Beyrouth (Lakkis, 1983). (10) Adriatic Sea (Ganoturska) (Vucetic, 1957). (11) and (12): Adriatic Sea, extreme values (Komizi and Roviny) (Gamulin, 1954). (13) Gulf of Gascogne (Arbault & Lacroix Boutin, 1970). (14) Moroccan shelf waters (Belfquieh, 1980). (15): Georges Bank (Riley & Bumpus, 1946). (16): Block Island Sound (Deevey, 1952). (17): Long Island Sound (Deevey, 1956).

in two eastern Atlantic neritic areas: the continental shelf of the Gulf of Gascogne (Arbault and Lacroix-Boutin, 1970) and the shelf waters of the Moroccan region (Belfquieh, 1980). These areas have been extensively investigated during several years in a broad geographic coverage. Higher values are obtained at several stations near the shore, but when annual means are considered, it appears that these two regions are poorer than the Mediterranean neritic areas. Similarly, a comparative study of Vives et al. (1975) shows also that the zooplankton of the Alboran Sea is evidently richer than the zooplankton of the Cadiz bay. On the contrary, the western Atlantic shelf waters generally have standing stocks twice or three times higher than in the Mediterranean, depending on the more or less sheltered character of the sampling stations (Fig. 3).

Oceanic area

In the surface layer of the offshore waters (0-200 m), the

published quantitative results concern the plankton collected during several cruises, for different seasons and different areas. The main results are given by Boucher and Thiriot (1972); Razouls and Razouls (1978) and Nival et al. (1975). From these observations, it appears that the biomass of zooplankton of the northern part of the central basin increases rapidly during the spring bloom (from 1.1 to 15.2 $mg \cdot m^{-3}$). This maximum is followed by low values (0.4 $mg \cdot m^{-3}$ in May, then 0.2 $mg \cdot m^{-3}$ in September. In the west Corsica-Sardinia region, a value of 4.5 mg is found in February, contrasting with the value 1.8 $mg \cdot m^{-3}$ in the Sicilio-Tunisian strait recorded at the same time. In the Alboran Sea, in May-June, the plankton is richer than in the southern part of the central basin (1.6 mg instead of 0.8 $mg \cdot m^{-3}$).

During the "Bannock" cruise in the Tyrrhenian Sea, in autumn, Vives (1967) noted that the southern part was about twice as rich as the northern part (average values being 0.58 and 0.27 $mg \cdot m^{-3}$ respectively, corresponding to 660 and 250 $individuals \cdot m^{-3}$, from fig. 3). During the same period and using a similar methodology, Ehrhardt (1967) compared the number of zooplankton individuals collected from the Sardinia channel (1275 to 1375 $ind.m^{-3}$) and from the south Tyrrhenian Sea (400 to 739 $ind.m^{-3}$) and attributed these differences to a stronger influence of the Atlantic current in the Sardinia channel region. Soenen's data (1969) are interesting to consider because they relate to an extended area, south of the 40th parallel of the central basin, which was investigated during three seasonal cruises (autumn, winter and spring). The average number of individuals was 180, 606 and 560 for these three seasons. Geographical differences appeared between the western stations (transect Majorque - Cap Bengut) and the Sardinia channel area. They were attributed by the author to an enrichment by the Atlantic current, more marked in the western part of the basin.

From all these data, it seems that a seasonal abundance cycle exists for the oceanic plankton, with stronger variations in the northern part of the occidental basin; and that the standing stocks of the different regions show sensible differences within seasons. The Tyrrhenian Sea, and particularly its northern part, is poor in comparison with the south part of the Central basin and the Alboran Sea.

In the offshore waters of the south-eastern Mediterranean, Porumb et al. (1981) indicated seasonal standing stocks varying between 14.2 $mg \cdot m^{-3}$ (summer) and 23.7 $mg \cdot m^{-3}$ (winter). Converted to dry weight (17% wet weight), these values (2.4 and 4 $mg \cdot m^{-3}$) and the corresponding number of individuals per m^3 (211 to 373) are in the same range as the several values found in the western basin.

The plankton abundance of the North Atlantic Ocean was analyzed by Be et al. (1971) from numerous samples collected during ten years.

Low displacement volumes (0.01 to 0.05 $ml \cdot m^{-3}$) were found over a large extent in the temperate latitudes. Data from the "Yakutat" cruise indicate dry weight from 1 to 2 $mg \cdot m^{-3}$ in almost all the stations between the 15th degree of longitude off Spain and Morocco and the 60th degree, near Bermuda. These values do not indicate a richness for the oceanic Atlantic waters significantly higher than many values found in the Mediterranean offshore area.

Very little data are available for deep water fauna. Furnestin (1968), in a station from the central part of the western basin, and Ianora and Scotto di Carlo (1981), in the Tyrrhenian Sea, found a rapid decrease of the biomass and abundance of plankton between 0 and 200 m, then a tendency to reach low and uniform values in the deeper layers. This type of distribution would be expected to be the same in the whole basin, due to the great homogeneity of the deep plankton, according to Razouls and Razouls (1978) and Vives (1979).

QUALITATIVE ASPECT

This analysis is limited to the mesoplankton: the microzooplankton (50 to 200 µm) is still poortly known, most of the sampling systems used for collecting plankton until now being unsuitable for its capture. The macroplankton has rarely been studied as a whole but useful information concerning the macroplankton of the western Mediterranean can be found in Casanova (1977) and Franqueville (1971). As strong seasonal variations can affect the community structure, only annual studies with quantitative data have been considered for the Neritic area (see the author's references in Table 2 and 3).

Despite the variety of zoological groups present in the zooplankton, the bulk of the neritic communities consists in a few of them: copepods, cladocerans, appendicularians, chaetognaths and some meroplankton elements, crustacean larvae particularly. On an annual basis, these categories account for 85 to 95% of the number of metazoans (Table 2). Copepods are usually dominant, with 55 to 95% of the total number. Appendicularians and cladocerans can reach 19%, and crustacean larvae, 6% of the community. Chaetognaths never exceed 3%. Some geographic differences appear in the proportion of cladocerans, which is higher at Malaga and Alger, two areas strongly influenced by the lower salinity waters of the Atlantic current and for appendicularians which are more numerous at Castellōn and at Monaco.

In the offshore area, the dominance of copepods is less marked, and the number of cladocerans and appendicularians decreases sharply. On the contrary, heteropods, pteropods and ostracods are more frequent

Table 2. Percentages of the main components of the neritic communities of the western Mediterranean (based on annual means)

Locality	Malaga (1)	Castellón (2)	Marseille (3)	Monaco (4)	Ligurian coast (5)	Napoli (6)	Calvi (7)	Alger (8)
Copepods	60	55	81	74	83.5	77	94.6	68
Cladocerans	19	7	4	7	0.9	0.9	2.9	12
Appendicularians	5.5	19	5	12.4	5.9	3.7	0.4	+
Chaetognaths	+	2.6	1	0.4	2.9	1.1	+	2
Crustacean larvae	5.3	+	6	0.4	+	6.2	-	-

(1) Rodriguez, 1979; (2) Vives, 1966; (3) Gaudy, 1971a; (4) Gilat et al., 1965; (5) Carli and Sertorio, 1964; (6) Hure and Scotto di Carlo, 1968; (7) Dauby, 1980; (8) Bernard, 1955.

Table 3. Percentages of the main copepod species of the Neritic Mediterranean communities (based on annual values)

Locality	Malaga (1)	Castellon (2)	Marseille (3)	Villefranche (4)	Monaco Cap Martin (5)
Clausocalanus arcuicornis	16	35	18.7	38.4	32.1
C. furcatus					34
Paracalanus parvus	10	21.8	12.3	0.7	
Centropages typicus(15)	2.6	7.6	23.5	0.7	12
Temora stylifera	1.5	9.6	11.5	1.5	+
Acartia clausi	67	0.4	12.6	38.3	21
Oithona spp.	+	9	+	1.2	12.6
Oncaea spp.	+	+	+	+	+
Corycaeus spp.	+	1.9	+	+	2.1

Table 3 (Cont.)

Locality	Ligurian coasts (6)	(7)	(8)	Naples (9)	Calvi (10)	Alger (11)	Adriatic Dubrovnic (12)	Athens (13)	Lebanon coasts (14)
Clausocalanus arcuicornis	32.1	34.1	42.5	7	(32.6)	26.2	12.7	6.4	4
C. furcatus									
Paracalanus parvus	5.8		1.3	13.2	5.8	15.2	3.4	0.7	40
Centropages typicus (15)	6.9	16.6	15.7	7	10.6	3.6	2.4	39	8
Temora stylifera	1.5	17.5	8.9	5.3	+	3.7	7.6	3.2	5
Acartia clausi	0.1	8.8	8.4	14.4	28.2	1.5	0.5	53.1	5
Oithona spp.	15.3	+	+	8.4	+	20.9	0.3	3.5	3
Oncaea spp.	+	+		3.4	+	9.3	+	1.8	+
Corycaeus spp.				+	+	+	+	3.1	+

Data cited or calculated from values from (1) Rodriguez, 1979); Vives, 1966 (3); Gaudy, 1971; (4) Seguin, 1981; (5) Gilat et al., 1965; (6) Carli & Sertorio, 1964; (7) Della Croce, 1952; (8) Della Croce, 1959; (9) Hure & Scotto di Carlo, 1968; (10) Dauby, 1980; (11) Bernard, 1955; (12) Hure & Scotto di Carlo, 1968; (13) Moraitou-Apostolopoulou, 1974; (14) Lakkis, 1983; Centropagus typicus is replaced by C. chierchiae in Malaga and by C. kroÿeri in Lebanon.

than in the neritic area. The thaliaceans are sometimes relatively abundant in the central part of the basin.

Throughout the year, the cladocerans display an irregular cycle of abundance, with generally a strong summer maximum (Podon, Evadne spp.) or an autumnal peak (Penilia avirostris). The appendicularians are principally Oikopleura longicauda, during summer and Fritillaria pellucida or F. borealis during winter. The chaetognaths are dominated by Sagitta serratodentata, S. bipunctata and S. setosa (this latter, mainly in diluted waters). S. enflata seems to be related to the influence of Atlantic waters (Furnestin, 1979).

The copepods constitute the main part of the zooplankton community. As they have been more intensively studied than other groups, the following discussion will be mainly based on them. In the neritic community, they include about 50 to 60 species, but only few of them are characteristic of this community according to their number and their frequency throughout the year: Clausocalanus arcuicornis, C. furcatus, Paracalanus parvus, Centropages typicus or C. chierchiae, Temora stylifera, Acartia clausi, Oithona helgolandica, O. nana and a few small species of Oncaea, Corycaeus and Euterpina genera which generally display low abundance.

The first five general cited contribute for 75 to 95% of the copepods total number (Table 3, Fig. 4). Among them, Clausocalanus and Paracalanus (sometimes considered together in some studies) are generally the main components. Acartia clausi is present everywhere, but its relative abundance can vary considerably, being narrowly linked to the proximity of the coastline (see below) and depends on the sampling site.

All these copepods display considerable seasonal variations in abundance. Some of them, such as Paracalanus parvus and Centropages typicus are very numerous during spring and summer, while others display high numbers of individuals in late summer or autumn (Temora stylifera). Compared to other species, Clausocalanus arcuicornis is well-represented during the winter period. Generally, for each species, several peaks of abundance succeed each other throughout the year with a spring maximum more or less delayed until the beginning of summer, an autumn peak, and generally a mid-summer peak. Winter (December-January) is a period of scarcity for most species.

The detailed analysis of the biological cycle of the principal species shows that these different peaks correspond to the successful development of some of the 5 to 7 annual generations which succeed each other throughout the year (Gaudy, 1971; Razouls, 1972). These successive generations correspond to a synchronizing effect of the phytoplankton cycle on the egg laying (Gaudy 1971d). The algal cycle is itself regulated by the discontinuous intake of nutrients in the surface marine environment and by the seasonal changes of

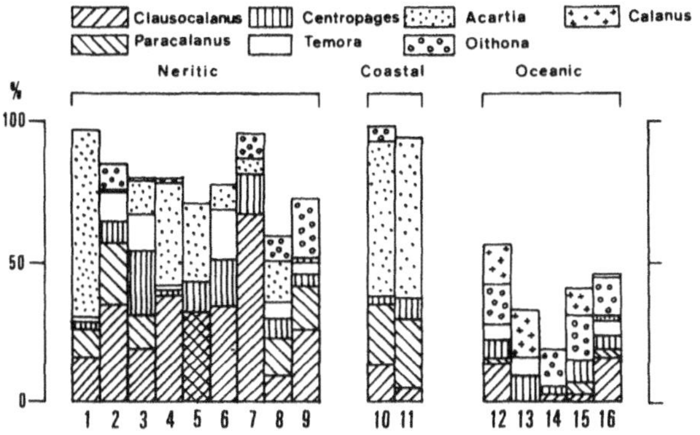

Fig. 4 Abundance of the main copepod genera in percentage of total copepods number in some neritic, coastal and oceanic Mediterranean waters (western basin). (1) Malaga (Rodriguez, 1979). (2) Castellón (Vives, 1966). (3) Marseille (Gaudy, 1971a). (4) Villefranche (Seguin, 1981). (5) Monaco Cap-Martin (Gilat et al., 1965). (6) Ligurian coasts (Della Croce, 1952). (7) Calvi (Dauby, 1980). (8) Naples (Hure & Scotto di Carlo, 1968). (9) Alger (Bernard, 1955). (10) Marseille harbour (Patriti, 1976). (11) Genova harbour (Zunini Sertorio et al., 1977). (12) South of the 40th parallel (Soenen, 1969). (13) South of the 40th paralel (Boucher & Thiriot, 1972). (14) North of the 40th parallel (Boucher & Thiriot, 1972). (15) South of the Tyrrhenian Sea (Ehrhardt, 1967). (16) Off Naples (Hure and Scotto di Carlo, 1968).

water stability (Sournia, 1973). The quantity of adults produced results from the factors acting on the larval mortality (temperature and abundance of algal cells suitable in size during larval development). Adults from these different generations display some somatic variations (size, respective proportion of body parts) related to physical factors, mainly temperature and trophic conditions existing during the growth (Gaudy, 1971a; Razouls, 1972; Deevey, 1960).

Besides these dominant species and the presence of some other ones, less abundant but frequent in the neritic area, several species from outside origin can maintain themselves in the neritic at different moments of the year. Some of them come from subsurface or deep waters. They are carried up to the surface by upwellings (Duran, 1955; Vives, 1971; Gaudy, 1971a) which occur frequently along the northern coasts of the Mediterranean. These species can also reach the surface neritic waters when the thermal barrier is broken during the winter homothermal period. Among them, Calanus helgolandicus, which is a species common in the neritic zone from the Atlantic Ocean and adjacent seas, appears in the Mediterranean shelf waters only during winter or occasionally the rest of the year with upwelling. Its abundance in the western Mediterranean is generally low, except in Marseille where two generations can develop (Gaudy, 1971b). The rest of the year this species inhabits deep waters, up to 1000 to 1500 m, according to Hure and Scotto di Carlo (1968), in the Naples region. Other species from offshore regions, can be carried occasionally to coastal areas by superficial currents. Several are considered as indicator species for the Atlantic current, because their frequency and relative abundance decrease progressively in the Alboran Sea, then along the Algerian coasts, the West Corsica-Sardinia region and the Tyrrhenian and Ligurian Seas. They can be found as far as the Gulf of Marseille (Gaudy, 1963). One of the most typical of these species is Temora longicornis, a species abundant in the neritic and littoral communities of the Atlantic Ocean and North Sea, and which appears only scarcely in the western Mediterranean. Nevertheless, the previously-established lists of Atlantic indicator species probably have to be shortened: some recent works have shown that several of them such as Centropages violaceus, were normal inhabitants of the eastern basin. As it is not excluded that local populations of these species could be enriched by new immigrants, it is difficult to test the validity of these indicator species.

A peculiar facies of the neritic community is found in the littoral areas where changes of physical or chemical factors such as temperature, salinity, turbidity, pollutants, are more marked in time and space than in the neritic zone sensus stricto. Under these conditions, the less tolerant species are eliminated: Temora stylifera, then Clausocalanus spp., Centropages typicus and Oithona helgolandica, whereas a few other species can swarm, taking advantage of the rich food conditions usual in such environments. Thus, the zooplankton populations, very rich in individuals, display poor specific diversity: only few species characterize this facies. The copepod populations are dominated by Acartia species; A clausi is accompanied, then replaced if the salinity is lower, by other euryhaline forms found only near the coast, such as A. discaudata var. medit., A. italica, A. enzoi, A. josephina and A. latisetosa. Some other neritic copepods are also present when the salinity conditions are not too low (Paracalanus parvus, Oithona nana and Euterpina acutifrons). The chaetognath Sagitta setosa, the cladoceran Podon polyphemoides and several meroplanktonic larvae, particularly nauplii of Balanus amphitrite and lamellibranch larvae complete the general feature of this type of plankton, which seems also to characterize harbour environments (Rodriguez, 1982; Patriti, 1976; Della Croce, et al., 1972; Zunini-Sertorio et al., 1977) and brackish or estuarine areas (Fatemi, 1938; Benon et al., 1976).

In the oceanic area, from the works of Mazza (1967), Soenen (1969), Ehrhardt (1967), Vives (1967) and Hure and Scotto di Carlo (1968), the bulk of the copepod fauna includes some species found also in the neritic zone, such as Clausocalanus arcuicornis,

Oithona helgolandica and O. plumifera (this latter being of minor abundance in the coastal community). Temora stylifera can also be found far from the coast. On the other hand, Acartia clausi disappears and Paracalanus parvus is scarce. Contrary to their large dominance in the shelf waters, these species account for less the half of the total number of copepods (fig. 4), the rest being an assemblage of various species notably Calanus minor, Pleuromamma gracilis and several species of Oncaeidae, Corycaeidae and Microsetellidae. Other species such as Euchaeta marina, Lucicutia flavicornis, Candacia spp. and Sapphirina spp., although less numerous are nevertheless characteristic forms of the oceanic environment.

The structure of the upper layer oceanic community is complicated by the frequency of deep species migrating towards the surface or carried upward by vertical currents (Mazza, 1967). Of course, the transition between neritic and oceanic faunas is progressive: for example, various neritic species can be found in the Tyrrhenian and Ligurian Seas with rather high abundance due to the relative proximity of the coasts (Della Croce, 1959; Soenen, 1969; Vives, 1967). Near Naples Hure and Scotto di Carlo (1968) showed the transition between these types of fauna: the abundance of Centropages typicus decreases rapidly in the two most external stations, but Clausocalanus arcuicornis remains relatively constant; on the other hand, Pleuromamma gracilis, Lucicutia flavicornis, Haloptilus longicornis, Oithona plumifera and Corycella rostrata display higher abundance in the offshore station.

The deep water plankton has been studied throughout an annual cycle only once, in the Naples region, this study being restricted to copepods (Hure and Scotto di Carlo, 1968). These authors observed low seasonal variations, the recruitment of larvae being continuous in deep water copepods, according to Mazza (1967). Geographical variations appear also to be very limited, according to Casanova (1970) for the western Mediterranean and to Hure and Scotto di Carlo (1968) for the whole Mediterranean Sea, a similar fauna existing everywhere. For example, in the plankton collected during the Atlantis cruises, of the 69 species of copepods collected in the eastern basin, and 64 in the western basin, 61 are common to both regions (Scotto di Carlo and Ianora, 1982). Nevertheless, some differences in the faunal composition of copepods appear between the Tyrrhenian and the Ionian Seas due to the currents of the Sicilia strait and of the Sicilo-Tunisian sill (Vaissière and Seguin, 1980). The specific variety of the deep plankton is high, but only a few species are dominant (Casanova, 1970). In the Tyrrhenian Sea, 4 species of copepods account for 72% of the total copepod assemblage between 1000 and 1500 m and only three species contribute for about the same percentage between 1500 and 2000 m (Scotto di Carlo et al., 1975). Casanova and Hure and Scotto di Carlo agree on the arbitrary character of separating meso and bathyplanktonic communities in the Mediterranean: few species are strictly restricted to deep waters. On the contrary, mesopelagic species, which can reach the surface or subsurface layers during their vertical migrations, are also caught up to 1000 or 2000 m depth. There is no real bathypelagic community, but a fauna with a large vertical range richer at 300-500 depth than in deeper waters, where it cohabits with some species having their optimum of abundance at these levels. This type of structure is very different from the Atlantic one, as for example described by Deevey and Brooks (1971), between surface and 2000 m, in the Sargasso Sea.

BIOGEOGRAPHICAL ASPECT

The degree of uniqueness of the Mediterranean pelagic fauna compared to the Atlantic plankton is low when it is examined from a zoogeographical point of view (Furnestin, 1979). Most species are the same as in the Atlantic. They are, for the most part (86% of copepods) from Indo-Pacific origin, according to the definition of Sewell (1948) for the pelagic fauna. The other species originate from the Atlantic itself or from the Arctic Ocean. This fauna would have entered the Tethys more or less easily, from the east or from the west, with probably several complete renewals following the strong climatic changes occuring during the Quaternary period. Despite these changes, some species seem to have succeeded in surviving in some "refuge" areas, such as the Adriatic Sea or the deep waters. For example, Pseudocalanus elongatus populations from the Adriatic Sea are considered as boreal relict forms, and Eucalanus subcrassus and Rhincalanus cornutus, which occur in offshore areas, would be tropical relict forms (Furnestin, 1979).

Despite the peculiar physical or chemical characteristics of its waters, the Mediterranean zooplankton has few endemic species. For copepods, most of them appear to have their origin in the Adriatic Sea. In the occidental basin, littoral waters are the most favourable to speciation, which appears principally in copepods of Acartia genera (Crisafi, 1974). In the offshore area, which differs from the Atlantic one by its higher temperature and transparency of its waters, some morphological variations can be found in several macroplankton species, such as the eye size or the neotenic character of the copulation appendages in crustaceans, or the biometry of the pteropod species. Thus several oceanic or deep water plankton species have Mediterranean subspecies or varieties, without mixing of fauna with their Atlantic homologues via Gibraltar (Casanova, 1977).

The general features of the neritic copepod assemblage of the western basin and of the eastern Atlantic are very close, even though the relative abundance of species may be different. Referring to the works of Belfquieh (1980) for the Moroccan area and of Beaudouin (1975) for the Gulf of Gascogne, the main difference among the dominant species, is the greater proportion of Calanus helgonandicus in both Atlantic sectors. In the Moroccan area, Centropages chierchiae replaces C. typicus. In the Gascogne gulf, Temora longicornis replaces T. stylifera, and Pseudocalanus elongatus, a normal inhabitant of the northern neritic communities, is frequent. Thus, the Mediterranean copepod fauna of the neritic areas displays an intermediate character between north and south Atlantic influences, as shown by the mixing of subtropical species such as Temora stylifera and temperate species such as Centropages typicus. Some other species, numerous in the Atlantic communities, such as the tropical form Centropages chierchiae and the temperate species Pseudocalanus elongatus and Temora longicornis are restricted to waters strongly influenced by the Atlantic current (Alboran Sea more particularly: Vives et al., 1980).

The Atlantic and Mediterranean deep faunal assemblages show much marked differences, because many Atlantic bathypelagic species are unable to cross the Gibraltar sill, as much as the deep countercurrent runs from east to west. Other species cannot adapt to the deep water temperature conditions of the Mediterranean, which are very different from the Atlantic ones for the same depth. This could explain the impoverishment of the deep water fauna (only 60 copepod species are Mediterranean deep water inhabitants instead of 350 deep species in the Atlantic Ocean, according to Sewell (1948). On the other hand, because the absence of a low temperature barrier, the whole water column can be occupied by migrating mesopelagic species.

CONCLUSION

At the end of this brief analysis, the most salient points can be emphasized.

- The neritic zooplankton of the western Mediterranean shows a great homogeneity, either quantitatively (with the exception of the Tyrrhenian coast, which seems less productive), or qualitatively, the same bulk of few species occurring in all the studied areas, where they display similar seasonal variations. The neritic Mediterranean communities do not seem to be quantitatively poorer than the eastern Atlantic neritic communities. By its specific composition, the copepod fauna shows an intermediate character between temperate and tropical Atlantic influence.
- The epipelagic oceanic fauna shows mean standing stocks equivalent to the biomass value of most offshore Atlantic temperate regions. Geographic and seasonal variations occur in relation to the complicated hydrology of the western basin (Atlantic current, divergence areas, etc...).
- The deep fauna displays a greater uniqueness when it is compared

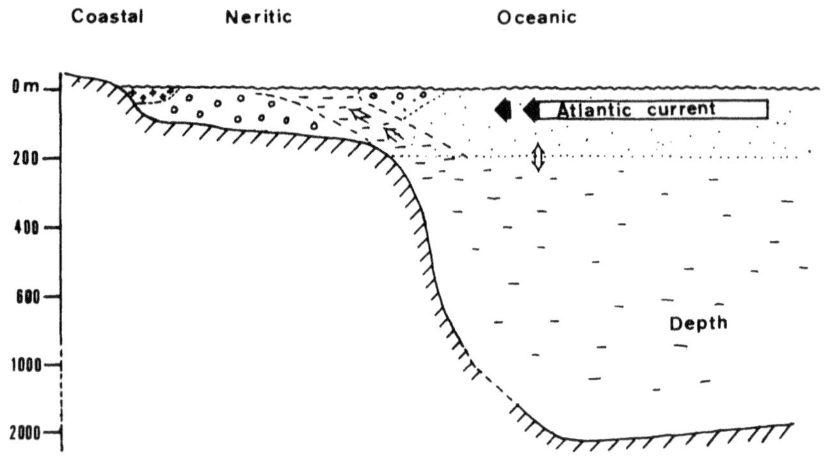

Fig. 5 Interrelations between the zooplankton communities of the western Mediterranean. Coastal facies = crosses; neritic community = open circles; upper waters oceanic community = dashes; the white arrows indicate upwellings and vertical movements of migration mesozooplankton; the black arrow shows the possible transport of Atlantic indicator species by surface currents.

to the Atlantic one: it is poorer qualitatively and quantitatively, and most of the species are mesopelagic forms, largely distributed in the whole water column.

- The uniqueness of the Mediterranean fauna is low from a zoogeographical point of view, most of its species being the same than in the Atlantic. There are few endemic species, and the speciation phenomenon seems restricted mainly to littoral waters,
- In the Mediterranean western basin, frequent interactions occur between neritic, epi-oceanic and deep plankton. They are summarized on fig. 5. They result from the homothermal conditions of deep waters, and from the hydrodynamic movements occurring in divergence areas or near the coast (upwellings, favourished by the narrowness of the shelf, or intake by surface currents of offshore fauna in the neritic area). A part of this fauna could be introduced into the Mediterranean by the Atlantic current.

SUMMARY

1. The abundance of zooplankton is analyzed in the neritic, offshore and deep waters of the western Mediterranean. Except for the deep fauna, which seems poorer, the standing stocks are similar to the eastern Atlantic ones.

2. Few species constitute the bulk of the neritic community. They display large seasonal variations related to successive generations. Some species from deeper waters can maintain in the community during upwellings or homothermal period. A special facies with low specific diversity and large abundance inhabits the brackish or eutrophic littoral areas.

3. The oceanic fauna contains, besides several species common with the neritic area, a varied assemblage of species with low abundance.

4. The deep water fauna is mainly constituted by mesopelagic species. Only few species account for the bulk of this community which seems uniform all over the Mediterranean.

5. On a zoogeographical approach, the Mediterranean plankton is very close to the Atlantic one. There are few endemic species. The deep water plankton is impoverished, due to the barrier effect of the Gibraltar sill, and the special temperature conditions of Mediterranean deep waters. The surface plankton shows an intermediate character between temperate and tropical influence.

REFERENCES

Arbault, S., and Lacroix-Boutin, N., 1970, Quatre ans de mésures volumetriques de plancton total dans le Golfe de Gascogne (1964-1967), Rev. trav. Inst. Pêches marit., 34:59.

Arellano Lenox, C., 1970, Recherches sur la biomasse zooplanctonique (région du Golfe du Lion et du Golfe de Marseille), Thèse Doct. 3ème cycle, Univ. Aix-Marseille.

Be, A.W.H., Forns, J.M., and Roels, O.A., 1971, Plancton abundance on the North Atlantic Ocean, in: "Fertility of the Sea," Vol. 1, J.D. Costlow, ed., Gordon and Breach Science, New York.

Beaudouin, J., 1975, Copépodes du plateau continental du Golfe de Gascogne en 1971 et 1972, Rev. trav. Inst. Pêches marit., 39:121.

Belfquieh, M., 1980, Les copépodes du plateau continental atlanto-marocain. Un cycle annuel dans les zones d'upwelling. Thèse Doct. 3ème cycle, Univ. Aix-Marseille.

Benon, P., Balnc, F., Bourgade, B., Charpy, L., Kantin, R., Kerambrun, P., Leveau, M., Romano, J.C., and Sautriot, D., 1976, Golfe de Fos, Impact de la Pollution, Bull. Observatoire de la Mer, 3, suppl. 1:1.

Bernard, F., 1950, Note sur la distribution du plancton au large de la Baie d'Alger, Arch. Zool. exp. gen., 86:103.

Bernard, M., 1955, Etude préliminaire quantitative de la répartition saisonnière du zooplancton de la Baie d'Alger. 1ère année 1950-51, Bull. Inst. Oceanogr. Monaco, 1065-26.

Boucher, J., and Thiriot, A., 1972, Zooplancton et micronecton estivaux des 200 premiers mètres en Méditerranée occidentale, Mar. Biol., 15:47.

Caminas, J.A., 1981, Distribution spatiale et temporelle de la biomass zooplanctonique superficielle dans le secteur nord-occidental de la Mer d'Alboran, Rapp. Comm. int. Mer Medit., 27:125.

Carrada, G., Hopkins, T.S., Bonaduce, G., Janora, A., Marino, D., Modigh, M., Ribera d'Alcala, M., and Scotto di Carlo, B., 1980, Variability in the hydrographic and biological features of the Gulf of Naples, P.S.Z.N.I. Marine Ecology, 1:105.

Carli, A.M., and Sertorio, T., 1964, Ricerche planctologiche Italiane dell' Anno Geofisico Internazionale 1957-58 I. Osservazioni sullo zooplancton raccolto nel Mere Ligure, Arch. Ocean. Limnol., 13:107.

Casanova, J.P., 1970, Essai de classement bathymétrique des formes zooplanctoniques en Méditerrannée, Rev. trav. Inst. Pêches marit., 34:45.

Casanova, J.P., 1977, La faune pélagique profonde (zooplancton et micronecton) de la Province Atlanto-Méditerranéenne. Aspects taxonomiques, biologique et zoogéographique, Thèse Doct. es Sciences, Univ. Aix-Marseille.

Crisafi, P., 1974, Inquinamento et speciazione: Acartia josephina e A. enzoi (Copepoda, Calanoida) specie nuove del Mar Mediterraneo, Boll. Pesca Piscic. Idrobiol., 29:5.

Dauby, P., 1980, Cycle annuel du zooplancton de surface de la baie de Calvi (Corse). Biomasse totale et plancton copépodien, Oceanol. Acta, 3:403.

Deevey, G.B., 1952, Quantity and composition of the zooplancton of Block Island Sound, Bull. Bingham oceanogr. coll., 13:120.

Deevey, G.B., 1956, Oceanography of Long Island Sound. V. Zooplankton, Bull. Bingham oceanogr. coll., 15:113.

Deevey, G.B., 1960, Relative effects of temperature and food on seasonal variations in length of marine copepods in some Eastern American and Western European waters, Bull. Bingham oceanogr. coll., 17:54.

Deevey, G.B., and Brooks, A.L., 1971, The annual cycle in quantity and composition of the zooplankton of the Sargasso Sea off Bermuda. II. The surface to 2,000m, Limn.Ocean., 16:927.

Della Croce, N., 1952, Variazioni stagionali dello zooplancton di superficie pescato a Punta del Mesco (La Spezia) tra il 15/10/49 e il 30/3/51, Boll. Mus. Ist. Biol. Univ. Genova, 24:87.

Della Croce, N., 1959, Copepodi pelagici raccolti nelle crociere talassografiche dell "Robusto" nel Mar Ligure ed alto Tirreno, Boll. Mus. Ist. Biol. Univ. Genova, 29:29.

Della Croce, N., Salemi-Picone, P., and Zunini-Sertorio, T., 1972, Popolamenti planctonici nel Mar Ligure. II Convegno "Côte d' Azur-Riviera dei Fiori": Pollutions et Aménagements, Genova, 28-30 Marzo 1972.

Della Croce, N., Fabiano, M., and Zunini Sertorio, T., 1979, Biomassa planctonica, sali nutritivi, parametri idrobiologici (Chiavari): ottobre 1977-ottobre 1978, Univ. Genova, Cattedra di Idrobiologia e Piscicoltura, Rapporto tecnico No. 8.

Duran, M., 1955, El zooplancton de Castellón, Observaciones y problems, II Reunion sobre Productividad y Pesquaras, 2:52.
Ehrhardt, J.P., 1967, Contribution a l'étude du plancton superficiel et subsuperficiel du Canal de Sardaigne et de la Mer sud-Tyrrhénienne, Cah. oceanogr., 19:657.
El Maghraby, A.M., and Halim, Y., 1965, A quantitative and qualitative study of the plankton of Alexandria waters, Hydrobiologia, 25:221.
Fatemi, M., 1938, Les variations saisonnières du plancton de l'étang de Thau a l'embouchure du Canal de Sète, Thèse Doct. es Sciences, Univ. Montpellier.
Fenaux, R., and Quelart, G., 1977, Evolution de la biomasse zooplanctonique dans les eaux superficielles de la baie de Villefranche-sur-Mer en 1972, Rapp. Comm. int. Mer Medit., 24:159.
Franqueville, C., 1971, Macroplancton profond (invertebrés) de la Méditerranée nord-occidentale, Tethys, 3:11.
Furnestin, M.L., 1968, Le Zooplancton de la Méditerranée (Bassin occidental). Essai de synthèse, J. Cons. perm. int. Explor. Mer, 32:25.
Furnestin, M.L., 1979, Aspects of the zoogeography of the Mediterranean plankton, in: "Zoogeography and diversity in Plankton," Vander Spoel and Pierrot-Bults, eds., Bunge Scientific Publ., Utrecht.
Gamulin, T., 1954, La ponte et les aires de ponte de la sardine (Sardina pilchardus) dans l'Adriatique de 1947 à 1950, Izvjesca, 4:64.
Gaudy, R., 1963, Sur la présence à Marseille d'espèces planctoniques indicatrices d'eaux d'origine atlantique, Rapp. Comm. int. Mer Medit., 20:363.
Gaudy, R., 1971a, Contribution à l'étude du cycle biologique des copépodes pélagiques du Golfe de Marseille. 1. L'environnement physique et biotique et la composition de la population de copépodes, Tethys, 3:921.
Gaudy, R., 1971b, Contribution a l'étude du cycle biologique des copépodes pélagiques du Golfe de Marseille. II. Etude du cycle biologique de quelques espèces caractéristiques, Tethys, 4:175.
Gaudy, R., 1971c, L'allongement antennulaire chez Centropages typicus; signification adaptive et utilité biométrique, Rapp. Comm. int. Mer Medit., 20:363.
Gaudy, R., 1971d, Etude expérimentale de la ponte chez trois espèces de copépodes pélagiques (Centropages typicus, Acartia clausi et Temora stylifera), Mar. Biol., 9:65.
Gilat, E., Kane, J.E., and J.C. Martin, 1965, Study of an ecosystem in the coastal waters of the Ligurian Sea. II Surface zooplankton, Bull. Inst. oceanogr. Monaco, 65(1353):56.
Hure, J., and Scotto di Carlo, B., 1968, Comparazione tra lo zooplancton del Golfo di Napoli e dell'Adriatico meridionale presso Dubrovnik. I. Copepoda, Pubbl., Staz.zool. Napoli, 36:21.

Ianora, A., and Scotto di Carlo, B., 1981, Vertical distribution of Mediterranean deep sea copepods, Rapp. Comm. int. Mer Medit., 27:165.

Jespersen, P., 1923, On the quantity of macroplankton in the Meditterranean and in Atlantic, Rep. Dana oceanogr. Exped. Medit., 3:1.

Lakkis, S., 1983, Contribution à la connaissance de l'écosystème néritique des eaux libanaises (Méditerranée orientale), Thèse Doct. es Sciences, Univ. Paris VI.

Mazza, J., 1967, Les copépodes pélagiques en Méditerranée occidentale (Développement post-larvaire, Biologie, Ecologie), Thèse Doct. es Sciences, Univ. Aix-Marseille.

Moraitou-Apostolopoulou, M., 1974, An ecological approach to the systematic study of planktonic copepods in a polluted area (Saronic Gulf, Greece), Boll. Pesca Piscic. Idrobiol., 29:29.

Moraitou-Apostolopoulou, M., 1981, The annual cycle of zooplankton in Elefsis Bay (Greece), Rapp. Comm. int. Mer Medit., 27:105.

Nival, P., Nival, S., and Thiriot, A., 1975, Influence des conditions hivernales sur les productions phyto- et zooplanctoniques en Méditerranée Nord Occidentale. V. Biomasse et Production zooplanctonique, Mar. Biol., 31:249.

Patriti, G., 1976, Structure et fonctionnement des systèmes planctoniques dans l'écosystème des zones fermées et semi-fermées des ports de Marseille (dystrophie et pollution), Thèse Doct. es Sciences, Univ. Aix-Marseille.

Porumb, F., Oncium, T., Ialina, E., and Petran, A., 1981, Sur les quantités de zooplancton dans les eaux de la Méditerranée du sud, Rapp. Comm. int. Mer Medit., 27:119.

Razouls, C., 1972, Estimation de la Production secondaire (Copépodes pélagiques) dans une Province néritique méditerranéenne (Golfe du Lion), Thèse Doct. es Sciences, Univ. Paris VI.

Razouls, C. and Razouls, S., 1978, Valeur énergétique de l'échelon secondaire pélagique dans différentes aires maritimes, Vie Milieu, 28)29:39.

Riley, G.A. and Bumpus, D.F., 1946, Phytoplankton-zooplankton relationships on Georges Bank, J. marine Res., 6:33.

Rodriguez, J., 1979, Zooplancton de la bahia de Malaga. Approximación al conocimiento de una communidad planctonica neritica en el Mar de Alboran, Thèse Doct. Univ. Malaga.

Rodriguez, V., (in press), Ecologie du plancton dans le port de Malaga. II. L'associé congénérique d'Acartia (Copepoda Calanoida), Rapp. Comm. int. Mer Medit., Cannes 2-11 December, 1982.

Scotto di Carlo, B., Hure, J., and Miralto, A., 1975, Bathypelagic copepods collected off the island of Ponza (Mediterranean Sea) in June 1973 and June 1974, Pubbl. Staz. zool. Napoli, 39:176.

Scotto di Carlo, B., and Ianora, A., (in press), Atlantis II cruises: Copepod assemblage in deep Mediterranean waters, Rapp. Comm. int. Mer Medit., Cannes 2-11 December, 1982.

Seguin, G., 1981, Dynamique des copépodes pélagiques de Villefranche-sur Mer a partir de prélèvements quotidiens (année 1972). Bilan quantitatif et écologique, Oceanol. Act., 4:405.

Sewell, R.B.S., 1948, The free swimming planktonic Copepoda. Geographical distribution, Res. John Murray Exped., 8(3), 592 pp.

Soenen, M., 1969, Contribution a l'étude du zooplancton superficiel et profond du bassin sud-occidental méditerranéen, Thèse Doct. 3ème cycle, Univ. Aix-Marseille.

Sournia, A., 1973, La production primaire planctonique en Méditerranée. Essai de mise à jour in:"Bull. Etude en Commun de la Méditerranée," No. special 5, Monaco.

Taniguchi, A., 1973, Phytoplankton-zooplankton relationships in the western Pacific Ocean and adjacent seas, Mar. Biol., 21:115.

UNESCO, 1968, Zooplankton sampling, Monogr. Oceanogr. Methodol., 2, UNESCO Press, Paris.

UNESCO, 1976, Zooplankton fixation and preservation, Monogr. Oceanogr. Methodol., 4, UNESCO press, Paris.

Vaissière, R., and Seguin, G., 1980, Etude préliminaire de peuplements de copépodes (juillet 1977) en relation avec l'hydrologie des mers Tyrrhénienne et Ionienne, Oceanol. Act., 3:17.

Vives, F., 1965, Sur les copépodes néritiques (Calanidae) de la Méditerranée occidentale, Rapp. Comm. int. Mer Medit., 17:547.

Vives, F., 1966, Zooplancton neritico de las aguas de Castellón, Invest. Pesq., 30:49.

Vives, F., 1967, Los copepodos planctonicos del Mar Tirreno en septiembre y octobre de 1963, Invest. Pesqu., 31(3):539.

Vives, F., 1971, L'affleurement d'eau sur la côte catalane et les indicateurs biologiques (Copépodes), Invest. Pesq., 35(1):161.

Vives, F., 1979, La population des copépodes de la Méditerranée occidentale pendant le mois d'octobre 1976, Rapp. Comm. int. Mer Medit., 25/26, p. 135.

Vives, F., Santamaria, G., and Trepat, I., 1975, El zooplancton de los alrededores del Estrecho de Gibraltar en junio-julio de 1972, Res. Exp. cient. B/O Cornide, 4:7.

Vives, F., Rodriguez, V., and Rodriguez, J., 1980, Présence de Pseudocalanus elongatus Boeck (Copepoda, Calanoida) en Mer d' Alboran, in: "Journées Etud. System. Biogeogr. Medit.," Cagliari, C.I.E.S.M.:355.

Vucetic, T., 1957, Zooplankton investigation in the sea water "Malo Jezero" and "Veliko Jezero" in Island of Mljet (1951-1953), Acta Adriatica, 4:51.

Yannopoulos, C., 1976, The annual variations of the zooplankton standing stock in the Saronikos Gulf, Aegean Sea. Quantitative results during 72-74, Rapp. Comm. int. Mer Medit., 23: 107.

Zunini-Sertorio, T., Picone, P., and Della Croce, N., 1977, Ecologia y biologia dei porti del Mar Ligure e alto Tirreno. 5. Composizione dello zooplancton del Porto di Genova, Rapp. Comm. int. Mer. Medit., 24:107.

THE ZOOPLANKTON COMMUNITIES OF THE EASTERN MEDITERRANEAN (LEVANTINE BASIN, AEGEAN SEA); INFLUENCE OF MAN-MADE FACTORS

M. Moraitou-Apostolopoulou

Zoological Laboratory, University of Athens
Athens 15 771, Greece

INTRODUCTION

The Eastern Mediterranean is a semienclosed basin of the Mediterranean Sea. This relatively small area connects with three different biogeographical areas, the Western Mediterranean and the Atlantic Ocean through the strait of Sicily, the Black Sea through the Bosphorous and the Red Sea and the Indian Ocean through the Suez Canal.

The establishment of the fauna and flora of the Eastern Mediterranean is the result of important geological changes in the Mediterranean Sea. The living forms of the area have been enriched by migration of species through the pathways of the Strait of Sicily, the Suez Canal and Bosphorous. To this mixture of faunas are added the endemic forms.

Although civilisation has developed around the Mediterranean Sea for over 5 000 years, only 100 years ago did human activities begin to introduce recognizable environmental and biological changes in the Eastern Mediterranean. Recently, the effects of human activities have reached a dramatic magnitude. The opening of the Suez Canal in 1869 made possible the exchange of organisms between the Mediterranean and the Red Sea.

The construction of the Aswan High Dam in 1965 caused important changes in the hydrographic conditions of the south part of the Eastern Mediterranean because of the reduction of freshwater and nutrient inputs. Apart from other changes in the natural resources of the area, the construction of the Aswan High Dam made easier the migration of Red Sea forms to the Mediterranean.

Dams have also recently been constructed on different rivers of the Black Sea, influencing hydrological conditions of the area and consequently the exchanges of water and organisms between the Black Sea and the Mediterranean.

The problem of pollution in the area is also increasing. Urban and industrial waste disposal and river-born contaminants are affecting an ever increasing portion of the coastal region. Moreover the Eastern Mediterranean is the main route of oil transport from the Middle East to Europe. Huge amounts of oil have already been released or spilled in the sea.

The large scale ecological experiments of the opening of the Suez Canal and the construction of the Aswan High Dam,added to all the above-mentioned events,constitute the Eastern Mediterranean as a natural Laboratory and a unique biotope for ecosystem study, and particularly for evaluation of the impact of human activities on the environment.

Because of their importance in the marine food web, the study of zooplankton communities seems to be particularly useful for the evaluation of ecosystem conditions, even though their passive transportation does not permit the formation of clearly-defined units as is possible for benthic systems.

Unfortunately, extensive data on the zooplankton communities of the Eastern Mediterranean, and particularly recent ones, are lacking; the problems of characterisation of the ecosystem structure and function, and the determination of the effects of man therefore become very difficult.

HYDROGRAPHY

The Levantine Basin, and to a lesser degree the Aegean Sea, are characterized by high surface temperatures (Table 1). In the deeper layers, the annual range of temperature is progressively diminished. Below 250 meters, the temperature in the North Aegean Sea remains practically unchanged throughout the year (13.5 to 14.0°C). Due to intense evaporation and lack of important freshwater input and also to restricted rainy season, the Eastern Mediterranean is characterized by high salinities. The hydrography of the area is locally influenced by the penetration of lower salinity waters: the Atlantic current, the Black Sea current and the outflow of Nile (till some years ago) and other smaller rivers.

The Atlantic current (salinity 37 o/oo) flows along the north african coast. In summer this water submerges at Cyrenaica, while in winter it flows as far as Palestine and the south coast of Turkey with gradually increasing salinity (Gulf of Sirte 38.3o/oo; coasts

Table 1. Surface temperatures (°C) of some coastal areas of the Eastern Mediterranean.

	Egypt	Lebanon	S. Aegean	N. Aegean
summer	29.0	30.0	28.0	25.5(max)
winter	16.5	17.0	12.5	11.04(min)

of Lebanon 38.8o/oo; coasts of Israel 39o/oo). During winter the cooling of surface waters makes them submerge and thus intermediate water is formed and flows to the west.

The northeast part of the Aegean Sea receives a surface current from the Black Sea through the Marmara Sea and the Bosphorus. This water is characterized by very low salinity (about 20o/oo). The Black Sea current and the freshwater discharge from the rivers along the north greek coast cause an important decrease of salinity in the shallow, northern part of the Aegean. During summer and in stations near the north coast of Greece, salinities as low as 30.88o/oo have been measured (Kiortsis and col., 1969).

A subsurface countercurrent traverses the Sea of Marmara, going in the opposite direction to the Black Sea current and transporting waters of 35 to 38o/oo salinity from the Aegean to the Black Sea.

Some years ago the surface current from the Black Sea was much more important than the countercurrent from the Aegean, due to the large quantities of freshwater from the major rivers (Danube, Dnepr, Dnestr). After the construction of dams in these rivers, the outflow of fresh water has been significantly reduced, resulting in a corresponding reduction of the waters flowing to the Aegean Sea. Till 1965, and for many centuries, the periodic rise and fall of the Nile transported huge quantities of fresh water and nutrients to the coasts of Egypt, resulting in a strong lowering of salinities and causing high productivity of the sea water during the flood period. After the construction of the High Dam significant changes in the hydrographic conditions have occurred. From August to September, formerly the season of Nile outflow, high surface salinities (39 to 39.4o/oo) are observed. The surface layer is underlain by a layer of reduced salinity (38.0 to 38.7o/oo) due to the presence of the Atlantic current (Emara et al., 1973). Water exchanges between the Mediterranean and the Red Sea take place through the Suez Canal. For more than 9 months (October to July) the current in the canal flows northwards, while during summer (July to October) it is reversed. Since 1966, with the completion of the Aswan High Dam, there is evidence that the northward current is reduced in velocity during summer and is probably no longer reversed (Gerges, 1976).

CHARACTERISTICS OF THE ZOOPLANKTON COMMUNITIES OF THE LEVANTINE BASIN

a) Coast of Egypt

Till some years ago, the huge amounts of nutrients transported during the Nile flood created a very fertile area around the Nile mouth. The distribution and ecology of plankton communities before the construction of the Aswan Dam was studied by Dowidar (1965) and Dowidar and El-Maghraby (1971a,b). The total zooplankton population formed two maxima, a spring maximum and an autumn one. The autumn maximum was much more important and coincided with the Nile flood. The nutrient enrichment caused an immediate and considerable phytoplankton bloom (more than 10 million cells/l), much higher than any other Mediterranean region (Halim, 1960). The phytoplankton bloom was followed by an intense development of zooplankton. The inshore zooplankton communities were more rich in individuals (up to 250,000 ind/m^3) than the offshore (up to 130,000 ind/m^3). The zooplankton population was dominated all year round by small-sized copepods characteristic of surface warm waters (mean 65% in the inshore and 76% in the offshore). Next to the copepods were the tintinnids (8.3 and 10.5% respectively) and appendicularians (1.3 and 2.2%). The planktonic larvae became abundant only in inshore waters (31 in inshore, 6% in offshore).

Of the 84 species of copepods recorded only 18 formed the great bulk of the copepod community. The most important species were: *Paracalanus parvus*, *Euterpina acutifrons*, *Oithona nana*, *O. plumifera*, *Centropages kroÿeri*, *Isias clavipes*, *Clausocalanus arcuicornis*, *Acartia sp*. The first three species were very abundant only in neritic waters and were strongly reduced in the offshore stations. Some of the dominant species were permanent (present in adult or larval stages) in the plankton throughout the year (*Paracalanus parvus*, *Oithona nana*, *Euterpina acutifrons*, *Clausocalanus arcuicornis*), others were present only during part of the year (*Centropages kroÿeri*, *Acartia negligens*, *A. latisetosa*) while others were characterized by very brief period of occurrence (*Isias clavipes*, *Acartia clausi*, *Temora stylifera*, *Oncaea sp*., *Corycaeus sp*., *Corycella sp*.).

Medusae became abundant in zooplankton from June to May and from early October to November. The principal species were: *Liriope tetraphylla*, *Pandea conica*, *Pantachogon haeckeli*. From all medusae species only *Liriope tetraphylla* and *Obelia sp*. were found in inshore waters.

Siphonophora became rather frequent only in the offshore stations. *Chelophyes appendiculata* was the most common followed by *Muggiaea kochi*.

Chaetognatha (Sagitta bipunctata, S. serratodentata, S. hexaptera, S. enflata) formed two maxima, the first in June (up to 1875 ind./m³), the second in October (up to 1500 ind./m³).

Cladocerans, not significant numerically, were represented by 5 species: Evadne nordmani (the most common species of the genus occurring from March to autumn), E. spinifera, E. tergestina (both thermophilic occurring during summer), Podon polyphemoides (from early spring to early autumn) and Penilia avirostris which was represented only by few specimens.

Appendicularia were found almost all year round, formed two peaks (one in late spring to early summer, the other from late summer to early autumn) and the principal species (Oikopleura longicauda, O. dioica) became abundant only in inshore stations.

A characteristic element of plankton in the egyptian coasts was Noctiluca scintillans whose presence was almost confined to the flood season (late August to October or November). Noctiluca reached its maximum (up to 18.600 ind./m³) in September. According to Halim et al. (1974) Sagitta friderici and Oikopleura sp., abundant forms along the egyptian shores during the period of the Nile flood, can also be characterized as indicators of the Nile outflow, while the abundance of Appendicularia sicula is among the characteristic features of the S.E. part of the Levantine Basin.

Data on the zooplankton communities of egyptian coasts after the construction of the Aswan High Dam are limited. An important decrease in zooplankton abundance of 1966 compared with the values of previous years was noticed by El-Maghraby and Dowidar (1973) and Dovidar and El-Maghraby (1973): 22.400 ind./m³ in 1962 to 9.900 in autumn at the same collection site (Abu-Qir bay). Regarding the qualitative aspects of the populations, the general pattern was the same as before the construction of the High Dam. Copepods dominated the population (average 77%). Planktonic larvae (10.8%) followed in abundance; appendicularians and radiolarians were also found in high numbers.

Following the damming of the Nile, the whole food chain was affected. As early as 1966 the total catch for all fish species had dropped to about 30% of the 1962 value. Sardinella which feeds on copepods and diatoms was more dramatically affected. Previous research had shown a correlation between plankton biomass and Sardinella fisheries.

Morcos and El-Rayiss (1973) have noticed a vertical gradient in nutrient concentration and a general nutrient increase during winter. With the observed rise in surface salinity and surface density it is not unlikely that winter convection from the deeper layers will gradually become more effective. According to Halim

(1976) a partial recovery in the biological productivity is not to be excluded.

There are very limited data concerning the impact of pollution on zooplanktonic communities on the coast of Egypt, although recent observations indicate that marine ecosystems are affected by pollution. Toma et al. (1980) found high concentrations of heavy metals in the Nile continental shelf sediments. Also the coastal waters near Alexandria show high values of apparent oxygen utilisation (El-Hehyawi and El-Tabbakh (1980). El-Sharkavi (1980) concludes that there is a significant risk of contracting typhoid from bathing in the polluted waters of Alexandrian swimming beaches. Increased accumulation of floating tar balls resulting from a heavy shipping traffic have been noticed in the area of Alexandria (Aboul-Dahal and Halim, 1980; Wahby, 1980). Finally El-Sokkary (1980) found high levels of Hg in fishes in populations of *Solea vulgaris*.

Recently El-Hehyawi (1978) noticed a significant drop in zooplankton (hyponeuston) abundance in the area surrounding the petroleum wells. Furthermore, the mixture of sea water and drainage water, with high concentrations of industrial wastes and hydrocarbons, in Abu-Qir Bay was also found to cause a sharp decline in the abundance of fish eggs and a virtual disappearance of meroplanktonic forms.

b) Coast of Israel

In the inshore and offshore waters of the coast of Israel the zooplankton communities form two main annual peaks in April-May and in December (Kimor and Berdugo, 1967). In earlier years the arrival of the Nile flood along the coast of Israel stimulated a later summer peak.

During the period of peaks, the great bulk of zooplankton is formed by the copepods (up to 85% of the total zooplankton standing stock and up to 940 ind./m^3). The December peak however consists also of large populations of other zooplanktonic groups, chiefly chaetognaths, pteropods and larval stages of decapods. In the offshore waters, pteropods and cladocerans constitute elements of major importance. Previously the drop of the salinity of the sea water caused by the arrival of the Nile flood resulted in an almost immediate increase in the abundance of the cladoceran *Podon polyphemoides* to a state of mass development (Komarovsky, 1953; Oren and Komarovsky, 1961).

Among the copepods, calanoids dominate the planktonic community both in coastal and in offshore areas. In the neritic stations the more abundant copepods are: *Paracalanus parvus*, *Centropages kroÿeri*, *Isias clavipes*, *Acartia clausi*, *Euterpina acutifrons*, while in the oceanic, Eucheta marina and Calanus gracilis. Fairly abundant in both environments are: Clausocalanus furcatus, Temora stylifera, Centropages violaceus, Acartia negligens. As a result of pronounced seasonal cycles the composition of copepods presents sharp differences from one season to another. The spring peak consists mainly of Paracalanus parvus, Centropages kroyëri, Isias clavipes and Acartia clausi. The December peak was characterized by Paracalanus parvus, Temora stylifera, Corycaeus sp. and Oncaea sp. Due to the homothermal conditions in this period of the year, some deep water species are added to the community (Lucicutia flavicornis, Euaetideus giesbrechti).

Three species of cladocerans have been identified on the coast of Israel: Evadne spinifera, E. tergestina and Podon polyphemoides. They are all thermophilic, so the cladoceran peak is formed in July (up to 454 ind./m³). Chaetognaths were represented mainly by two species Sagitta friderici and S. enflata abundant in neritic waters (up to 44 ind./m³). Two molluscs Creseis and Limacina are quite common in the coastal waters (up to 200 ind./m³). Decapoda reach up to 125 ind./m³ and consist mainly of larval stages of benthic species and Lucifer.

Cases of abrupt change in the composition and/or abundance of the zooplankton community either through predation by carnivorous zooplankton or fishes (Sardinella) or competition within the ecosystem have been referred to by Komarovsky (1959) and Kimor (1965).

c) Coast of Lebanon

The coast of Lebanon is characterized by a very narrow continental shelf. During the warm period (April-November) a net stratification of the water masses is observed, while during winter the combination of currents and winds leads to the creation of isothermal conditions. Nutritive salts brought to the surface layers result in a phytoplankton bloom accompanied by a rather strong zooplankton development.

The zooplankton communities of the Lebanese coast have been studied since 1969 by Lakkis (1971a,b, 1973, 1976a,b, 1977, 1980, 1981) and Lakkis and Zeidane (in press a,b). The zooplankton of the Lebanese coast presents high diversity and pronounced seasonal quantitative variations. 330 zooplankton species have been identified, copepods (109 species) forming the 65% of the total biomass. The other groups are much less abundant (crustacean larvae 13%, cladocerans 4%, salps and appendicularians 4%, chaetognaths 3%, medusae and siphonophores 2%).

From the quantitative point of view the zooplankton is rich: 3.000 ind./m³ in spring when the annual maximum is formed, dropping to 1.00 ind./m³ during the autumn minimum.

Eleven species of copepods form the quasi-totality of copepod biomass: Acartia clausi, A. latisetosa, A. longiremis, Calocalanus styliremis, Centropages kroÿeri, Clausocalanus furcatus, Cl. arcuicornis, Cl. mastifophorus, Paracalanus parvus, P. crassirostris, Temora stylifera, Corycaeus clausi, C. flaccus, Corycella rostrata, Euterpina acutifrons, Oithona nana, O. plumifera, Oncaea conifera, O. mediterranea. 25 species of siphonophores have been identified, 5 physonects and 20 calycophores. The commonest physonects are Agalma, Halistema and Nanomia, while the most abundant calycophores are Abylopsis, Chelophies, Eudoxoides, Hippopodius, Lensia, Muggiaea. The cladocerans comprise five species: Evadne spinifera and E. tergestina forming the 90% of the population, Podon polyphemoides (7%), P. intermedius (2%) and Evadne nordmani (1%). The chaetognaths, well diversified, include 10 species, Sagitta friderici is the dominant followed by S. enflata. S. megalophthalma is a mesopelagic form recorded for the first time in the Eastern Mediterranean. 12 species of medusae have been identified. Between the 5 euphausiids species Euphausia brevis is the most common while between the 7 species of amphipods Lestrigonus schizogeneios and L. latissima.

Three stages of ecological succession have been identified in the annual cycle of zooplankton populations:

a) During winter (December-March) the temperature and salinity minima are observed. The plankton biomass is low and a high diversity is noted. During this period the zooplankton is dominated by predators: chaetognaths, molluscs, medusae and bathypelagic copepods (Pleuromamma, Eucheta, Candacia).

b) During spring a phytoplankton bloom is observed, followed by the development of herbivores: copepods (Paracalanus, Temora, Centropages, Acartia, Clausocalanus, Euterpina), appendicularians, and various nauplii and larvae. The zooplankton biomass is high and the diversity low.

c) The thermocline, formed during summer (July-September) and resultant water stratification inhibit the vertical migration of plankton to the surface.

Between the various sources of pollution on the coast of Lebanon, domestic sewage seems to be the most important. More than ten open sewers discharge untreated sewage into the ocean in the area of Beirut. This untreated domestic sewage results in extremely high concentration of dissolved nitrogen along the coastal area of Beirut to about 10 kilometers southwards. This nutrient enrichment causes phytoplankton standing crops within the polluted area 8 to 10 times higher than in unpolluted areas further south or elsewhere in the Eastern Mediterranean (Hardy and Jubayli, 1974).

The study of zooplankton at different coastal stations in the area of Beirut subjected to different kinds of pollution gives some useful information for the influence of pollution on marine ecosystems. In a general way, a reduction in the variety of zooplankton at the polluted stations has been observed. Furthermore, the zooplankton abundance is lowered with the exception of the area near Beirut harbour. Some species seem equally distributed in the polluted and non-polluted stations (A. longiremis, A. discaudata, Euterpina acutifrons, Evadne spinifera, Sagitta friderici, Oithona nana, Centropages kroÿeri, Lucifer sp.). Some others seem to avoid polluted waters (Paracalanus parvus, Temora stylifera, Corycaeus flaccus, Thalia democratica, Sagitta enflata, S. serratodentata, Oikopleura sp., Obelia sp. different fish eggs, siphonophores and Medusae (Lakkis and Kouyoumjian, 1974; Lakkis and Abboud, 1976).

d) Offshore waters of the Levantine Basin

The plankton of the offshore waters of the Levantine Basin collected from an extensive net of stations throughout the Basin, during one cruise in 1964 (summer) and three in 1968 (May-June, September, October-November) was studied by Israelian scientists (Kimor and Berdugo, 1967; Pasteur et al., 1976).

The quantitative estimation of zooplankton based on displacement volumes and numerical abundance showed similar trends: zooplankton showed higher densities during spring-early summer than those of September and November-December.

The main feature is the predominance of large populations of copepods. This group, formed mainly of calanoid copepods, constituted up to 85% of the total zooplankton standing crop. In some cases, cyclopoids, represented by one or several species of Corycaeus, were dominant. The numerical abundance of copepods, high in the May-June period, was lowered in September. This decrease continued to an ever greater extent in December following the general trend of total plankton biomass pattern. Clausocalanus furcatus was the dominant species of copepods especially abundant in May-June followed by Temora stylifera, also an abundant species and Centropages violaceus and Acartia negligens both common but not abundant. Paracalanus parvus was abundant only in the 1964 cruise. The two additional groups of importance as biomass formers were pteropods and cladocerans (only during summer). Pteropods were represented mainly by Limacina which constituted an important element of the zooplankton in the offshore waters of the Levantine Basin. During May-June Limacina formed 30 to 40% of the zooplankton numbers in the area of Rhodes. According to Furnestin (1960) Limacina and Creseis acicula an also abundant species in the Eastern Mediterranean are favoured by the high salinity conditions.

Cladocerans were more abundant during the May-June cruise (up to

21 rarely 35 to 75%) dropping in September to 7%. Two species of cladocerans have been found, Evadne spinifera the dominant form and E. tergestina. Chaetognaths constituted a group of minor importance (up to 4% in May-June). The main species was Sagitta enflata. Decapods comprised mainly larval stages of sergestids and Lucifer (up to 12 ind/1.000 m^3). Appendicularians were only occasionally numerous. The same was true for mysidaceans and siphonophores which became abundant only at individual stations.

INFLUENCE OF THE RED SEA

The construction of the Suez Canal made possible the exchange of faunas between the Mediterranean and the Red Sea. For a long period these Lessepsian migrations (Por, 1969,1971) remained insignificant because of the presence of two natural obstacles; a topographical one: the very shallow depth of the canal and a hydrological one: the pronounced differences in salinity between the gulf of Suez and the Mediterranean at the area of Nile outflow.

Recently, the excavation of the canal and the damming of the Nile have facilitated exchanges between the two seas. The salinity of the Bitter Lakes has been significantly lowered (from 51.84 o/oo in 1924 to 44 o/oo in 1971) and the salinity in the Mediterranean coast has reached the normal values of the area: 39 o/oo). Furthermore, the speed of the current to the Mediterranean has increased. It seems therefore quite possible that species of some ecological tolerance could be transported from the Red Sea to the Mediterranean. This would result in an enrichment of the fauna of the latter region which presents a low variety of forms compared with that of the Red Sea.

Transportation of holoplanktonic animals is easier than that of planktonic larvae which are usually more sensitive to salinity changes than adults.

Determination of the immigrants from the Red Sea in not easy. Both areas comprise for the most part tropical and subtropical species and the majority of the common species are cosmopolitan or of a wide distribution. However, there exist several Red Sea species which have not been mentioned from the Mediterranean. Several examples of Lessepsian migrations have been referred to in the literature and specimens of Red Sea species have been captured as far away as the coast of Lebanon and the south Aegean Sea. According to Furnestin (1979) the proofs of these migrations are of unequal value and only the recent records of Indo-Pacific species which are observed at the same time in the Red Sea, the Canal and in various places of the Levantine Basin seem to bear witness to the present relations between the Red Sea and the Mediterranean.

According to Halim (1970) 5 species of dinoflagellates have migrated from the Red Sea and 7 of tintinnids (Tintinnopsis aperta, T. dadayi, T. gracilis, T. mortensenii, T. tocantinesis, Metacylus annulifera, Codonelopsis bulbulosus) according to Dowidar (1974). Schmidt (1972,1973) states that at least two species of hydromedusae (Laodicea fijiana, Euphysora bigelowi) have been transported through the Suez Canal and he considers that as the number of hydromedusae species is much higher in the Red Sea than in the Western Mediterranean, other species must also have been introduced. According to the same author an important obstacle for the settlement of planktonic hydrozoans in the Eastern Mediterranean is that hard substrata, in which hydropolyps usually live,are rare. Sagitta neglecta a typical Red Sea species was found by Guergues and Halim (1973) in the waters near Alexandria.

Among copepods the following have been considered as immigrants from the Red Sea: Paracalanus crassirostris found at a distance of 20-40 miles from the Suez Canal (Shmeleva, 1973), Calanopia elliptica and C. media captured along the coasts of Israel and Lebanon (Berdugo, 1968; Lakkis, 1976b), Arietellus pavoninus in the south Aegean (Moraitou-Apostolopoulou, 1969), Eucheta concinna (Casanova, 1973), Acartia centrura (Berdugo, 1974), Labidocera madurae, L. detruncata, Acartia fossae, Arietellus aculeatus, Centropages ponticus, Sagitta bierii, Paracytaeis vetuna (Lakkis, 1976b), Salpa cylindrica (Godeaux, 1960).

The transportation of species from the Red Sea to the Mediterranean seems consequently limited, although recent systematic investigations are lacking.

Enrichment of the Mediterranean populations may also occur both for species which have penetrated and settled to the Mediterranean and for species which are common in both areas. Some species such as Paracalanus aculeatus and Calocalanus pavo, found in fairly large numbers in the area of Alexandria, may indicate their probable enrichment from the Red Sea (Dowidar and El-Maghraby, 1971a). According to Steinitz (1968) introduction of individuals of species already present in the involved area is immigration,at least from the numerical point of view.

The passage in the direction Mediterranean-Red Sea is very restricted, mainly because the southward current is very limited in duration and probably since recently the northward current is not longer reversed.

AEGEAN SEA

The Aegean Sea is topographically divided into two basins by approximately the 38^{0} parallel, the north and the south Aegean Sea.

The hydrological and topographical conditions differ strongly between the two basins. The north Aegean Sea is characterized by an extreme continental shelf, favouring the proliferation of neritic and mero-planktonic forms, while in the south Aegean the continental shelf is limited except in the area around the Cyclades) and the area has the characteristics of a pelagic zone. In the north Aegean are deversed the outflows of some important rivers (Nestos, Axios, Evros, Strymon) which lower the salinity locally. Furthermore, the current through Dardanelles causes a decrease of salinity in the northeast Aegean.

The Aegean Sea can be characterized as an oligotrophic area. With the exception of the area around the mouth of rivers, where dense zooplankton communities are formed, the plankton biomass remains low: 500 to 1.000 ind/m^3 or 0.5 to 2.9 mg/m^3. According to Pavlova (1966) the Aegean Sea is 12 to 18 times inferior to the Black Sea, in secondary productivity, 4 times to the Adriatic Sea and similar to that of Tyrrhenian and Libyan Seas.

The zooplankton communities are well diversified. For example 120 copepod species have been found by Pavlova (1966) and 114 in mainly superficial samples (Moraitou-Apostolopoulou, 1972). The appendicularians show also high diversity: 29 species have been identified (Fenaux, 1974).

The zooplanktonic communities of the Aegean Sea are clearly dominated by copepods. They usually constitute the 75-95% of the total number of zooplanktonic organisms (Moraitou-Apostolopoulou, 1972,1974a,b). The dominant species are: a) north Aegean Sea: *Temora stylifera*, *Centropages typicus*, *Calanus minor*, *Corycaeus giesbrechti*, *Centropages violaceus*, *Corycaeus ovalis*, *Pleuromamma gracilis*, *Clausocalanus furcatus*, *Cl. arcuicornis*, *Acartia clausi*. b) South Aegean Sea: *Temora stylifera*, *Clausocalanus arcuicornis*, *Oithona plumifera*, *Calanus minor*, *Clausocalanus furcatus*, *Acartia negligens*, *Calocalanus pavo*, *Corycella rostrata*, *Corycaeus typicus*, *C. giesbrechti*. Many differences have been observed between the copepod fauna of the two basins of the Aegean Sea. These differences are qualitative (9 species have been found only in the north Aegean) and, especially, quantitative. Almost all important species show pronounced quantitative differences between the two basins. The north Aegean Sea presents the characteristics of a semi-enclosed and ecologically-isolated area while the fauna of the south Aegean, in its general trends, shows a transitional character between the fauna of the north Aegean and that of the Western and Eastern Mediterranean. The main characteristic of the zooplanktonic communities of the north Aegean is the extreme abundance of *Temora stylifera* which, near the north coasts and in the vicinity of the rivers, forms very dense populations (up to 400 ind/m^3).

Two species *Corycaeus giesbrechti* and *Candacia aethiopica*, rare in the Mediterranean, are also found in large numbers in the north

Aegean. Other forms such as Clausocalanus arcuicornis, Oithona plumifera and Mecynocera clausi, which are abundant throughout the Mediterranean, are found in restricted numbers in the north basin. Particularly striking is the paucity in numbers of Paracalanus parvus, a dominant form both in the Western and Eastern Mediterranean and also in the Black Sea, although the ecological conditions (shallow waters, lower salinities) are favourable. This form is also scarce in the south Aegean. Centropages typicus, a form which is very abundant in the Western Mediterranean and progressively diminishes (Adriatic) or disappears (Levantine Basin) to the east, is found in very large numbers in the north Aegean.

Between the transitional characters of the plankton fauna of the south Aegean one can mention the increase in numbers of Clausocalanus arcuicornis, Cl. furcatus, Oithona plumifera, Mecynocera clausi and the decrease of Temora stylifera, Corycaeus giesbrechti, Candacia aethiopica, while the numerical decrease of Centropages typicus is a resemblance to the Eastern Mediterranean fauna.

In some features the plankton of the Aegean resembles that of the western Mediterranean: Centropages kroÿeri a dominant copepod in the Levantine Basin is found in restricted numbers in both the Western Mediterranean and the Aegean Sea.

Four species of cladocerans have been found in the Aegean Sea (Moraitou-Apostolopoulou, 1973,1974c): Evadne spinifera (the most important), E. tergestina, Penilia avirostris, Podon intermedius. With the exception of Podon intermedius all species are thermophilic so this group becomes abundant only during summer. The percentage of cladocerans to the total zooplankton number is low: 1 to 5%. Penilia avirostris forms dense populations on the north coast of Greece and around the mouths of rivers. This form which has not been found in the Levantine Basin, is practically, absent from the south Aegean.

10 species of chaetognaths have been identified from the Aegean Sea (Ghirardelli and Rottini, 1979), Sagitta minima and S. enflata being the most abundant forming together 44% of the chaetognath population. Kronitta subtilis (2.53%) is a form rare in the Western but well-represented in the Eastern Mediterranean. 9 species of mysids have been identified from the Aegean Sea. Four of them Anchialina oculata, A. agilis, Siriella thompsoni and S. norvegica form 92% of the total number (Hatzakis, 1974). Hyperia schizogeneios is the commonest of the 14 amphipod species found in the Aegean, while isopods are represented by only two species (Euridice truncata, Idothea metallica) (Veini and Kiortsis, 1974). Euphausids, encountered mainly in pelagic stations and nocturnal samples, are represented by 4 species, Euphausia krohni being the most abundant (Kiortsis and col., 1969). Although siphonophores are considered as a group of decreasing importance from the Western to the Eastern

Mediterranean, 22 species have been identified in the Aegean Sea (Malouchou-Grimba, personal communication). Bassia bassensis is the commonest followed by Eudoxoides spiralis.

The only species of medusae regularly found in the Aegean is Liriope tetraphylla. Cotylorhiza multituberculata occurs in swarms during summer and early automn. The commonest pteropod in the surface waters of the Aegean is Creseis acicula but also swarms of Cymbulia peroni and to a lesser extent of Firoloida desmaresti are also found. Thalia democratica is the dominant form among salps, sometimes encountered in large numbers, while Oikopleura longicaudata is the most common among appendicularians (Kiortsis and col., 1969).

INFLUENCE OF THE BLACK SEA

The Mediterranean is connected with the Black Sea through the Bosphorous strait and the Marmara Sea. The existence of currents facilitates the exchange of planktonic organisms between the two areas. Pronounced differences in the environmental factors, chiefly salinity, render the settlement of organisms from one region in the other difficult.

The planktonic fauna of the Black Sea is a very impoverished atlantomediterranean fauna enriched by some freshwater forms.

Five Black Sea species have been found by Pavlova (1966) in the Aegean Sea: Paracalanus parvus, Pontella mediterranea, Centropages kroÿeri, Oithona similis, O. nana. To this catalog must be added Calanus helgolandicus, Acartia clausi and Anomalocera pattersoni which have also been found in the Aegean Sea (Moraitou-Apostolopoulou, 1972). Acartia clausi, Paracalanus parvus and Calanus helgolandicus are very abundant in the Black Sea. In the Aegean they are found in large numbers in the northeast part, diminishing strongly to the south. The same pattern characterizes also the distribution of Pontella mediterranea. An enrichment of the Aegean Sea populations of these species is not to be excluded, although the numerical decrease of the two former could also be due to their neritic character. Anomalocera pattersoni is the only species localised exclusively in the north basin where it is a rare form, while it is very rare or absent in the Levantine Basin. Demir (1959) found large numbers of Pontella mediterranea and Anomalocera pattersoni on the northeast coast of the Aegean during spring and early summer and concludes that these species are transported from the Aegean to the Black Sea. The fact that, during this period of the year, the current from the Black Sea to the Mediterranean is more intense and that this current is subsuperficial while both species are hyponeustonic forms, leads to the conclusion that a transportation of Black Sea individuals to the Aegean is more plausible.

The transportation of planktonic animals from the Mediterranean to the Black Sea through the Sea of Marmara presents several difficulties. Of the transported only those individuals survive that are subjected to a preliminary acclimation in the surface fresh waters of the Sea of Marmara (salinity 20.5 to 30 o/oo). Due to construction of dams on several Black Sea rivers a change in the hydrological conditions has occurred, resulting in an increase in strength of current from the Aegean. Many investigators have noticed an increase in the number of mediterranean organisms penetrating in the Black Sea. Thus some planktonic organisms (Clausocalanus arcuicornis, Calocalanus pavo, C. pavoninus, Oncaea mediterranea, O. minuta, O. dentipes, Lucicutia flavicornis, Microsetella rosea, Euterpina acutifrons, Corycella rostrata, Corycaeus typicus, C. flaccus, C. clausi and Eudoxoides spiralis) seem recently able to endure the migration into the Black Sea (Bogdanova, 1973; Pavlova, 1964). In spite of the increased transportation of mediterranean planktonic animals to the Black Sea caused by man-made factors which leads to a certain "mediterranisation" of the Black Sea, the influence of the Mediterranean fauna on the Black Sea remains low, since some very abundant planktonic forms of the Aegean as Temora stylifera, Centropages typicus, Calanus minor and Corycaeus giesbrechti have not yet been found in the Black Sea.

THE ZOOPLANKTON COMMUNITIES OF SOME SPECIFIC ENVIRONMENTS OF THE AEGEAN SEA

In gulfs and closed bays of the Aegean Sea the zooplankton communities deviate strongly from those of open sea conditions. A characteristic example of this type of ecosystem is Evoikos Gulf. This gulf is a long and narrow channel between the coasts of Attiki peninsula and Evia island. In this neritic and protected environment, although copepods remain the main zooplankton group, forming the 11% (June) to 98% (February) of the total zooplankton numbers, another group, the cladocerans, become particularly abundant (Moraitou-Apostolopoulou and Kiortsis, 1977; Moraitou-Apostolopoulou and Vournazou, in press) forming from 0.6 to 89% of the total zooplankton numbers with a mean of 45%).

Although the copepod variety remains high, 83 species identified from only horizontal samples, two species clearly dominate the copepod community: Acartia clausi which forms 41.5% of the total copepod numbers and to a lesser degree Temora stylifera, the leading form of the Aegean, forming the 14.3%. The other copepod species are found in reduced numbers and only some of them (Centropages typicus, Calanus helgolandicus, Clausocalanus furcatus, Ctenocalanus vanus) attain quantitative importance.

All six mediterranean cladoceran species were found in Evoikos gulf. By far the two most abundant were Evadne spinifera and Penilia avirostris.

Due mainly to the high production of cladocerans, the mean zooplankton biomass of Evoikos gulf (4.5 mg/m³ according to Yannopoulos and Yannopoulos, 1976) is significantly higher than that of the Aegean Sea.

Between the west coast of the Attikis penninsula and the east coasts of Peloponnesus is formed Saronicos gulf (gulf of Athens). This gulf is an extension of the south Aegean Sea. Due to the presence of turbulent movements, nutrients and detritus are raised from the bottom, resulting in an enrichment of the sea water and creating eutrophic conditions in the gulf. Water rich in organic material, originating from Elefsis Bay also contributes to the eutrophication of the gulf. An important quantitative increase of the zooplankton biomass compared with that of the Aegean Sea was observed in the Saronicos gulf. The dominant planktonic group all year round is the copepods (Moraitou-Apostolopoulou, 1974) forming the 66% of the total zooplankton numbers. This group is well diversified: 83 species identified from only superficial samples. Forms abundant in the Aegean such as Temora stylifera and Corycaeus giesbrechti are reduced in numbers in the gulf, while Acartia clausi becomes the dominant form (53%) followed by Centropages typicus.

All six cladoceran species were also found in Saronicos gulf, Evadne spinifera being the most abundant. The percentage of cladocerans to the total zooplankton numbers ranges from 8% (March) to 70% (August).

The study of the composition and vertical distribution of the zooplankton community of the surface layer (0-100 cm) of Saronicos gulf performed by a slightly-modified Zaitsev hyponeuston net, proved that the hyponeuston of Aegean Sea has not the importance attributed by Zaitsev (1971) to this layer of the Black Sea. The superficial water layer (0-20 cm) contains less planktonic organisms than the deeper microlayers. The euneustonic (permanently neustonic) forms are rare, the majority of species having a tendency to concentrate in deeper microlayers (Moraitou-Apostolopoulou, 1978a,b).

The north east part of the Saronicos gulf, Elefsis Bay, offers a characteristic example of the influence of man-made factors on the marine ecosystems. Elefsis Bay is bordered by the most heavily-populated and industrialized area of Greece. At the east part of this bay, at Keratsini, is the main sewage outfall of the cities of Athens and Piraeus. Industrial waste and hydrocarbon products discarded from industries contribute to the pollution of the bay. High concentrations of heavy metals have been measured in the sediments of Elefsis Bay (Grimanis et al., 1976), while the average level of dissolved zinc in the waters of Elefsis Bay was found to be four times higher than that of unpolluted regions of the Aegean Sea (Scoulos, 1980). Higher levels of 12 trace metals have been found in Acartia clausi a very abundant copepod of the bay (Zafiropoulos and Grimanis, 1976).

Nutrient values of the bay are 20 times higher than those of the Saronicos gulf (Coachman et al., 1976). As a result of nutrient enrichment, strongly eutrophic conditions characterize the area.

Very pronounced differentiation of the zooplanktonic communities have been noted in Elefsis Bay: very high biomass values (25-96, 1 mg/m^3 according to Yannopoulos and Yannopoulos (1976), during winter). Sampling performed on a year-round basis at beweekly intervals simultaneously: a) inside Elefsis Bay, and b) in a non-polluted area of Saronicos 25 km southwards (a 180 μ net was used, while previous sampling was done with a 200 μ net) revealed strong differences, both qualitative and quantitative, between the two populations (Moraitou-Apostolopoulou, 1981).

The Elefsis Bay population proved to be much more abundant (annual average 9.224 ind./m^3) than that of south Saronicos (average 1.287 ind./m^3). Furthermore the south Saronicos population presented a higher species diversity (e.g. 42 copepod species identified) than the Elefsis (16 species identified). Copepods and cladocerans form the virtual totality (about 96%) of the two populations. The list of dominant species and their abundance differs markedly between the two stations:

South Saronicos		Elefsis Bay	
Dominant Forms	Abundance (ind./m^3)	Dominant Forms	Abundance (ind./m^3)
Acartia clausi	435	Oithona nana	2.864
Clausocalanus furcatus	160	Penilia avirostris	1.926
Paracalanus parvus	127	Evadne nordmani	1.755
Oithona nana	126	Acartia clausi	1.488
Podon intermedius	97	Paracalanus parvus	1.305
Oncaea media	81	Evadne tergestina	1.105

The zooplankton annual cycle in Elefsis was remarkable (maximum abundance occurring in the period July-September followed by consequent decreases and increases in the population size during the rest of the year). In the south Saronicos gulf it was characterized by a period of lower abundance (May-October) followed by an increase (November-March). To find out if the pollution conditions have some impact on the physiology of planktonic animals living in polluted areas, comparative experiments have been conducted with specimens of Acartia living a) inside Elefsis and b) in a non-polluted area of Saronicos gulf. The results suggest some physiological differences between the two populations. Thus the pollution-adapted population of Acartia demonstrated higher ingestion and filtration rates than the population of the clean area (Moraitou-Apostolopoulou and Verriopoulos, 1980). Furthermore, tests of thermal tolerance proved that the specimens of Elefsis are more resistant both to the upper and lower temperature limits. Finally, the pollution-adapted Acartia survived longer under laboratory conditions (Moraitou-Apostolopoulou and Verriopoulos, 1981a,b).

In an attempt at better understanding of the adaptation of Acartia to pollution conditions, the effects of two metals (Cu,Cd) on specimens of the two populations have been studied comparatively. In acute toxicity tests for both metals the pollution-adapted Acartia proved more resistant, the LC5048h (concentration of a toxicant which kills the 50% of the test animals after 48 hours of exposure) values being significantly higher than the corresponding values of the non-pollution-adapted specimens. In subacute toxicity tests, Acartia specimens from the Elefsis Bay also demonstrated more resistance to both metals. They survived longer at low concentrations of both metals; their ingestion rate, egg production and respiration rate were also less affected than that of the outer Saronicos animals (Moraitou-Apostolopoulou, 1978c, Moraitou-Apostolopoulou and Verriopoulos, 1979, Moraitou-Apostolopoulou et al., 1979a,b).

CONCLUSIONS

Knowledge of the zooplankton communities of the Eastern Mediterranean is scarce and fragmentary; systematic data from most parts of the area are lacking. The paucity of recent information and the lack of standardised sampling methods permit only some rough conclusions to be drawn about the present status of the zooplankton communities of the Aegean Sea and the Levantine Basin.

The Eastern Mediterranean has long being considered as an oligotrophic area. It has repeatedly been stated that the Mediterranean is characterised by low levels of nutrients and consequently of productivity and that this poverty increases from west to east. However recent data do not seem to support this view. In the neritic waters of the Western Mediterranean the average number of zooplankters per cubic meter fluctuates from 1.000 to 1.200. Comparison of the quantitative data between the Western Mediterranean and other areas proves that the Mediterranean values exceed the values from the gulf of Gascogne and from the atlanticomoroccan neritic areas, but are lower than the values recorded along the west atlantic coasts (Gaudy, this book).

Similar values (500 to 1.000 for the Aegean, 1.000 to 3.000 for the coast of Lebanon, 1.100 for the coast of Israel) have also been noted for the Eastern Mediterranean (Table 2). According to Pavlova (1966) 806 ind./m^3 and 12.1 mg/m^3 have been measured in the

Table 2. Structure of the zooplankton community of some areas of the Eastern Mediterranean

Area	Standing stock (ind./m^3)	Copepods	Cladocerans	Planctonic larvae	Pteropods	Chaeto-gnaths	Appendicu-larians	Medusae	Main species
Coasts of Egypt									PP,Ea,On,Op,Ck,Ic, Ca,Asp,As,Ns,Sf, Osp
1) before the Aswan Dam									
- offshore waters	up to 130,000	76%	N.i.	6%		1%	2.2%	A	
- inshore waters	up to 250,000	65%	N.i.	31%			1.3%		
2) after the Aswan Dam	up to 9,900	77%	N.i.	10.8%					
Coasts of Israel	100 to 1,000	85%	A		A				Pp,Ck,Ic,Ac,Ea,Cf, Ts,Cv,An,Csp,Lsp
Coasts of Lebanon	1,000 to 3,000	65%	4%	13%		3%	4%	2%	Ac,Al,Ala,Cs,Ck,Cf, Ca,Cm,Pp,Ts,Cc,Cfl, Op,Cr
Offshore waters of the Levantine Basin		85%	3-21%		7-40%	up to 40%	Oc.i.		Cf,Ts,Cv,An,Pp,Lsp, Es
North Aegean Sea	500 to 1,000	85%	2-5%	C.	N.i.	C	C	Oc.i	Ts,Ct,Cm,Cg,Cv,Co, Pg,Cf,Ca,Ac,Cac,Cp
South Aegean Sea	500 to 1,000	85%	1-3%	N.i.	Oc.i	C	C	Oc.i	Ts,Ca,Op,Cm,Cf,An, Cp,Cr,Ct,Cg
Evoikos gulf (Greece)	4,5 mg/m^3	45%	1-90%	8%	N.i.	2.4%	7.1%	2.4%	Ac,Ts,Ct,Ch,Cf,Ctv, Es,Pa
Saronicos gulf(Greece)	1,287	66%	8-70%		N.i.	N.i.	Oc.i.	Oc.i.	Ac,Ct,Cf,Pp,On,Ome, Pi
Elefsis Bay(Greece)	9,224	77%	20%		N.i.	N.i.	N.i.	N.i.	On,Ac,Pp,Pa,En,Et

(continued)

Table 2. Footnotes

A = abundant, N.i. = not important, Oc.i. = occasionally important, C = common.

Pp=*Paracalanus parvus*, Ea=*Euterpina acutifrons*, On=*Oithona nana*, Op=*O. plumifera*, Ck=*Centropages kroÿeri*, Ic=*Isias clavipes*, Ca=*Claucalanus arcuicornis*, Asp=*Acartia sp.*, As=*Appendicularia sicula*, Ns=*Noctiluca scintillans*, Sf=*Sagitta friderici*, Osp=*Oikopleura sp.*, Ac=*Acartia clausi*, Cf=*Clausocalanus furcatus*, Ts=*Temora stylifera*, Cv=*Centropages violaceus*, An=*Acartia negligens*, Csp=*Creseis sp.*, Lsp=*Limacina sp.*, Al=*Acartia longiremis*, Ala=*A. latisetosa*, Cs=*Calocalanus styliremis*, Cm=*Clausocalanus mastigophorus*, Cc=*Corycaeus clausi*, Cfl=*Corycaeus flaccus*, Cr=*Corycella rostrata*, Es=*Evadne spinifera*, Ct=*Centropages typicus*, Co=*Corycaeus ovalis*, Pg=*Pleuromamma gracilis*, Cp=*Cymbulia peroni*, Cm=*Calanus minor*, Cp=*Calocalanus pavo*, Cty=*Corycaeus typicus*, Cg=*C. giesbrechti*, Ch=*Calanus helgolandicus*, Ctv=*Ctenocalanus vanus*, Pa=*Penilia avirostris*, Ome=*Oncaea media*, Et=*Evadne tergestina*, En=*E. nordmani*, Pi=*Podon intermedius*, Cac=*Creseis acicula*.

Aegean and 1.207 and 16.9 in the Levantine Basin, while the Adriatic Sea (1.701 and 53) and the Black Sea (10.000 and 257) are richer areas. Detailed data on the standing stocks for the whole Mediterranean Basin are given by Scotto di Carlo (1983).

Due to pronounced seasonal cycles of the most important zooplankton species (mainly due to temperature fluctuations) strong quantitative differences are noted in the zooplankton abundance from one season to another. A principal maximum is formed in spring, usually accompanied by a less important autumn or early winter peak. In Elefsis Bay (Saronicos gulf, Greece) where pollution conditions prevail the zooplankton annual cycle shows remarkable modifications.

Reports of the Dana expedition (1908-10) revealed a sharp decrease in zooplankton variety from west to east. The same view was supported recently by Godeaux (1974) and Alvarino (1974) based on the study of thaliaceans and siphonophores. However data from other planktonic taxa are not in accordance with this view and prove that the plankton of the Eastern Mediterranean is well diversified. For example 132 copepod species have been found in the egyptian coasts, 141 in the gulf of Sirte and the Levantine Basin (Delalo, 1966), 120 in the Aegean Sea and 109 at the coasts of Lebanon.

The majority of zooplankton species are small-sized atlanto-mediterranean forms known from the Western Mediterranean, while some species have migrated from the Black Sea and, especially, from the Red Sea. Some species can be characterized as Red Sea immigrants and are confined to the Eastern Mediterranean.

By far, the most important component of the zooplankton communities is the copepods (mainly calanoids) which form the main bulk (65 to 95%) of the zooplankton. The second planktonic group, in numerical importance, are the cladocerans which in some cases (summer, protected, or eutrophicated environments) become more numerous than copepods. Appendicularians and chaetognaths are, usually, the most numerous groups of the secondary constituents of zooplankton, while pteropods in offshore waters and benthic larvae in neritic environments attain also quantitative importance.

Although copepods display a noticeable variety of species, a restricted number of species are dominant forms. This group of species (Table 2) remains practically the same throughout most of the studied area (Levantine Basin, south Aegean) and differs slightly from that of the Western Mediterranean.

The extreme environmental conditions which characterize the Eastern Mediterranean (high temperatures and salinity) seem to favour the proliferation of some planktonic forms such as Appendicularia sicula, Creseis acicula, Cymbulia peroni, Centropages violaceus, C. kroÿeri. On the contrary, some other species (Mecynocera clausi, Oithona plumifera, Clausocalanus arcuicornis, C. furcatus) seem to undergo a numerical reduction compared with their extreme abundance in the Western Basin. Characteristic also is the disappearance of Centropages typicus, a form very abundant in the Western Mediterranean, the north Aegean and the Saronicos gulf, from the Levantine Basin.

Cladocerans present an irregular occurrence. The high salinity of the area does not favour their proliferation. Only in coastal waters, and particularly in enclosed protected environments (e.g. Evoikos gulf) this group forms dense populations. Some cladocerans (Penilia avirostris, Evadne tergestina, E. nordmani) were found in large numbers in polluted areas. Penilia avirostris was practically absent from the Levantine Basin and the south Aegean.

Many copepod species among those included in the list of Furnestin (1968) as Atlantic current indicators are found in the Levantine Basin and the Aegean Sea. The use of most of them as Atlantic current indicators is denied by the Eastern Mediterranean researchers. It seems that most species transported by the Atlantic current have gradually adapted to the Mediterranean environment and are settled in this region. However the simultaneous transportation of individuals of these species by the Atlantic current is not to be excluded.

Strong differences have been observed between the plankton fauna of the two Basins of the Aegean Sea. The north Aegean presents several peculiarities and a tendency of ecological isolation (e.g. the extreme abundance of Temora stylifera and the high numbers of Corycaeus giesbrechti and Candacia aethiopica and the reduced numbers of Paracalanus parvus, Oithona plumifera and Mecynocera clausi). The fauna of the south Aegean resembles that of the Western and Eastern Mediterranean and often presents a transitional character between the fauna of the north Aegean and that of the rest of the Mediterranean.

With the construction of the Aswan High Dam, fresh water and nutrients from the Nile are no longer emptied into the mediterranean coast of Egypt. An immediate and important decrease of the productivity in Egyptian waters has been observed, but, at least, a partial recovery is expected. Changes in the seasonal cycle of zooplankton were also felt after the construction of the Aswan Dam, on the coast of Lebanon: the autumn bloom due formerly to the Nile flood is no more formed; the same is true for the swarms of Podon polyphemoides.

The increase of salinity of mediterranean waters after the construction of the High Dam has reduced the environmental differences between the Red Sea and the Mediterranean facilitating thus the exchange of living forms. Although there exist recent references of plankton forms being transported, a systematic investigation would certainly increase considerably the catalog of Red Sea species in the Mediterranean. In any case the numbers of planktonic immigrants seem to be significantly less than that of other groups such as fishes.

The construction of dams in various Black Sea rivers has changed the hydrological conditions in the Black Sea. A decrease in the velocity of the superficial current going from the Black Sea to the Mediterranean is noted, accompanied by an increase of the subsuperficial current going from the Aegean to the Black Sea. Thus, a progressive "mediterranisation" of the Black Sea is going on.

The effects of pollution in the Eastern Mediterranean are mainly felt at the coastal waters in the vicinity of the big cities. A typical case of mainly organic pollution is noted in Elefsis Bay, Saronicos gulf. An important increase of plankton biomass is observed accompanied by a reduction of species variety. Some, few, forms (Acartia clausi, Oithona nana, Penilia avirostris) clearly predominate and are found in very high numbers and form the almost exclusive component of the zooplankton community. In addition some planktonic animals belonging to Elefsis Bay populations present significant differentiations in various physiological processes and higher resistance to various heavy metals.

SUMMARY

Although the Eastern Mediterranean is considered to be a very poor area, a comparison of quantitative plankton data proves that similar zooplanktonic biomass values are found in the Eastern and the Western Basin of the Mediterranean. Furthermore no significant differences have been noted in the variety of forms for the most important zooplankton groups between the Eastern and Western Mediterranean.

The majority of zooplankton species are atlantomediterranean forms, while some species have migrated from the Red Sea and are found in the Levantine Basin and the south Aegean.

The most important component of zooplanktonic communities are the copepods, followed by the cladocerans, while appendicularians, chaetognaths, benthic larvae (in neritic areas) and pteropods (in offshore waters) are the most important of the secondary constituents.

Some planktonic forms favoured by the environmental conditions of the area are found in increased numbers compared with their abundance in the Western Basin, while others are found in decreased numbers.

The north Aegean Sea presents some peculiarities in its planktonic fauna and a tendency to ecological isolation while the south Aegean usually presents the character of a transitional area.

The construction of the Aswan High Dam has caused a drop in the productivity of the egyptian Mediterranean waters and a change of zooplankton seasonal cycles was felt as far away as the costs of Lebanon. The construction of the Aswan Dam has also facilitated the migration of Red Sea forms to the Mediterranean. The construction of dams in many Black Sea rivers has provoked a change in the hydrological conditions favouring the process of "mediterranisation" of the Black Sea.

The effects of pollution in the Western Mediterranean are mainly felt in the coastal waters in the vicinity of the big cities. Apart from changes in the quantitative and qualitative composition of zooplankton, pollution causes also some physiological differentiation in the zooplanktonic animals.

REFERENCES

Aboul-Dahal, O., and Halim, Y., 1980, Oil pollution of the marine environment in the areas of Alexandria, Ves Journées Etud. Pollutions Cagliari, C.I.E.S.M.:201.

Alvarino, A., 1974, Distribution of siphonophores in the regions adjacent to the Suez and Panama Canals, Fish. Bull., 72:527.

Berdugo, V., 1968, Sur la présence dans la Mediterranée orientale de deux espèces du genre Calanopia (Copepoda, Calanoida), Rapp.Comm.int. Mer Medit., 19, 3:445.

Berdugo, V., 1974, On the occurrence of Acartia centrura (Copepoda, Calanoida) in the neritic waters of the Eastern Mediterranean, Rapp. Comm. int. Mer Medit., 22:85.
Bogdanova, A.K., 1973, Hydrological conditions of penetration of Mediterranean Sea organisms into the Black Sea, Rapp. Comm. int. Mer Medit., 21:507.
Casanova, J.P., 1973, Pénétration du copèpode. Euchaeta concinna Dana en Méditerranée orientale par le canal de Suez, Rapp. Comm. int. Mer Medit., 21:513.
Coachman, L.K., Hopkins, T.S., and Dugdale, R., 1976, Water masses of the Saronicos gulf in winter, Acta Adriatica, XVIII:129.
Delalo, E.P., 1966, Le zooplancton de la partie est de la Méditerranée (mers Levantine et Syrte) in: "Recherches sur le plancton des mers du sud," Ed. Nauka.
Demir, M., 1959, Pontellidae and Parapontellidae (Copepoda) from the Southern Black, Marmara, and North East Aegean Sea, Publ. Hydrobiol. Res. Inst. Univ. Istambul, Ser. B, IV:17.
Dowidar, N.M., 1965, Distribution and ecology of the neritic plankton in the region of Alexandria, Egypt, Ph.D. Thesis, Alexandria University.
Dowidar, N.M., 1974, Tintinnids from the Suez Canal, Rapp. Comm. int. Mer Medit. 22:123.
Dowidar, N.M., and El-Maghraby, A.M., 1971a, The neritic zooplankton of the SE Mediterranean at Alexandria. I. Distribution and ecology of the zooplankton organisms with special reference to Copepoda, Bull. Inst. Oceanogr. Fish, I:227.
Dowidar, N.M., and El-Maghraby, A.M., 1971b, The neritic zooplankton of the southeastern Mediterranean at Alexandria. II. Considerations of the total zooplankton community, Bull. Inst. Oceanogr. Fish, I:277.
El-Hehyawi, M.L., 1978, New data on the distribution of pollutants and their effect on some hyponeuston constituents in the S.E. Mediterranean, IVes Journées Etud. Pollutions, Antalya, C.I.E.S.M.:301.
El-Hehyawi, M.L., and El-Tabbakh, T.I., 1980, The effects of pollution upon the apparent oxygen utilisation, the N/P ratios and the fish production in the SE Mediterranean waters adjacent to Alexandria, Ves Journées Etud. Pollutions C.I.E.S.M. Cagliari:1035.
El-Maghraby, A.M., and Dowidar, N.M., 1973, Observations on the zooplankton community in the Egyptian Mediterranean waters, Rapp. Comm. int. Mer Medit., 21:527.
El-Sharkawi, F., 1980, The relation between the state of pollution on Alexandria swimming beaches and the occurrence of typhoid among bathers, Ves Journées Etud. Pollutions, Cagliari, C.I.E.S.M.:189.
El-Sokkary, I.H., 1980, Mercury accumulation in fish from Mediterranean coastal area of Alexandria, Egypt, Ves Journées Etud. Pollutions C.I.E.S.M., Cagliari:493.

Emara, H., Halim, Y., and Morcos, S.A., 1973, Oxygen, phosphate and oxydizable organic matter in the Mediterranean waters along the Egyptian Coast, Rapp. Comm. int. Mer Medit., 21:345.

Fenaux, R., 1974, Sur les appendiculaires du golfe Saronique (Grece), Rapp. Comm. int. Mer Medit., 22:101.

Furnestin, M.L., 1960, Zooplancton du golfe du Lion et de la côte orientale de Corse, Revue Trav. Inst. (scient. tech). Pêch. marit. 24:153.

Furnestin, M.L., 1968, Le zooplancton de la Méditerranée (bassin occidentale) Essai de synthèse, J. Cons. Int. Explr. Mer, 32:25.

Furnestin, M.L., 1979, Aspects of the zoogeography of the Mediterranean plankton, in: "Zoogeography and diversity in plankton," S. van der Spoel and A.C. Pierrot-Bults, eds., Arnold, London.

Gaudy, R., (this book), Features and peculiarities of zooplankton communities from the western Mediterranean.

Gerges, M., 1976, The damming of the Nile river and its effects on the hydrographic conditions and circulation pattern in the southeastern Mediterranean and the Suez canal. Acta Adriatica, XVIII:179.

Ghirardelli, E., and Rottini, L., 1979, Ghétognathes récoltés dans la mer Egée occidentale et la mer Ionienne orientale, Rapp. Comm. int. Mer Medit. 25/26:153.

Godeaux, J., 1960, Tuniciers pélagiques du golfe d'Eylath, Bull. Sea Fish. Res. Stat. Haifa 29 (Contr. Knowledge Red Sea 18):9.

Grimanis, A., Vassilaki-Grimani, M., and Griggs, G.B., 1976, Pollution studies of trace elements in sediments from the upper Saronicos gulf, Greece, in: Proc. int. Conf. on modern trends in activation analysis," München, Sept. 1976.

Guergues, S.K., and Halim, Y., 1973, Chétognathes du plancton d'Alexandria. II. Un spécimen mûr de Sagitta neglecta Aida en Méditerranée, Rapp. Comm. int. Mer Medit., 21:497.

Halim, Y., 1960, Observations on the Nile bloom of phytoplankton in the Mediterranean, J. Cons., 26:57.

Halim, Y., 1970, Microplancton des eaux egyptiennes. III espèces Indopacifiques ou érythréennes a l'extreme nord du Canal de Suez, Journées Etud. Planctonol. Rapp. Comm. int. Mer Medit.: 57.

Halim, Y., 1976, Marine Biological studies in egyptian Mediterranean waters. A review, Acta Adriatica, XVIII:29.

Halim, Y., Aboul-Ezz, S., and Guergues, S.K., 1974, Appendiculaires de la mer d'Alexandrie et des eaux avoisinantes, Rapp. Comm. int. Mer Medit., 22:113.

Hardy, J.T., and Jubayli, Z., 1974, Phytoplankton standing crop and sewage nutrient enrichment along the central coast of Lebanon, IIes Journees Etud. Pollutions, Monaco, C.I.E.S.M.:149.

Hatzakis, N., 1974, Mysidacés trouvés en mer Egée (partie sud) de 1966-1971, Rapp. Comm. int. Mer Medit., 22:81.

Kimor, B., 1965, Fluctuations in the abundance and composition of the plankton communities off the Mediterranean coast of Israel in relation to the Sardinella aurita fishery during the years

1955-1960, Proc. Gen. Fish. Coun. Medit., 64:1.
Kimor, B., and Berdugo, V., 1967, Cruise to the Eastern Mediterranean, Cyprus 03 Plankton reports, Sea Fish. Res. St. Haifa, Bul. No. 45:6.
Kiortsis, V. and collaborators, 1969, Marine Fauna of the Aegean Sea. Surface zooplankton from the north Aegean Sea. Report of Naval Research, Oceanic Biology branch 1-82.
Komarovsky, B., 1953, A cladoceran from the plankton as a possible indicator for the presence of the Nile flood off the coasts of Israel, Nature, 171:937.
Komarovsky, B., 1959, A study of the food of Sardinella aurita off the Mediterranean coast of Israel during a peak season (May-June, 1958), Proc. Gen. Fish. Coun. Medit., 5:311.
Lakkis, S., 1971a, Contribution à l'étude du zooplancton des eaux libanaises, Mar. Biol. II, 138.
Lakkis, S., 1971b, Variations saisonnières en abondance des copépodes pélagiques dans le Bassin Levantine à Beyrouth, Thalassia Yugoslavica, 7:159.
Lakkis, S., 1973, Note préliminaire sur la présence et la répartition des copépodes dans les eaux superficielles libanaises, Rapp. Comm int. Mer Medit., 21:459.
Lakkis, S., 1976a, Consideration on the distribution of pelagic copepods in the Eastern Mediterranean off the coasts of Lebanon, Acta Adriatica, XVIII:39.
Lakkis, S., 1976b, Sur la présence dans les eaux Libanaises de quelques copépodes d'origine indo-pacifique, Rapp. Comm. int. Mer Medit., 23:83.
Lakkis, S., 1977, Les chétognathes des eaux libanaises. Observations faunistiques et écologiques, Rapp. Comm. int. Mer Medit. 24:137.
Lakkis, S., 1980, Le plancton des eaux libanaises (Méditerranée orientale). Caractéristiques biogeographiques, Journées Etud. System. Biogéogr. Cagliari, C.I.E.S.M.:59.
Lakkis, S., 1981, Les cladocères des eaux Libanaises. Observations faunistiques et écologiques. Rapp. Comm. int. Mer Médit., 27:155.
Lakkis, S., and Kouyoumjian, H., 1974, Observations sur la composition du zooplancton aux embouchures d'effluents urbains des eaux de Beyrouth, Rapp. Comm. int. Mer Medit., 22:107.
Lakkis, S., and Abboud, M., 1976, Zooplankton et pollution du secteur Libanais en Méditerranée orientale, Rapp. Comm. int. Mer Medit., 23:79.
Lakkis, S., and Zeidane, R. (in press)a, Diversité spécifique, dominance et successions écologiques chez les copépodes des eaux libanaises, Rapp. Comm. int. Mer Medit. Cannes, 1 82.
Lakkis, S., and Zeidane, R., (in press)b, Caractéristiques écologiques et dynamiques du zooplancton des eaux côtières Libanaises, Rapp. Comm. int. Mer Medit. Cannes, 1982.
Moraitou-Apostolopoulou, M., 1969, Sur la présence en mer Egée d'Arietellus pavoninus copépode pélagique cité pour la première

fois en Mediterranée. Biologia Gallo-Hellen., II:189.
Moraitou-Apostolopoulou, M., 1972, Occurrence and fluctuation of the pelagic copepods of the Aegean Sea with some notes on their ecology, Hell. Oceanology and Limnology, XI:325.
Moraitou-Apostolopoulou, M., 1973, The cladocerans of the Aegean Sea: Occurrence and seasonal variation, Mar. Biol., 20:137.
Moraitou-Apostolopoulou, M., 1974a, Caractères écologiques des copépodes de la mer Egée, Rapp. Comm. Int. Mer Medit., 22:89.
Moraitou-Apostolopoulou, M., 1974b, An ecological approach to the systematic study of planktonic copepods in a polluted area, Boll. Pesca, Piscic, Idrobiol., 29:29.
Moraitou-Apostolopoulou, M., 1974c, Répartition des cladocères en mer Egée, Rapp. Comm. Int. Mer Medit., 22:87.
Moraitou-Apostolopoulou, M., 1978a, Le zooplancton superficiel (0-100 cm) du golfe Saronique, Grèce: a) aspect general de la communauté zooplanctonique, Thalassographica, 2:155.
Moraitou-Apostolopoulou, M., 1978b, Contribution à la systématique et écologie du zooplancton de surface (0-100 cm) dans une zone polluée, Biol. Gallo-Hellen. VII:201.
Moraitou-Apostolopoulou, M., 1978c, Acute toxicity of copper to Acartia clausi (Copepoda, Calanoida), Mar. Poll. Bull., 9:278.
Moraitou-Apostolopoulou, M., 1981, The annual cycle of zooplankton in Elefsis bay (Greece), Rapp. Comm. Int. Mer Medit., 27:198.
Moraitou-Apostolopoulou, M., and Kiortsis, V., 1977, The Cladocerans of the north Evoikos gulf, Thalassographica 1:205.
Moraitou-Apostolopoulou, M., and Verriopoulos, G., 1979, Some effects of sublethal concentrations of copper on the marine copepod Acartia clausi (an experimental study), Mar. Poll. Bull. 10:88.
Moraitou-Apostolopoulou, M., Verriopoulos, G., and Palla, P., 1979a, Temperature and adaptation to pollution as factors influencing the acute toxicity of cadmium to the planktonic copepod Acartia clausi, Téthys, 9:97.
Moraitou-Apostolopoulou, M., Verriopoulos, G., and Lentzou, P., 1979b, Effects of sublethal concentrations of cadmium as possible indicators of cadmium pollution for two populations of Acartia clausi (Copepoda) living at two differently polluted areas, Bull. Environm. Contam. Toxicol., 23:642.
Moraitou-Apostolopoulou, M., and Verriopoulos, G., 1980, Feeding activity of three annual generations of unperturbed and pollution impacted Acartia clausi (Copepoda) populations in Saronicos gulf, Greece, Rev. Int. Oceanogr. Med., LV:29.
Moraitou-Apostolopoulou, M., and Verriopoulos, G., 1981a, The longevity of three annual generations of unperturbed and pollution impacted Acartia clausi (Copepoda) populations in Saronicos gulf (Greece), Hydrobiologia 77:9.
Moraitou-Apostolopoulou, M., and Verriopoulos, G., 1981b, Thermal tolerance of two populations of Acartia clausi (Copepoda) living at differently polluted areas, Hydrobiologia 77:3.
Moraitou-Apostolopoulou, M., and Vournazou, M., in press, A systematic and ecological study of planktonic copepods at the

north Evvoikos gulf, Boll. Pesca Piscic. Idrobiol.

Morcos, S.A., and El-Rayiss, O.A., 1973, The Levantine intermediate water, oxygen and nutrients off Alexandria, Thalassia Jugoslavica, 9:13.

Oren, O.H., and Komarovsky, B., 1961, The influence of the Nile flood on the shore waters of Israel, Rapp. Comm. int. Mer Medit., 16:655.

Pasteur, R., Berdugo, V., and Kimor, B., 1976, The abundance, composition and seasonal distribution of epizooplankton in coastal and offshore waters of the Eastern Mediterranean, Acta Adriatica, XVIII:53.

Pavlova, E.V., 1966, Composition et répartition du zooplancton en mer Egée, in: Recherches sur le plancton des mers du Sud, Ed. Nauka (in Russian).

Por, F.D., 1969, The Ganuellidae (Copepoda, Harpacticoida) in the waters around the Siani Peninsula and the problem of 'Lessepsian' migration of this family, Israel. J. Zool., 18:169.

Por, F.D., 1971, One hundred years of Suez Canal century of Lessepsian migration: Retrospect and view points, Syst. Zool., 20: 138.

Schmidt, H.E., 1972, The distribution of Hydroida (Hydrozoa-Coelenterata) in the Red Sea and the Eastern Mediterranean, 17 Congr. int. Zool. Monte Carlo:8.

Schmidt, H.E., 1973, Biogeographical problems of the Red Sea area exemplified by Hydroids, in: "The biology of the Indian Ocean," B. Zeitzschel and S.E. Gerlach, eds., Springer Verlag, Berlin.

Scotto di Carlo, B., and Ianora, A., 1983, Standing rocks and species composition of Mediterranean zooplankton, in: "Quantitative analysis and simulation of Mediterranean coastal Ecosystems: The Gulf of Naples, a case study," Unesco Reports in Marine Science, No. 2.

Scoulos, M., 1980, Dissolved and particulate zinc in a polluted Mediterranean Bay, Ves Journées Etud. Pollutions C.I.E.S.M., Cagliari:483.

Shmeleva, A.A., 1973, New data on planktonic fauna from the east of the Mediterranean Sea, Rapp. Comm. int. Mer Medit., 21:537.

Steinitz, H., 1968, A tentative list of immigrants via the Suez Canal, Israel J. Zool., 13:166.

Toma, S.A., Saal, M.A.H., Salama, M.S., and Halim, Y., 1980, The distribution of some adsorbed elements on the Nile continental shelf sediments, Ves Journées Etud. Pollutions, Cagliari C.I.E.S.M.:377.

Veini, M., and Kiortsis, V., 1974, Amphipodes et isopodes planctoniques de la mer Egee, Rapp. Comm. int. Mer Medit., 22:93.

Wahby, S.D., 1980, Pollution by petroleum hydrocarbons along Alexandria coasts, IVes Journées, Etud. Pollutions, Antalya, C.I.E.S.M.:93.

Yannopoulos, C., and Yannopoulos, A., 1976, Zooplankton biomass in the Saronicos gulf, winter 1972-73, Acta Adriatica, XVIII: 327.

Zafiropoulos, D., and Grimanis, A.P., 1976, Trace elements in Acartia clausi from Elefsis Bay of the upper Saronikos gulf, Greece, IIIes Journées Etud. Pollutions Split, C.I.E.S.M.:63.

Zaitsev, Y., 1971, "Marine neustonology," Transl. from Rus.: Israel program for scientific translations.

PROCESSES OF DIFFERENTIATION BETWEEN MEDITERRANEAN POPULATIONS OF THE SUPER-SPECIES *Tisbe clodiensis* Battaglia and Fava (1968) (COPEPODA)

Giancarlo Fava[1] and Bruno Battaglia[2]

[1]Istituto Biologia del Mare C.N.R., Venezia

[2]Istituto Biologia Animale, Università di Padova

INTRODUCTION

Evolutionary research deals mainly with the mechanisms that promote the formation of new species. This is not only because speciation is viewed as the central event of the whole evolutionary process, but also because it is the step which, under certain conditions, is more amenable to experiment.

General theories describing the mechanisms of speciation have been presented in recent years (Dobzhansky, 1970; Mayr, 1970; White, 1978) with the support of many examples taken from both animal and plant kingdoms. But there are a few experimental studies providing an analysis of the genetical and ecological factors active in the early stages of speciation.

The analysis of evolving species, if not the only level at which the evolutionary process can be fruitfully investigated (Abren, Grobois and Beardmore, 1982), might provide an answer to several unsolved questions, as pointed out also by Nevo (1982). At this stage it is possible to obtain a more precise estimate of the levels of genetic diversification, in relation to the influence of different ecological factors, which are involved in the process of formation of new species, and of the rate at which the phenomenon takes place. In studies on the genetic bases of speciation, marine organisms for reasons that have been recently discussed by Battaglia (1982), have received little attention in recent years; investigations have been made on copepods, isopods and polychaetes (Bocquet, 1976; Solignac, 1978; Menzies, 1972; Åkeson, 1978).

Among copepods, those belonging to the genus Tisbe have been studied for several years by our research group. These are benthic copepods living on algal substrates along coastal shallow waters. Several species penetrate estuaries and coastal brackish lagoons.

These organisms offer excellent opportunities for genetic and evolutionary research. They are widely distributed. Some species present naturally-occurring, visible polymorphisms under genetic control, which are involved in the determination of fitness. Finally, they can be easily grown under a variety of experimental conditions.

Two species, T. clodiensis and T. holothuriae, have been most extensively studied during recent years. Both possess a wide distribution and can be considered cosmopolitan. They occur together in many localities and are able to colonize brackish lagoons. But while T. clodiensis is present only in areas close to the sea, where short-term salinity fluctuations, although present, are rather reduced, T. holothuriae can be found in the inner parts of the lagoons and is common also in areas, such as estuaries and deltas, where the salinity reaches quite low values and where fluctuations of more than 10 units in a few hours can be observed (Fava and Volkmann, 1975; 1977; Fava, 1980). Thus, this species can occupy a variety of ecological niches and appears to be highly tolerant toward variable and extreme environments. Its distribution is practically continuous along the coasts, in contrast with the less tolerant T. clodiensis, whose distribution is rather discontinuous.

Both species have been found to be members of two distinct groups of sibling species, that is both come from ancestors which have undergone speciation events recently (Volkmann - Rocco, 1972; Volkmann-Rocco and Battaglia, 1972).

Several crosses have been carried out between various European populations of T. holothuriae and between European and North American populations. All the crosses proved to be perfectly fertile and in the following F1, clear signs of heterosis have been found (Battaglia and Volkmann-Rocco, 1973). This strongly suggests that the species is not undergoing any process of splitting into different incipient species, and that it is in a stage of "evolutionary equilibrium".

A similar series of laboratory crosses involving mainly mediterranean populations of T. clodiensis led to different results (Volkmann et al., 1978). In some cases perfect fertility was obtained. In others the production of adult offspring was dependent upon the geographic origin of the female and of the male, but in these instances hybrids showed a slower development, with a strong alteration of the sex ratio and a reduction of the number of offspring per female. The last kind of crosses gave no offspring, or dying larvae, in both directions.

The highest degrees of isolation have been shown by the Venice strains, when the male was used, and by Arcachon populations when the females were involved.

The conventional taxonomic characters did not show any obvious difference, also when completely intersterile populations were compared. A first study on the biochemical affinity showed that the degree of distance is intermediate between those usually shown by geographic races and by sibling species (Bisol, 1976).

Thus *Tisbe clodiensis* appears to be a superspecies representing a cluster of semispecies in *statu nascendi* several of which being present in the Mediterranean.

The investigations discussed so far indicate also that the role played by geographic distance was not determinant in promoting differentiation. This conclusion was further supported by recent crosses between the populations of Grado and Trieste (Fava, unpublished data), which are geographically very close but reproductively isolated.

The animals from Trieste showed some visible difference in the polymorphism which characterizes all the populations of the superspecies. It is a sex-limited polymorphism, visible in the female sex only, which consists in the presence or absence of a red-violet pigment in some segments of the second, third, and fourth pair of thoracic legs, the character "presence of pigment", *P*, being determined by a dominant allele. The pigment in females from Trieste, in about five per cent of the cases, showed the pattern which is typical for all other populations; but in the most frequent morph it had a more limited distribution. The same variant was found by Fava in the eastern coast of the Adriatic Sea, at Split.

Tisbe clodiensis may then be considered as a species in a phase of "evolutionary instability" and which is undergoing a process of incipient speciation.

As previously stated, this is an evolutionary situation that provides the possibility to test experimentally the relevance of various genotype-environment interactions in promoting the differentiation between geographic populations, which might eventually end in the formation of new species.

A series of studies have been planned considering several genetically controlled characters, biochemical, morphometrical physiological and colour polymorphisms, whose behaviour will be compared in populations presenting increasing levels of differentiation as to their ecological requirements and reproductive isolation.

The correspondent patterns shown by *T. holothuriae* will help to reveal if there is an identity, or not, between the processes of differentiation between geographic populations due to adaptation to local environments, and those promoting speciation.

The attention was focused, at first, on comparing closely-related populations, coming from geographically distinct but ecologically similar environments T. clodiensis samples were collected from comparable areas in the lagoons of Grado and Venezia, which in Roman times were part of a single lagoon from Grado to the Po river.

These two populations are reproductively compatible and particularly suitable for the study of the very early stages of differentiation.

Three kinds of analyses have been performed: 1) the behaviour of the visible polymorphism in experimental populations; 2) the effect of reduced salinities on some life-history traits; 3) the effect of reduced salinity on morphology.

EXPERIMENTAL POPULATIONS

Three kinds of populations have been analyzed:

A. Populations established with animals coming from the lagoon of Venezia.

B. Populations obtained from animals collected in the lagoon of Grado.

C. Populations whose founder animals came from crosses between strains from the two lagoons.

In all experiments 200 females and 100 males, both heterozygous, that is with $f_o(\underline{p}) = f_o(\underline{P}) = 0.5$, where introduced into bowls containing 2,000 ml of sea water with a salinity ranging between 34 and 36 per 1000 at a temperature of 18°C. The medium was renewed every two generations and samples of adult females from each population were counted and scored for colour polymorphism at intervals of about 24 days (one generation).

One year after, that is about fourteen generations after, the frequency of the recessive allele, $f(\underline{p})$ in the populations A, reached the value of 0.70; the paths followed by the populations suggested that eventually an equilibrium value would have been reached between 0.7 and 0.8, as previously obtained in similar experiments.

In the populations B an equilibrium frequency of $f(\underline{p}) = 0.40$ was reached within two-three generations. A similar behaviour was observed for populations C (Fava, in preparation).

The maintenance of equilibrium frequencies which are different from the starting ones suggests that the polymorphism is balanced in both populations, but the observed differences indicate that the adaptive values of the involved genotypes differ in the two strains.

From simple probabilistic considerations, it follows that Grado and Venezia populations differ in at least two loci. Such a reduced figure would be reliable only under the very restrictive assumptions that: 1) different alleles for both loci are nearly fixed in the two populations; 2) Grado alleles are dominant, and 3) there is no selection against Grado alleles.

Even if not unrealistic, these conditions appear rather unlikely, and probably the differences involve a higher number of loci.

Whatever the actual situation would be, the results strongly indicate that the structure of the gene sequence, probably representing a supergene, involved in the polychromatism of T. clodiensis is different in the two lagoon populations, even if no evident phenotypical modifications are present.

A further difference between lagoons is that the gene frequencies appear "fixed" at two quite separate values: in both cases only minor fluctuations have been observed, so that the polymorphism appears rigid with respect to environmental fluctuations.

EFFECTS OF LOW SALINITY ON LIFE HISTORY

As reported in the introductory remarks, the ecological observations made on several Adriatic lagoons and marine areas indicated for T. clodiensis a reduced tolerance of extreme ecological conditions, in particular of low levels of salinities with highly variable regimes.

This suggested the opportunity to conduct an experimental analysis of the effects of reduced salinities under laboratory-controlled conditions.

A first test was made with the experimental populations described above. Samples taken from the three types of populations were put into low salinity water, but no allele frequency changes were observed within four generations (Fava and Cervelli: unpublished).Even if not conclusive, this outcome indicates that the adaptedness of the polymorphism is not relevant with respect to salinity, and that the response to this environmental parameter is probably controlled by a different set of genes.

Another series of experiments was carried on to study the effects of reduced salinities on several life history parameters.

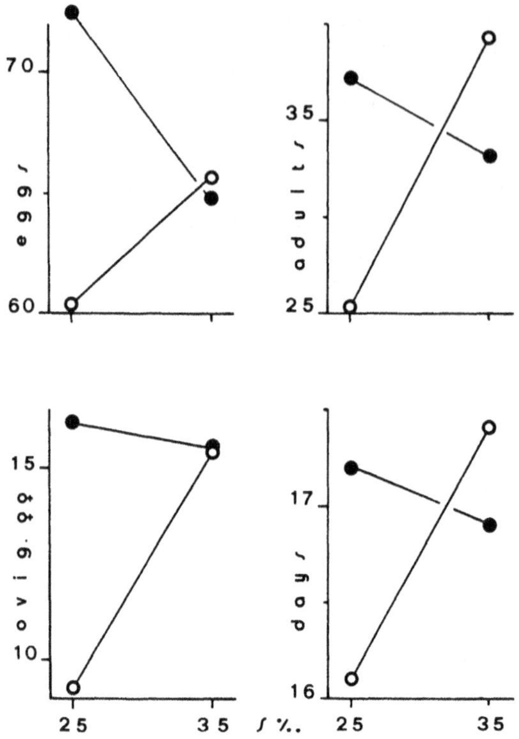

Fig. 1. Differential effect of salinity in four life history parameters. o–o Grado, ●–● Venezia populations. Salinities are in abscissa.

They were planned to test three salinities, that is 35, 25 and 20 per 1000, but the last value could not be further investigated because both T. clodiensis strains, although surviving, failed to reproduce at this salinity.

This confirmed the view that the species is less tolerant of low salinity than T. holothuriae, which was found to reproduce at salinities lower than 20 per 1000, even in populations coming from relatively stable areas like the gulf of Marseille (Gaudy et al., 1982).

Both strains were maintained under identical conditions in the laboratory for more than seven years, and were equally preadapted for one generation to the experimental salinity. Thus, differences in the behaviour of the two strains could be viewed mainly as reflecting differences at a genetic level.

Among the biological parameters which have been considered, four gave significantly different results depending upon their geographic origin (Fig. 1).

The average number of eggs produced per female, the number of adults per female and the number of ovigerous females per family decreased from 35 to 25 per 1000 salinity in the animals from Grado, while they showed more or less pronounced increment in those from Venezia. The developmental time became shorter for the former while it did not give significantly different values for the Venezia strain.

Further experiments, involving selection, confirmed the possibility that this different behaviour is due to differences in the genetic structure of the two populations (Fava et al., in press).

Some preliminary results are available for the hybridgs, which appeared intermediate for some biological parameters, or more like the one or the other parent, for others; a similar behaviour indicates that the differences observed should be ascribed to the effect of several different loci.

MORPHOLOGICAL EFFECTS OF LOW SALINITY

A third aspect that has been taken into account when comparing the populations of T. clodiensis from the two lagoons was the impact of low salinity on a series of morphological characters.

Most probably, many of the genes controlling the variability of quantitative characters are different from those involved in the processes discussed above, being related to biochemical and physiological paths which are not the same. Thus, as a first approximation, they may provide a third, independent estimate of the presence of a genetic differentiation between Grado and Venezia populations.

The animals to be measured came from the adults obtained with the experiments described in the previous section. Measurements have been limited to the females, which, because of their larger size, are easier to handle. In fact, in these organisms size imposes some limitations in the study of quantitative characters, and the choice is necessarily restricted. The majority of estimates were confined to the lengths of some body parts, that is segments of the first antenna, segments of the endopodites of the first and second pairs of thoracic legs and abdominal somites, whose ends can be determined with the required precision, and to counts of the number of spinelets in the second pair of thoracic legs. Total body lengths and volumes cannot be easily estimated.

Females from Venezia showed no differences when grown at 35 or 25 per 1000 salinity, except for the length of the second segment of the antenna, whose size increased at the lower salinity.

The situation was quite different for the females coming from the lagoon of Grado. For all the measurements reported in Fig. 2,

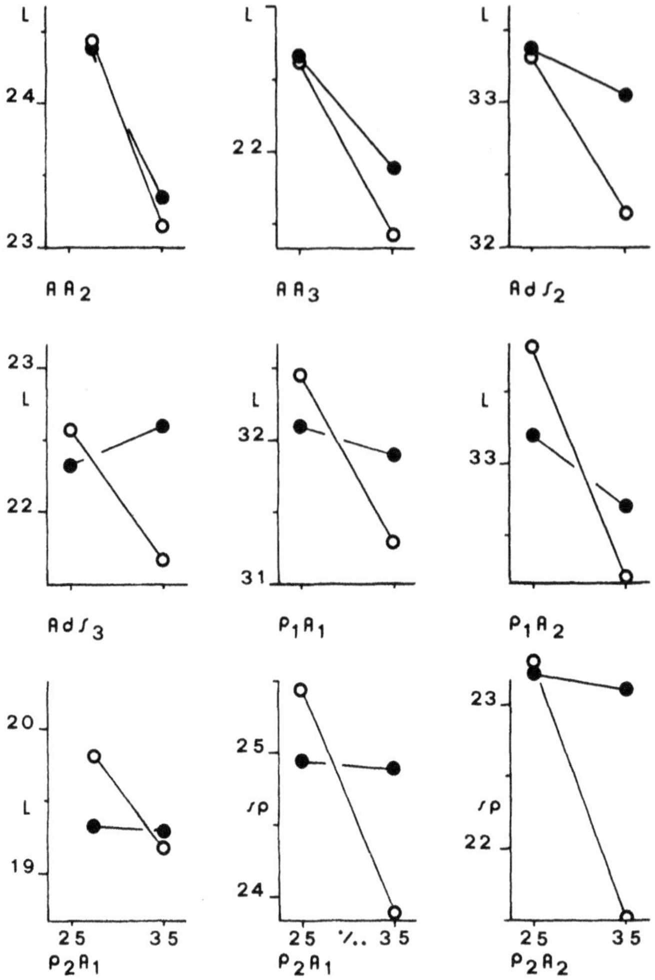

Fig. 2. Morphological effects of salinity. Measures of length (l) are in arbitrary units. (sp.) = number of spines. AA_2, AA_3: 2nd and 3rd segment antenna. AdS_2, AdS_3: 2nd and 3rd abdominal somite. P_1A_2, A_1A_2: 1st and 2nd segment in the first pair of thoracic legs. P_2A_1, P_2A_1: 1st and 2nd segment in the second thoracic legs. Salinities are in abscissa. o—o Grado, ●—● Venezia populations.

values were significantly higher at the salinity of 25 per 1000.

The fourth segment of the antenna and the fourth and fifth abdominal somite did not show any appreciable variation between 35 and 25 per 1000. A more detailed analysis of these data will be given elsewhere (Fava et al., in preparation).

Thus, the morphological effects of low salinity were also rather different in the two populations. It is not possible, at the present stage of the research, to assess precisely how much genetic differentiation is demonstrated by such results. Heritability estimates for the analyzed characters are not yet available for T. clodiensis. The values obtained for T. holothuriae (Fava and Lazzaretto, in preparation), ranging from .3 to .6, suggest that a great amount of variability can be genetically controlled, but this aspect needs further investigations.

Nevertheless, the existence of differences for many characters, in strains which underwent the same treatment for many generations, and which were exposed to the same tests, strongly suggests the existence of a diversification at the genetic level.

DISCUSSION

Field observations showed that Tisbe clodiensis is absent from areas with highly variable regimes of low salinity, such as the inner part of the lagoon of Venezia and the embayments in the delta of the Po river. A similar discontinuity in the distribution of this species was observed in marine areas close to Porto Garibaldi and Porto Corsini, south of the Po, which are strongly influenced by this river.

The results of some of the experiments reported above provide further support to the hypothesis that T. clodiensis is a species particularly sensitive to salinity variations, and that this environmental variable can be effective in promoting the reduction of the gene flow between natural populations, facilitating diversification.

The behaviour of T. holothuriae is rather different; each population can cope with a wide range of salinities, as shown by Gaudy et al. (1982).

Some physiological races, characterized by a different tolerance of somotic shocks, have been found (Battaglia, 1967). But, unlike T. clodiensis, they came from distant geographic areas with different ecological characteristics. Moreover the tolerance of low salinities appears higher in T. holothuriae than in T. clodiensis.

The different evolutionary paths shown by these two species may partly depend upon these different tolerances of this important ecological parameter.

The two populations of T. clodiensis apparently did not differ much in their ability to survive at a given low salinity, and both failed to reproduce at 20 per 1000. Nevertheless they showed a differential effect of reduced salinity on some population parameters directly related to fitness. The dissimilar response at the morphological level probably has no direct relation to fitness *per se*, rather an indirect one. The higher or lower ability to control body size might reflect a differential ability to maintain the body "milieu" at the optimal conditions.

These differences indicate that the genes controlling the biochemical and physiological mechanisms of adaptation to varying salinities have different structures in the two populations.

Considering the short geographic distance between the lagoons of Grado and Venezia and the fact that their salinity regimes do not differ strongly, the results obtained suggest the possibility that substantial changes in the genetic structure of natural populations of *Tisbe clodiensis* may be induced by relatively small differences in the ecological factors they experience.

But salinity is evidently not the only parameter that promoted the differentiation between the two lagoon populations. The group of genes that are directly involved in the determination of fitness, and whose action can be revealed thanks to the marker locus that controls the colour polymorphism, apparently is not affected by salinity variations. Nevertheless, as was shown by studying the experimental populations, they differ, and the fitness of colour phenotypes from Grado is quite far from that of the corresponding phenotypes from Venezia. Therefore other environmental parameters, which at present cannot be identified, are active in producing genetic differentiation between natural populations of *Tisbe clodiensis*.

Battaglia (1982) remarked, discussing the results of crossbreeding experiments, that Arcachon and Venezia populations appeared to be the more isolated populations, and that they came from similar brackish lagoon habitats, located in comparable alluvial basins. This suggested the hypothesis that "higher levels of differentiation and reproductive incompatibility are features of common occurrence in populations of *Tisbe clodiensis* inhabiting these areas".

The results obtained comparing the strains from the Grado and Venezia lagoons seem to be in agreement with this hypothesis. In this case there are no signs of reproductive incompatibility, probably due to the fact that gene flow between the two populations has been reduced only in recent times, with the fragmentation of the ancestral lagoon.

Further evidence that ecological factors, more than distance *per se*, are important in promoting differentiation and isolation is given by the population of *T. clodiensis* from Trieste. Its colour morphs are different and it is reproductively isolated from that of Grado, although geographically very close, being less than 30 kilometers apart. The similar strain found in Split suggests that it is distributed all along the eastern coast of Adriatic Sea. This coast is ecologically very different from the western one, in its physicochemical and biological parameters as well.

The populations of Grado and Venezia represent a very early stage in the process of evolutionary differentiation. The fact that they present a substantial amount of genetic diversification, in spite of the similarity of their environments, and the quite recent partial separation, indicate that T. clodiensis is a species whose populations possess a genetic structure conferring a rather limited flexibility with respect to the variations of the environment.

Thus, the average genotype seems to confer on its carriers a higher specialization to fit local conditions. The species appears able to colonize marginal areas, such as lagoons, mainly through a very few colonizers with a preadapted genotype and/or throughout a rearrangement of the genetic structure. In both cases the resulting subpopulations appear gentically differentiated, more specialized and particularly adapted to the "average environment", and evolving rigid polymorphisms.

On the other hand, T. holothuriae appears to possess rather an "all purpose" genetic constitution, conferring on its carriers a high individual flexibility. So this species can cope with a variety of environments without undergoing a substantial process of genetic reorganization.

These differences in the genetic structure of the two species which determine two distinct kinds of genotype-environment interactions may provide an explanation for the opposite "evolutionary statuses" which are shown by these two species of Tisbe.

A genetic mechanism that apparently fits the present data is that discussed by Carson (1959). But, before going on in suggesting a detailed description of the most likely mechanisms that determine the divergences among populations in T. clodiensis, it is necessary to obtain a better knowledge of the levels of genetic variability characterizing various natural populations, and to include comparisons between strains representing deeper levels of differentiation, considering also reproductively-isolated populations, such as those from Grado and Trieste, and populations showing intermediate levels of compatibility.

SUMMARY

A relatively small number of studies on speciation has been performed on marine animals, and little is known about the genotype-environment interactions which may promote speciation in the sea.

Good opportunities for this kind of studies are provided by some species of marine benthic copepods of the genus Tisbe, in particular *T. holothuriae* Humes and the polymorphic *T. clodiensis*. Both are cosmopolitan, with a wide ecological distribution, and can be reared under a variety of experimental conditions. The first species is particularly tolerant of highly variable regimes of low salinity, so that its distribution along the coasts is almost continuous, but the second presents several discontinuities close to estuarine areas.

Crosses between North American and European populations were perfectly fertile for *T. holothuriae*, but several degrees of incompatibility were found in *T. clodiensis*, which appears to be subdivided into a number of semispecies.

The very early stages of differentiation have been studied in *T. clodiensis* utilizing interfertile populations from the lagoons of Grado and Venezia.

Comparisons between experimental populations established with animals from the two lagoons and with their hybrids revealed the existence of a differentiation in the genes involved in the polymorphism of this species and in their control of fitness.

The biological effects of low salinities also showed a different behaviour in the two strains, indicating the existence of a certain amount of differentiation in the genes controlling the physiological responses to this environmental factor. Low salinity has also been proved to affect differentially several morphological characters.

The considerable level of genetic differentiation that can be inferred for these populations of *T. clodiensis*, which are very similar from an ecological point of view, suggests that this superspecies, unlike *T. holothuriae*, possesses an "average" genetic structure which confers poor phenotypic flexibility.

The colonization of marginal environments, such as lagoons, may take place only through an *ad hoc* rearrangement of the genetic structure. The resulting subpopulations are then genetically differentiated, and more adapted to the "average environment" probably by evolving rigid polymorphisms.

REFERENCES

Abreu-Grobois, F.A., and Beardmore, J.A., 1982, Genetic differentiation and speciation in the brine shrimp *Artemia*, *in*:

"Mechanisms of speciation," C. Barigozzi, ed., Liss. Inc., New York.
Åkesson, B., 1978, A new Ophryotrocha species of the labronica group (Polychaeta, Dorvilleidae) revealed in crossbreeding experiments, in: "Marine Organisms: Genetics, Ecology and Evolution," B. Battaglia and J.A. Beardmore, eds., Plenum Pb. Co., New York.
Battaglia, B., 1967, Genetic aspects of benthic ecology in brackish waters, in: "Estuaries," American Association for the Advancement of Science.
Battaglia, B., 1982, Genetic variation and speciation events in marine copepods, in: "Mechanisms of speciation," C. Barigozzi, ed., Liss. Inc., New York.
Battaglia, B., and Volkmann-Rocco, B., 1973, Geographic and reproductive isolation in the marine harpacticoid copepod Tisbe, Mar. Biol., 19:156.
Bisol, P.M., 1976, Polimorfismi enzimatici ed affinitā tassonomiche in Tisbe (Copepoda, Harpacticoida), Accad. Naz. Lincei Rc., 60:864.
Bocquet, Ch., 1976, Les problēmes de l'espēce chez quelques Crustacēs: le genre Tisbe (Copēpodes, harpacticoides) et le complexe Jaera albifrons (Isopodes, Asellotes), in: "Les problēmes de l'espēce dans le règne animal," Ch. Bocquet, J. Gēnēramont and M. Lamotte eds., Sociētē Zoologique de France, Mémoire N. 38. Paris.
Carson, H.L., 1959, Genetic conditions which promote or retard the formation of species, Cold Spr. Harb. Symp. Quant. Biol., 24:87.
Dobzhansky, Th., 1970, Genetics of the evolutionary process. Columbia Univ. Press, New York.
Fava, G., 1980, Osservazioni ecologiche sulle specie di Tisbe (Copepoda, Harpacticoida) nella laguna di Marano, Atti. Ist. veneto Sci.,139:43.
Fava, G., (in preparation), Genetic Differentiation in experimental populations of the copepod Tisbe clodiensis from two Adriatic lagoons.
Fava, G., and Volkmann, B., 1975, Tisbe (Copepoda, Harpacticoida) species from the lagoon of Venice. I. Seasonal fluctuations and Ecology, Mar. Biol., 30:151.
Fava, G., and Volkmann, B., 1977, Tisbe (Copepoda, Harpacticoida) species from the estuarine embayment of Scardovari, Arch. Oceanogr. Limnol., 19:55.
Fava, G., and Lazzaretto, I., (in preparation), Ereditabilitā di alcuni caratteri quatitativi in una popolazione di Tisbe holothuriae (Copepoda, Harpacticoida).
Fava, G., Lazzaretto, I., and Martini, E., (in press), Effeti della riduzione di salinità in due diverse popolazioni lagunari di Tisbe clodiensis (Copepoda, Harpacticoida), Atti Ist. veneto Sci.

Fava, G., Lazzaretto, I., and Martini, E., (in preparation), Effeti morfologici della riduzione di salinitã in due popolazioni lagunari del copepode Tisbe clodiensis.

Gaudy, R., Guerin, J.P., and Moraitou-Apostolopoulou, M., 1982, Effect of temperature and salinity on the population dynamics of Tisbe holothuriae Humes (Copepoda, Harpacticoida) fed on two different diets, J. Exp. Mar. Biol. Ecol., 57:257.

Mayr, E., 1970, "Populations, Species and Evolution," Harvard University Press, Cambridge.

Menzies, L., 1972, Experimental interbreeding between geographically separated populations of the marine wood-boring isopod Limnaria tripunctata with preliminary indications of hybrid vigor, Mar. Biol., 17:149.

Nevo, E., 1982, Speciation in subterranean mammals, in: "Mechanisms of speciation," C. Barigozzi, ed., Liss Inc., New York.

Solignac, M., 1978, Genetics of ethological isolating mechanisms in the species complex Jaera albrifrons (Crustacea, Isopoda), in: "Marine Organisms: Genetics, Ecology and Evolution," B. Battaglia and J.A. Beardmore, eds., Plenum Pb. Corp., New York.

Volkmann-Rocco, B., 1972, Tisbe battagliai n.sp. a sibling species of Tisbe holothuriae Humes (Copepoda, Harpacticoida), Arch. Oceanogr. Limnol., 17:259.

Volkmann-Rocco, B., and Battaglia, B., 1972, A new case of sibling species in the genus Tisbe (Copepoda, Harpacticoida), in: "Vth Eur. Mar. Biol. Symp.," B. Battaglia, ed., Piccin, Padova.

Volkmann, B., Battaglia, B., and Varotto, V., 1978, A study of reproductive isolation within the super species Tisbe clodiensis (Copepoda, Harpacticoida), in: "Marine Organisms: Genetics, Ecology and Evolution," B. Battaglia and J.A. Beardmore, eds., Plenum Pb. Corp., New York.

White, M.J.D., 1978, "Modes of Speciation," W.H. Freeman and Company, San Francisco.

EVOLUTIONARY AND ZOOGEOGRAPHICAL REMARKS ON THE MEDITERRANEAN FAUNA OF BRACHYURAN CRABS

Carlos Almaça

Laboratório de Zoologia and Centro de Fauna Portuguesa (INIC) - Faculdade de Ciências, Rua da Escola Politécnica, 1294 Lisboa Codex, Portugal

INTRODUCTION

The crab fauna of the Mediterranean Sea appears to be the best-known in the world (Manning, 1981). Manning and Holthuis (1981) report more than 120 species from the Mediterranean, which means that about 40% of the East Atlantic known species are represented there. In other Atlantic areas the number of species (percentages relative to the known East Atlantic species) are as follows (see Manning and Holthuis, 1981 and Christiansen, 1982): tropical West Africa, 215 species (≈70%), Iberian Peninsula (Atlantic and Mediterranean coasts), 112 species (≈37%), northeast Atlantic north of British coasts, 60 species (≈20%), and Norwegian coast, 31 species (≈10%). In spite of the fact that these areas are different in size, the general picture of the above-mentioned numbers and percentages is a decreasing diversity from south to north.

The present Mediterranean crab fauna is, as is the fauna of any extended area, the result of different genetic mechanisms interacting with the geological evolution of the Mediterranean. Both genetic mechanisms and geological evolution will be considered although the last one very briefly, and only as far as the main zoogeographical consequences of the Mediterranean geological history are concerned.

According to a widespread classification of the mechanisms of faunal genesis (see, for instance, Banarescu, 1975), I think that the Mediterranean crab fauna has originated by: (a) autochthonous adaptive radiation, (b) repeated, or continuous, multiple colonization, and (c) adaptation to specialized habitats. The two former types of origin concern geographical areas, while the last one refers to particular environments. These mechanisms will subsequently be detailed and exemplified. First, however, some data on the geological evolution of the Mediterranean and a general picture of the present Mediterranean crab fauna will be presented.

MAJOR ZOOGEOGRAPHICAL EVENTS IN THE MEDITERRANEAN AREA

As a consequence of the extensive regression in the Oligocene and Lower Miocene, the connexions between the Mediterranean and the Indian Ocean were severed during the Miocene (Albrecht, 1982; Hsü, 1974; Vermeij, 1978). It seems that the communication to the Red Sea was lost before that to the Persian Gulf, since this last connexion, the Syrian Iranian Sea, or Fars Sea, retreated and was desiccated also as a consequence of the uplift of the Anatolian plateau, during the Pliocene (Banister, 1980).

From there on, and until a little more than a century ago, the only oceanic connexion of the Mediterranean Sea has been to the Atlantic Ocean. It seems, however, that during the Late Miocene Messinian stage, 7-5 million years ago, an important regression isolated the Mediterranean from the Atlantic, leading to the desiccation of the Mediterranean and to the total destruction of its marine faunas (Hsü, 1974; Hsü et al., 1973). Therefore, the post-Messinian marine faunas were very different from those of the Miocene. They resulted from the colonization by Atlantic species on and after the Atlantic water flooded the Mediterranean at the beginning of the Pliocene (Hsu, 1974).

As a result of this relatively recent colonization, and of Pleistocene extinctions during the glaciations, the Mediterranean, especially its eastern basin, is occupied by an impoverished biota (Vermeij, 1978). The regression during the glacial periods again limited the biotic exchanges through the Strait of Gibraltar. The glacial cooling's having been more pronounced in the western than in the eastern basin, subtropical species migrated from western to southern and eastern Mediterranean, while cold-tolerant Atlantic species invaded the western basin (Vermeij, 1978).

When the Suez Canal was opened in 1869, another phase began for the Mediterranean marine ecosystems: the immigration of Indo-West Pacific species.

THE PRESENT MEDITERRANEAN BRACHYURAN FAUNA

The roughly 125 brachyuran species inhabiting the Mediterranean Sea today are ranged in 23 families and 71 genera (table 1, based on Manning and Holthuis, 1981).

TABLE 1. Families and genera (alphabetically ranged) and number of species of the present Mediterranean Brachyuran fauna

FAMILY	GENUS	NUMBER OF SPECIES	FAMILY	GENUS	NUMBER OF SPECIES
Atelecyclidae	Atelecyclus	2	Majiidae	Acanthonyx	1
Calappidae	Calappa	1		Achaeus	2
Cancridae	Cancer	1		Anamathia	1
Corystidae	Corystes	1		Dorhynchus	1
Cymonomidae	Cymonomus	1		Ergasticus	1
Dorippidae	Ethusa	1		Eurynome	2
	Medorippe	1		Herbstia	1
Dromiidae	Dromia	1		Hyastenus	1
	Sternodromia	1		Inachus	5
Geryonidae	Geryon	1 or 2		Lissa	1
	Paragalene	1		Macropodia	5
Goneplacidae	Eucrate	1		Maja	3
	Goneplax	1		Pisa	6
Grapsidae	Brachynotus	3	Ocypodidae	Ocypode	1
	Euchyrograpsus	1	Palicidae	Palicus	1
	Pachygrapsus	3	Parthenopidae	Heterocrypta	1
	Planes	1		Parthenope	4
Homolidae	Homola	1	Pinnotheridae	Asthenognathus	1
	Paromola	1		Pinnotheres	3
Latreillidae	Latreillia	1	Pirimelidae	Pirimela	1
Leucosiidae	Ebalia	6		Sirpus	1
	Ilia	1	Portunidae	Bathynectes	2
	Ixa	1		Callinectes	1
	Leucosia	1		Carcinus	1
	Merocryptus	1		Charybdis	2
	Myra	1		Liocarcinus	9

(continued)

TABLE 1 (Cont.)

FAMILY	GENUS	NUMBER OF SPECIES	FAMILY	GENUS	NUMBER OF SPECIES
Portunidae (cont.)	Macropipus	1	Xanthidae	Atergatis	1
	Polybius	1		Eriphia	1
	Portumnus	2		Heteropanope	1
	Portunus	2 or 3		Microcassiope	1
	Thalamita	1		Monodaeus	2
	Xaiva	1		Paractaea	1
Raninidae	Notopus	1		Pilumnopeus	1
Thiidae	This	1		Pilumnus	5
				Rhithropanopeus	1
				Sphaerozius	1
				Xantho	3

According to Forest (1972b), the decapod Mediterranean marine fauna is composed of four distinct groups: (1) Subtropical-temperate East Atlantic elements, such as Monodaeus couchii (Couch, 1851) and Pisa tetraodon (Pennant, 1777); (2) Forms originated by East Atlantic species, such as Ebalia deshayesi Lucas, 1846 and E. edwardsii Costa, 1838, which seem to be derived from Ebalia tumefacta (Montagu, 1808), known from Norway to Spanish Sahara; (3) Mediterranean species without particular affinities to East Atlantic decapods, such as Xantho poressa (Olivi, 1792), which seems to be originally Mediterranean but extends to the Canary Islands, and Paragalene longicrura (Nardo, 1869), a Mediterranean endemic species; (4) Indo-Pacific immigrants which arrived in the Mediterranean through the Suez Canal.

Holthuis and Gottlieb (1958) consider two main components in the decapod fauna of the Mediterranean coast of Israel: the Atlantic and the Indo-West Pacific elements. The Atlantic element consists of 45 brachyuran species divided into three categories. The first category includes 40 Brachyura which inhabit the entire Mediterranean, generally penetrating into the East Atlantic and sometimes into the Black Sea. The second consists of 4 brachyuran species, Ocypode cursor (Linnaeus, 1758), Microcassiope minor (Dana, 1852), Maja goltziana D'Oliveira, 1888, and Pachygrapsus transversus (Gibbes, 1850), which are entirely, or almost entirely absent in the European part of the Mediterranean, occurring in the East Atlantic, along the North African coast and along the coasts of Israel and Syria. The third category is formed by Callinectes sapidus Rathbun, 1896, an American species unintentionally introduced into Israel waters.

At the time of Holthuis and Gottlieb's paper the Indo-West Pacific element was composed of 7 Brachyura, all of them recent immigrants which arrived in the Mediterranean by way of the Suez Canal. Other records of Indo-West Pacific brachyuran species immigrated into the Eastern Mediterranean were later referred to by Holthuis (1961) and Lewinsohn and Holthuis (1964).

The geographic origin of the Mediterranean brachyuran fauna having been thoroughly examined by the above mentioned authors (Forest, 1972b; Holthuis and Gottlieb, 1958), the mechanisms of faunal genesis and some of their implications will be now analyzed.

AUTOCHTHONOUS ADAPTIVE RADIATION

Between 1/5 and 1/3 of the Mediterranean Brachyura seem to have speciated in the Mediterranean Sea. Groups (2) and (3) referred to by Forest (1972b) (see the preceding section) correspond to these species. Indeed, their present geographic ranges are exclusively or almost exclusively restricted to the Mediterranean. The ancestors of the autochthonous temperate-tropical Mediterranean species have most probably been East Atlantic or eurythermic-temperate species, which colonized the Mediterranean in post-Messinian times and there gave rise to the present species or to their immediate ancestors. In any case, the present autochthonous brachyuran species will generally be younger than 5 million years.

Most of the autochthonous Mediterranean Brachyura seem to be derived from temperate-tropical East Atlantic species. This would have been the case for *Achaeus gracilis* Costa, 1839, *Inachus communissimus* Rizza, 1839, *Macropodia czerniavskii* (Brandt, 1880), *Monodaeus guinotae* Forest, 1976, *Parthenope angulifrons* Latreille, 1828, *Pilumnus aestuarii* Nardo, 1869, *P. villosissimus* (Rafinesque, 1814), *Pinnotheres marioni* Gourret, 1887, *Pisa corallina* (Risso, 1816), *P. muscosa* (Linnaeus, 1758), *Portumnus pestai* Forest, 1967, *Sirpus zariquieyi* Gordon, 1953, *Xantho granulicarpus* Forest, 1953, and *Xantho poressa*. Other autochthonous species, like *Carcinus aestuarii* Nardo, 1847, *Liocarcinus bolivari* (Zariquiey-Alvarez, 1948), *L. maculatus* (Risso, 1827), and *L. vernalis* (Risso, 1816), are most probably derived from eurythermic-temperate species. It is possible that the ancestors of *Carcinus* and *Liocarcinus* authochthonous species penetrated into the Mediterranean as a consequence of the Pleiostocene cooling. In this case, *Carcinus* and *Liocarcinus* autochthonous species would be younger than most of the temperate-tropical derivatives.

Littoral and sublittoral biotopes have provided the most suitable ecological niches for the speciation of Mediterranean autochthonous Brachyura. In fact, species like *Achaeus gracilis*, *Brachynotus foresti* Zariquiey-Alvarez, 1968, *B. gemmellari* (Rizza, 1839), *B. sexdentatus* (Risso, 1827), *Carcinus aestuarii*, *Ebalia deshayesi*, *E. edwardsii*, *Inachus communissimus*, *Liocarcinus bolivari*, *L. maculatus*, *L. vernalis*, *Lissa chiragra* (Fabricius, 1775), *Macropodia czerniavskii*, *Pachygrapsus maurus* (Lucas, 1846), *Parthenope angulifrons*, *Pilumnus aestuarii*, *P. villosissimus*, *Pisa corallina*, *P. muscosa*, *Portumnus pestai*, *Sirpus zariquieyi*, *Xantho granulicarpus*, and *X. poressa*, seem well adapted to littoral and/or sublittoral niches. A few other autochthonous species as *Monodaeus guinotae*, *Paragalene longicrura* and probably *Parthenope macrochelos*, seem, on the contrary, to represent adaptation to specialized deep water biotopes (Forest, 1972a,b).

Even assuming that the littoral and sublittoral Mediterranean fauna is better known than the deep water fauna, which is quite natural, the number of autochthonous littoral and sublittoral species seems to be significantly higher. A few different arguments could possibly explain this. One possible reason could be a higher diversity of littoral and sublittoral substrates, at least when the ancestoral Atlantic species colonized the Mediterranean Sea. In fact, Abele (1974) demonstrated that the number of substrates is the most important factor in determining the diversity of decapod Crustacea, probably because each species can make differential use of each substrate. The same author (Abele, 1974), did not verify any significant importance of the temperature range, salinity range, or tidal exposure.

Once an appreciable diversity was established in that way and/or any other, e.g. a selective colonization by larvae of littoral or sublittoral species, the diversifying selection due to the biotic environmental heterogeneity would increase the speciation rate (Almaça, 1982a). Day (1963) suggested for the benthic communities the existence of a correlation between the speciation rate and the complexity of the biotic environment, i.e. the greater the diversity is in species, the more opportunities exist for speciation. This was suggested particularly by the diversity of the carnivores of the benthic niche, which is, in itself, dependent on the newly-speciated forms added to the community. Such a progressive specialization demands the maintenance of genetic variability at a significant level (Almaça, 1982a). The diversifying selection operates towards that maintenance in natural populations (Mukai et al., 1982).

As for the modes of speciation of autochthonous Brachyura, it is evidently impossible to suggest a general pattern for all species. However, if we accept the Messinian desiccation of the Mediterranean, the founder principle must have played a role by marginal colonisation (Almaça, 1981, 1982) at least in the early post-Messinian times. A few examples taken among the autochthonous species can help to clear up the action of the speciation mechanisms in the formation of that autochthonous fauna.

According to Forest (1972b), the autochthonous Ebalia deshayesi and E. edwardsii exhibit morphological characters suggesting that they diverged from the boreal-warm temperate (Christiansen, 1969) or boreal (Ingle, 1980) East Atlantic E. tumefacta, a species whose geographical range extends from Norway and the Shetland Islands to the Spanish Sahara (Manning and Holthuis, 1981). E. deshayesi and E. edwardsii are both Mediterranean, the first one extending to the Canary Islands and Madeira (Manning and Holthuis, 1981). Some ecological attributes of the three species are summarized in table 2, and suggest: (1) a remarkable similitude in living depth between the putative parent and daughter species, and (2) an ecological divergence between the daughter species, which is shown by the types of bottom where they preferentially live, corresponding, quite probably, to different ecological niches for each species.

Other East Atlantic species of Ebalia also live in the Mediterranean, e.g. E. cranchii Leech, 1817, E. granulosa H. Milne Edwards, 1837, E. nux H. Milne Edwards, 1883, and E. tuberosa (Pennant, 1777), but they are there either rare or mainly deep water forms. One of them, however, E. tuberosa, occurs at similar depths as E. deshayesi and E. edwardsii, but it lives, in the Adriatic, on muddy or sandy bottoms (Zariquiey-Alvarez, 1968). In spite of

Table 2. Some ecological attributes of Ebalia tumefacta and its presumably derived Mediterranean species, E. deshayesi and E. edwardsii

	E. tumefacta	E. deshayesi	E. edwardsii
Bottom	Stony With pebbles, gravel, shell debris, sand More seldomly, sandy mud and clay[1]	With gravel Sandy mud and Zostera. Sand and Posidonia[3]	Shore Adriatic: with coral, commonly on Spondylus[3]
Depth	Scandinavian costs[1]: 5-12 to 60-100 m. British coasts[2]: 2-155 m Ria de Arosa (NW Spain)[3]: 20-54 m	5-100 m[3]	0-100 m[3]

[1] Christiansen, 1969

[2] Ingle, 1980

[3] Zariquiey-Alvarez, 1968

the even now impoverished Mediterranean biota (Vermeij, 1978), and the consequent, presumably weak competitive pressures, it seems that both autochthonous species, *E. deshayesi* and *E. edwardsii*, must have been selected to distinct ecological niches. In fact, they are sympatric, at least in some localities, e.g. Cadaqués (Spain), where *E. deshayesi* is the commoner species but *E. edwardsii* and *E. tuberosa* also occur, although the last one rarely (Zariquiey-Alvarez, 1968).

A more conspicuous example of ecological exclusion seems to be that of the Mediterranean species of *Brachynotus*. Three autochthonous species are recognized (Froglia and Manning, 1978): *B. foresti* and *B. sexdentatus*, intertidal, and *B. gemmellari*, sublittoral. A fourth species, *B. atlanticus* Forest, 1957, lives in the intertidal zone of the Atlantic adjacent to the Mediterranean, along the African coast. In table 3 the main ecological features of the Mediterranean species are shown, suggesting: (1) a vertical separation between *B. gemmellari*, and *B. foresti* and *B. sexdentatus*; (2) a strong similarity in the ecological attributes of *B. foresti* and *B. sexdentatus*.

According to Froglia and Manning (1978) the three species occupy

Table 3. Main ecological features of the autochthonous Mediterranean species of *Brachynotus*. Data from Froglia and Manning, 1978.

	B. foresti	*B. sexdentatus*	*B. gemmellari*
Depth	Shallow water	Shallow water	Sublittoral, 4-21 m
Ecological range	Mediolittoral zone. Generally in protected waters with little wave influence.	Mediolittoral zone	Usually limited to the lower part of the infralittoral zone
Bottom	Among algae or mussels growing on hard substrates. Apparently it may also dig in sand.	Apparently prefers soft bottoms. Usually found under stones but it may burrow in sand. On silty sand and mud with sea-weeds and *Caulerpa*.	Soft, muddy bottoms

different ecological habitats and usually do not live together. In fact, the three species are recorded from many localities in western and eastern Mediterranean and only in one locality (Sallambô, Tunisia) did the authors find *B. gemmellari* together with *B. sexdentatus*. Apparently, the competition between *B. foresti* and *B. sexdentatus*, which have very close ecological requirements, is avoided through geographical separation, while between those species and *B. gemmellari*, the avoidance of competition is achieved by habitats separation. Froglia and Manning (1978) remarked that the long, slender legs of *B. gemmellari* may be adaptive to the muddy bottoms where it lives.

Geographical exclusion of a higher degree concerns some other autochthonous species, e.g. *Carcinus aestuarii* and *Xantho granulicarpus*, which are vicarious to the eastern Atlantic species *C. maenas* (Linnaeus, 1758) and *X. incisus* (Leach, 1814), respectively. *X. granulicarpus* is exclusively mediterranean, while *C. aestuarii* extends to the Canary Islands.

C. maenas is a eurythermal, euryhaline and euryecious species living from Norway to Mauritania, which has been introduced in Australia and North America, probably in recent times (Almaça, 1962, 1963a). It exhibits an appreciable population variability, concerning biological, morphological and biometrical characters (Almaca, 1961a, 1972a, 1982b; Démeusy, 1958; Zariquiey-Alvarez, 1968). In spite of being a very common intertidal species (Almaça, 1982b; Christiansen, 1969; Ingle, 1980), *C. maenas* may occur to about 200 m (Christiansen, 1969). *C. aestuarii* is also an intertidal species, common as well in the brackish waters of littoral channels and lagoons and rarely found in the sea (Démeusy and Veillet, 1953). Démeusy (1958) admits that *Carcinus* is a Tethyan remnant, the ancestoral form of both recent species being Mediterranean. The isolation between the Mediterranean and the Atlantic would have provided the geographic barrier permitting the allopatric speciation of *C. aestuarii* and *C. maenas*.

It seems now that this speciation hypothesis may not conform with the Messinian extinction of the Mediterranean marine fauna. If the extinction was complete, *C. aestuarii* is the younger species, having resulted from the post-Messinian Mediterranean colonization by marginal populations of *C. maenas* or its immediate ancestor. At least one of the Pleistocene periods of isolation between the Atlantic and the Mediterranean would provide the barrier necessary to the genetic differentiation of *C. aestuarii*. In fact, no other mode of speciation than the allopatric one conforms to the highly tolerant and competitive populations of *C. maenas*.

A similar pattern of differentiation seems to have been followed by *X. granulicarpus*. This form probably evolved from the East Atlantic *X. incisus* or its ancestoral species by adaptation to Mediterranean conditions. There are, however, some doubts on the taxonomic status of granulicarpus, its specific or subspecific level not being clear (Almaça, 1959, 1961b, 1963b, 1972b).

X. incisus lives from the North Sea southward to Morocco, including the Azores, Madeira, Desertas, Canary Islands, and Cape Verde Islands, usually in shallow water, from the littoral to 30-40 m, having been recorded to 100 m (Manning and Holthuis, 1981). X. granulicarpus is a common Mediterranean sublittoral species which occurs from 0.5 to 7-10 m, but has been found to the depth of 100 m (Zariquiey-Alvarez, 1968). Another Mediterranean autochthonous species is X. poressa, which is very common and lives in shallow water on bottoms with gravel and stones (Zariquiey-Alvarez, 1968). X. poressa has extended its geographic range to the Canary Islands and Portugal. There is a third species still living in the Mediterranean, X. pilipes, which is known from Norway and the Shetland Islands southward to Angola. X. pilipes lives on hard, stony, and sandy bottom and bottoms with shell debris, from the shore to 110 m (Christiansen, 1969; Ingle, 1980). In the Mediterranean, Zariquiey-Alvarez (1968) refers to this species in shallow water and very common (Barcelona) or in depths between 20 and 30 m (Cadaquês).

It is possible that between X. granulicarpus and X. poressa a habitat exclusion has been established, since the first is a sublittoral form while X. poressa is essentially littoral. No data are available to predict the ecological relationships between X. granulicarpus and X. pilipes. As for X. incisus, the geographical exclusion between this species and X. granulicarpus, as well as the extended morphological similarity of both species, suggests the derivation of X. granulicarpus from X. incisus.

The above-examined examples, which are summarized in table 4, show how in the Mediterranean Brachyura, and likely in other groups, the greater the diversity is the more subtle and varied seem to be the ecological differences between the species. This means that speciation adds more and more specialized species to communities as Day (1963) formerly stated. With the increasing diversity the avoidance of competition becomes a progressively refined problem and the suitable space for expansion becomes more restricted.

Competition is crucial in determining the possible ways of speciation but once the new species is formed, i.e. once the ecological specialization is achieved and viable, further spreading of the new species will be limited by the available space suitable for that specialization, even if initial competitors are absent. This conforms to the "available space hypothesis", proposed by Anderson and Koopman (1981) from a different methodological and material basis. As a consequence, the smaller geographic ranges of species in more diverse faunas depend not only on interspecific competition but also on their specialized needs, which only a few habitats can support.

Table 4. Kinds of observed exclusion among Mediterranean living species of some genera

Genus	Number of Mediterranean living species	Kinds of observed exclusion among Mediterranean living species
Ebalia	6	Geographic(?) Habitat Ecological niche
Brachynotus	3	Geographic Habitat
Xantho	3	Habitat Ecological niche(?)
Carcinus	1	-

REPEATED OR CONTINUOUS MULTIPLE COLONIZATION

Besides the presumably autochthonous crabs, the major part of the present Mediterranean brachyuran fauna is the result of continuous or repeated colonization of species from the eastern and western Atlantic, and the Red Sea.

At least 3/5 of the species inhabiting now the Mediterranean are East Atlantic species, with a more or less enlarged geographic range, which colonized the Mediterranean Sea. According to their present range outside the Mediterranean these species can be classified into four different groups (see also Christiansen, 1969, 1982):

(1) Eurythermic temperate (≃23%)
(2) Warm Temperate (≃8%)
(3) Eurythermic temperate-tropical (≃31%)
(4) Warm temperate-tropical (≃37%).

Two somewhat arbitrary geographical references were taken to establish those groups: the Tropic of Cancer and the Peninsula of Brittany (France). Warm temperate species are those whose geographic range is comprised between the above references. If the geographic range extends northward of Brittany or southward of the Tropic the species are considered eurythermic temperate and temperate-tropical, respectively. Therefore, groups (1) to (4) are different in the extent of the geographic ranges of the species included in them, i.e., eurythermic temperate-tropical species always have larger ranges than warm temperate species, while eurythermic temperate or warm temperate-tropical generally, but not necessarily, have larger ranges than warm temperate species.

Some conclusions can be drawn from the grouping and percentages of species referred to in (1) to (4):

(a) No exclusively boreal (cold temperate) or exclusively tropical (East Atlantic), or arctic species colonized the Mediterranean Sea;
(b) The Mediterranean colonizing species have geographic ranges which extend over the warm temperate zone (all the species), the tropical zone (about 68% of the species), and the cold temperate zone (about 54% of the species);
(c) It can be concluded as a consequence of (a) and (b) that geographic contiguity has been an essential factor in Mediterranean colonization, and that in this colonization species that are also tropical are slightly advantaged over species that are also cold temperate.

East Atlantic species belonging to fifty different brachyuran genera have colonized the Mediterranean Sea. About 1/4 of these genera also differentiated into Mediterranean authchthonous species. These genera are: Achaeus, Ebalia, Inachus, Liocarcinus, Macropodia, Monodaeus, Pachygrapsus, Parthenope, Pilumnus, Pinnotheres, Pisa, Portumnus, and Xantho. Some problems about the probable or plausible ecological relationships among Mediterranean autochthonous and allochthonous species of a few of these genera have been dealt with before.

Many East Atlantic species living in areas contiguous to the Mediterranean have not colonized it. As for deep water species, Forest (1972b) explains this by the relatively small depth of the Gibraltar area (less than 300 m) and by the outflowing Mediterranean surface current which occupies the upper 100 m and impedes the penetration of planktonic larvae. This can explain the present situation, but not events occurring in the past when, after a regression period during which the Mediterranean was isolated from the Atlantic, oceanic water flooded the Mediterranean bringing a multitude of organisms. Whatever the reasons, an important group of deep water East Atlantic species, e.g., Cymonomus normani Lankester, 1903, Dicranodromia mahieuxii Milne Edwards, 1883, Ethusa microphtalma Smith, 1881, Ethusina alba (Filhol, 1884), E. talismani Milne Edwards and Bouvier, 1897, Homologenus rostratus (Milne Edwards, 1880), Progeryon paucidens Bouvier, 1922, and Rochinia carpenteri (Thomson, 1873), did not colonize the Mediterranean Sea. Other species are possibly kept out of the Mediterranean by competition with close forms which speciated in the Mediterranean or successfully colonized that sea. This could be the case of geographically contiguous East Atlantic species belonging to Acanthonyx, Brachynotus, Carcinus, Dromia, Ebalia, Ilia, Liocarcinus, Macropodia, Merocryptus, Parthenope, Pilumnus, Pinnotheres, Portunus, and Xantho. Examples of ecological exclusions referring to species of some of these genera were examined before.

Still others have not yet been discovered in the Mediterranean or did not yet have the opportunity of a successful colonization. In spite of the Mediterranean Brachyura's being a well known fauna, probably other species are still to be found as shown by the recent re-discovery of Parthenope expansa (Miers, 1879) in that sea (Pastore, 1975).

A few western Atlantic species were unintentionally introduced into the Mediterranean: Callinectes sapidus Rathbun, 1896, Planes minutus (Linnaeus, 1758), and Rhithropanopeus harrisii (Gould, 1841).

Callinectes sapidus is known from the Adriatic (Venice), Greece, Turkey, Israel, Egypt (Christiansen, 1969; Holthuis, 1961; Holthuis and Gottlieb, 1958), and Sicily. It is perfectly acclimatized in Turkey and Greece, being the object of small fisheries in both (Holthuis, 1961).

Planes minutus is a pelagic inhabitant of the Sargassum community. Occasionally it spreads on floating or swimming organisms such as marine turtles, other algae species, etc. It has been recorded from several localities in western and eastern Mediterranean (Christiansen, 1969; Holthuis and Gottlieb, 1958; Lewinsohn and Holthuis, 1964; Zariquiey-Alvarez, 1968).

Rhithropanopeus harrisii inhabits marine as well as brackish or nearly fresh water biotopes, having been introduced into the Black Sea and Sea of Azov (Bacescu, 1967; Christiansen, 1969).

Since the opening of the Suez Canal, in 1869, more than 120 Red Sea species colonized the eastern Mediterranean, whereas less than 10 Mediterranean species colonized the Red Sea (Agur and Safriel, 1981). About 12% of the species of Brachyura now living in the Mediterranean are Indo-West Pacific species which migrated through the canal. The nearly one-way migration from the Red Sea affected animal (fishes, molluscus, crabs) and algae species and was extensively discussed by Vermeij (1978). This author concluded that two factors may be primarily responsible for that migration: (1) the impoverishment of the eastern Mediterranean biota relative to the Red Sea biota; (2) the better adaptation of Red Sea species to the salinity regime of the Suez Canal (the waters of the canal, and especially those of the Bitter Lakes through which the canal runs, are hypersaline). According to Vermeij (1978), a competitive or predation-related superiority would have played a much less important role in that migration.

Briggs (1967), on the contrary, believes that the stable Indo-West Pacific ecosystems and high level of competition provide the suitable environment for the evolution of dominant species that can successfully invade the other regions. So, from the Indo-West Pacific, which has served as the evolutionary and distributional centre for the entire marine tropics, dominant species migrate across the open ocean to America, westward around the Cape of Good Hope into the Atlantic, and northward through the Suez Canal into the Mediterranean.

Utilizing a mathematical hydraulic model, Agur and Safriel (1981) concluded that the completion of a Mediterranean-bound passage of the Red Sea by free-drifting propagules is far faster and much more likely than a completion of a Red Sea-bound passage by Mediterranean propagules. Assuming that the chances of a Mediterranean-produced propagule encountering the northern entrance of the Suez Canal are roughly the same as those of a Red-Sea-produced propagule encountering the southern entrance of the canal, the probability of a Red Sea propagule being "absorbed" into the canal is 3 times higher than that of a Mediterranean propagule being "absorbed". Furthermore, once in the canal, the probability of successful dispersal is 16 and 75 times higher, when "two step" (the Great Bitter Lake as a "stepping stone") and "one jump" crossings are considered, respectively, for a Red Sea propagule to reach the Mediterranean, than for a Mediterranean propagule to reach the Red Sea.

Therefore, it seems that the action of various factors explains the nearly one-way migration through the Suez Canal and the successful colonization of eastern Mediterranean by Red Sea species. The distinct actions can be summarized in two main groups: (1) a high-pressure — low-pressure competitive gradient between a highly diversified Indo-West Pacific fauna and an eastern Mediterranean impoverished fauna; (2) salinity and current regimes facilitating the migration from the Red Sea to the Mediterranean more than the inverse migration.

The Mediterranean brachyuran fauna includes 15 species which are immigrants from the Red Sea. Until now these species are only known from the eastern Mediterranean. About 86.7% of the immigrants belong to genera which were not represented in the Mediterranean fauna:

Goneplacidae	*Eucrate crenata* de Haan, 1835
Leucosidae	*Ixa monodi* Holthuis and Gottlieb, 1956 *Leucosia signata* Paulson, 1875 *Myra fugax* (Fabricius, 1978)
Majidae	*Hyastenus hilgendorfi* De Man, 1887
Portunidae	*Charybdis helleri* Milne Edwards, 1867 *Charybdis longicollis* Leene, 1938 *Thalamita poissonii* (Audouin, 1826)

Raninidae	Notopus dorsipes (Linnaeus, 1758)
Xanthidae	Atergatis roseus (Rüppell, 1830) Heteropanope laevis (Dana, 1852) Pilumnopeus vauquelini (Audouin, 1826) Sphaerozius nitidus Stimpson, 1858

As for the successful colonization of the immigrant portunids, Vermeij (1978) suggested that they may be faster and more aggressive than the native Mediterranean species. Nevertheless, it can be accepted for the generality of the immigrant species that their success was a consequence of not meeting competitors when introduced into the eastern Mediterranean. In fact, as far as can be deduced from the taxonomic relationships among the immigrant and the eastern Mediterranean species, the first ones did not meet close competitors. Only two immigrant species, the portunid Portunus pelagicus (Linnaeus, 1758) and the xanthid Pilumnus hirsutus Stimpson, 1858 colonized the eastern Mediterranean in spite of existing closely related forms there. The eastern Atlantic-Mediterranean Portunus hastatus (Linnaeus, 1767) is a sublittoral species like P. pelagicus. P. pelagicus is very common on Israel coasts (Holthuis and Gottlieb, 1958) and some kind of niche exclusion probably exists in the sympatric area of P. pelagicus and P. hastatus. A similar type of exclusion will eventually exist among the immigrant Pilumnus hirsutus, and P. hirtellus (Linnaeus, 1761) and P. spinifer Milne Edwards, 1834 The three species are sublittoral and have been recorded from the eastern Mediterranean.

DISCUSSION

The present analysis was made considering the continuous or repeated post-Messinian colonization of the Mediterranean Sea and nothing seems to contradict the hypothesis of the Messinian desiccation, consequent destruction of the Mediterranean marine fauna and subsequent re-colonization by East Atlantic species. Inside the Mediterranean many of the East Atlantic post-Messinian immigrants speciated, originating an autochthonous contingent which partly spread over the adjacent Atlantic.

As stated before, many species living in the East Atlantic close to the Mediterranean did not colonize it. In the eastern Mediterranean (cf. Holthuis, 1961; Holthuis and Gottlieb, 1958; Lewinsohn and Holthuis, 1964) only about 70% of the Mediterranean species (autochthonous and allochthonous) with an East Atlantic origin are living today. All these species live also in the western Mediterranean (cf. Manning and Holthuis, 1981; Zariquiey-Alvarez, 1968). Moreover, many species present in the western Mediterranean have not yet been recorded from the eastern basin. All these facts suggest a relatively recent and gradual colonization of the Mediterranean by East Atlantic species.

On the other hand, the Mediterranean shows a clear receptivity to the admission of allochthonous species, mainly to tropical or partly tropical species, as can be demonstrated by the colonization and fast spreading of Indo-West Pacific immigrants into its eastern basin. But not only to those species: the temperate (warm and eurythermic) contingent is not negligible in the Mediterranean, as stated before. This too corroborates a relatively recent colonization: in old and evolved ecosystems available space is scarce and competition is usually too high to allow the fixation of so many and variously-adapted species.

One question remains to be answered: what are the chances, if any, for some pre-Messinian Mediterranean species or their derivatives to persist in the Mediterranean today ? As referred to before, Démeusy (1958) admitted that *Carcinus* is a Tethys remnant, the Mediterranean species being the most primitive. Other Brachyura have possibly been considered in that group: Forest (1972b) refers to *Paragalene longicrura* as a relict of an old Mediterranean fauna. It seems possible that, even if the Messinian desiccation was complete, some euryhaline species could have persisted in the Paratethys which, during the Middle and Upper Miocene, became isolated from the Tethys (Albrecht, 1982). In the course of that isolation the salinity of the Paratethys decreased, more markedly in its eastern part (Black Sea and Aralo-Caspian region) (Kosswig, 1955). But, immediately after the Messinian, during the Pontian, temporary connections were established between the central Paratethys (Pannonian Basin) and the Adriatic and Aegean regions (Albrecht, 1982). Therefore, the preservation of some Mediterranean pre-Messinian species and their subsequent return to the Mediterranean does not seem impossible.

SUMMARY

The Mediterranean brachyuran fauna is the result of three immigrant species flows, different in importance and geographic origin, and of an autochthonous contingent.

The most important flow is the East Atlantic temperate-tropical and temperate. The less important is the West Atlantic flow. Both have entered the Mediterranean by its western ocean connection (Gibraltar area).

The second most important flow, with an increasing importance, is the Indo-West Pacific one which attains the eastern Mediterranean through the Suez Canal.

The Brachyura which speciated inside the Mediterranean form an outgoing species flow which introduces into the adjacent Eastern Atlantic some autochthonous Mediterranean species.

Nothing contradicts, at least apparently, the hypothesis of the Messinian desiccation, consequent destruction of the Mediterranean marine fauna and subsequent re-colonization by East Atlantic species. On the contrary, the facts suggest a relatively recent and gradual colonization of the Mediterranean by the East Atlantic brachyuran crabs.

The possibility of a few pre-Messinian Mediterranean species, at least some euryhaline species, having been preserved in the central Paratethys and then, during the Pontian, recolonizing (or their derivatives) the Mediterranean is considered.

REFERENCES

Abele, L.G., 1974, Species diversity of decapod crustaceans in marine habitats, Ecology, 55:156.

Agur, Z., and Safriel, U.N., 1981, Why is the Mediterranean more readily colonized than the Red Sea, by organisms using the Suez Canal as a passageway ?, Oecologia (Berl), 49:359.

Albrecht, H., 1982, On the origin of the Mediterranean crayfishes, Quad. Lab. Tecnol. Pesca, 3:355.

Almaça, C., 1959, Sobre a variabilidade e a posição sistemática do Xantho incisus Leach da zona intercotidal do litoral português, Rev. Fac. Cien. Lisboa, 7:233.

Almaça, C., 1961a, Variabilidade de alguns caracteres usados na taxonomia do Gen. Carcinus Leach, Rev. Fac. Ciên. Lisboa, 8:137.

Almaça, C., 1961b, Notas sobre o Xantho incisus Leach de Parede (Portugal), Rev. Fac. Ciên. Lisboa, 8:155.

Almaça, C., 1962, Sur la distribution géographique du genre Carcinus Leach (Crust. Dec. Brach.), Rev. Fac. Ciên. Lisboa, 10:109.

Almaça, C., 1963a, Sur le problème de l'origine de Carcinus maenas (L.) du littoral américain, Rev. Fac. Ciên. Lisboa, 11:121.

Almaça, C., 1963b, Sur les crabes du genre Xantho Leach, 1815, du Muséum Zoologique de l'Université de Coimbra, Mem. Est. Mus. Zool. Univ. Coimbra, 281:5.

Almaça, C., 1972a, Note préliminaire sur les rapports entre quelques dimensions de la carapace de Carcinus maenas (L.) et C. mediterraneus Czerniavsky, Thalassia Jugoslavica, 8:49.

Almaça, C., 1972b, Le littoral portugais, zone d'intergradation entre Xantho incisus incisus (Leach) et X. incisus granulicarpus (Forest), Thalassia Jugoslavica, 8:59.

Almaça, C., 1981, Apport de la dynamique des populations aquatiques à la génétique évolutive, in: "Dynamique des populations et qualité de l'eau", H. Hoestlandt, ed., Gauthier-Villars, Paris.

Almaça, C., 1982a, Marine slides and allopatric speciation, in: "Marine slides and other mass movements," S. Saxov and J.K. Nieuwenhuis, ed., Plenum Press, New York and London.

Almaça, C., 1982b, Note sur la biologie des populations de Carcinus maenas (L.) de la zone intertidale du Portugal occidental, Quad. Lab. Tecnol. Pesca, 3:179.

Anderson, S., and Koopman, K.F., 1981, Does interspecific competition limit the sizes of ranges of species?, Amer. Mus. Nov., 2716:1.

Bacescu, M., 1967, Crustacea, Fauna Rep. Soc. Romănia, 4:1.

Banarescu, P., 1975, "Principles and problems of Zoogeography", U.S. Dept. Comm. and N.S.F., Washington.

Banister, K.E., 1980, The fishes of the Tigris and Euphrates rivers, Monog. Biol., 38:95.

Briggs, J.C., 1967, Dispersal of tropical marine shore animals: Coriolis parameters or competition?, Nature, 216:350.

Christiansen, M.E., 1969, "Marine Invertebrates of Scandinavia, No. 2, Crustacea Decapoda Brachyura," Universitets Forlaget, Oslo.

Christiansen, M.E., 1982, A review of the distribution of Crustacea Decapoda Brachyura in the Northeast Atlantic, Quad. Lab. Tecnol. Pesca, 3:347.

Day, J.H., 1963, The complexity of the biotic environment, in: "Speciation in the sea," The Systematic Association, ed., and publ., London.

Démeusy, N., 1958, Recherches sur la mue de puberté du Décapode Brachyoure Carcinus maenas Linné, Arch. Zool. Exp. Gen., 95:253.

Démeusy, N., and Veillet, A., 1953, Sur l'existene de deux populations de Carcinus maenas Pennant et sur les caractères morphologiques qui les distinguent, C.R. Acad. Sc. Paris, 236:1088.

Forest, J., 1972a, Une espèce nouvelle de Xanthidae des eaux bathyales de Méditerranée: Monodaeus guinotae sp. nov., Thalassia Jugoslavica, 8:63.

Forest, J., 1972b, ler Colloque de carcinologie méditerranéenne, remarques finales, Thalassia Jugoslavica, 8:143.

Froglia, C., and Manning, R.B., 1978, Brachynotus gemmellari (Rizza, 1839), the third Mediterranean species of the genus (Crustacea, Decapoda, Brachyura), Proc. Biol. Soc. Wash., 91:691.

Holthuis, L.B., 1961, Report on a collection of Crustacea Decapoda and Stomatopoda from Turkey and the Balkans, Zool. Verhand., 47:1.

Holthuis, L.B., and Gottlieb, E., 1958, An annotated list of the decapod Crustacea of the Mediterranean coast of Israel, with an appendix listing the Decapoda of the eastern Mediterranean, The Sea Fish. Res. Sta. Haifa, 18:1.

Hsü, K.J., 1974, The Miocene desiccation of the Mediterranean and its climatical and zoogeographical implications, Naturwissenschaften, 61:137.

Hsü, K.J., Ryan, W.B.F., and Cita, M.B., 1973, Late Miocene desiccation of the Mediterranean, Nature, 242:240.

Ingle, R.W., 1980, "British crabs", British Museum (N.H.) and Oxford University Press, London.

Kosswig, C., 1955, Zoogeography of the Near East, Syst. Zool., 4:49.

Lewinsohn, Ch., and Holthuis, L.B., 1964, New records of decapod Crustacea from the Mediterranean coast of Israel and the eastern Mediterranean, Zool. Meded., 40:45.

Manning, R.B., 1981, West African Brachyuran crabs, Nat. Geog. Soc. Res. Rep., 13:431.

Manning, R.B., and Holthuis, L.B., 1981, West African Brachyuran crabs (Crustacea: Decapoda), Smithsonian contrib. Zool., 306:1.

Mukai, T., Yamaguchu, O., Kusakabe, S., Tachida, H., Matsuda, M., Ichinose, M., and Yoshimaru, H., 1982, Lack of balancing selection for protein polymorphisms, in: "Molecular evolution, protein polymorphism and the neutral theory," M. Kimura, ed., Japan Sc. Soc. Press and Springer-Verlag, Tokyo and Berlin.

Pastore, M., 1975, Riscoperta di Parthenope expansa Miers in Mediterraneo, Mem. Biol. Mar. Ocean., 5:145.

Vermeij, G.J., 1978, "Biogeography and adaptation", Harvard Univ. Press, Cambridge, Massachussetts.

Zariquiey-Alvarez, R., 1968, Crustãceos Decãpodos Ibericos, Inv. Pesq., 32:1.

THE IMPACT OF THE LESSEPSIAN (SUEZ CANAL) FISH MIGRATION ON THE EASTERN MEDITERRANEAN ECOSYSTEM

A. Ben-Tuvia

Department of Zoology, The Hebrew University of Jerusalem, Jerusalem 91904, Israel

THE RATE OF IMMIGRATION

The number of Lessepsian (Suez Canal or Isthmic) immigrants (cross out "or isthmic") colonizing the eastern Mediterranean increases continuously with time. Ben-Tuvia (1978) reported 36 species of fishes. One of them, however, namely Carcharhinus brevipinna Müller and Henle, is a circumtropical species and its presence in central and eastern Mediterranean could be a result of the Atlantic connections (Garrick, 1982). Therefore it has now been excluded from the list of the Red Sea immigrants. An additional 6 species have been found since (Mouneimné, 1977,1979; Golani and Ben-Tuvia, 1982; Ben-Tuvia, 1983a,1983b; Ben-Tuvia and Golani, in preparation). In total there are 41 Indo-Pacific species that evidently reached the Mediterranean Sea via the Suez Canal or via earlier canals connecting the Red Sea with the Nile River (Isthmic immigration).

The first immigrant namely Atherinops lacunosus (Table 1) was recorded by Tillier (1902) off Alexandria, 33 years after the opening of the Suez canal in 1869. Nine additional fishes were recorded up to 1939, an average of one fish per 7 years. In the period 1940-1949 the number of newly recorded immigrants increased to one fish per two years. Between 1950 and 1983, 27 new immigrants were found, on average one fish per 1.2 years, the incidences being dispersed more or less evenly over the 34-year period.

Among the stream of new colonizers at least three species, namely Sillago sihama, Pempheris vanicolensis and Oxyurichthys papuensis appeared almost immediately as large populations. Thus it is evident that the process of colonization by the Red Sea fishes is continuous without any sign of decline. It might double the

Table 1. List of Red Sea fishes in the Levant Basin

Dasyatidae	1.	*Himantura uarnak* (Forsskal, 1775)
Clupeidae	2.	*Dussumieria acuta* (Valenciennes, 1847)
	3.	*Etrumeus teres* (DeKay, 1842)
	4.	*Herklotsichthys punctatus* (Rüppell, 1837)
	5.	*Spratelloides delicatulus* (Bennett, 1831)
Myraenosicidae	6.	*Muraenesox cinereus* (Forsskal, 1775)
Synodontidae	7.	*Saurida undosquamis* (Richardson, 1848)
Ariidae	8.	*Arius thalassinus* (Rüppell, 1835)
Exocoetidae	9.	*Paraexocoetus mento* (Valenciennes, 1846)
Belonidae	10.	*Tylosurus choram* (Rüppell, 1837)
Hemiramphidae	11.	*Hemiramphus far* (Forsskal, 1775)
Cyprinodontidae	12.	*Aphanius dispar* (Ruppell, 1828)
Atherinidae	13.	*Atherinops lacunosus* (Bloch and Schneider, 1801) (Syn. *Pranesus pinguis* Lacepêde)
Holocentridae	14.	*Sargocentron ruber* (Forsskal, 1775)
Scorpaenidae	15.	*Sebastapistes nuchalis* (Günther, 1874)
Platycephalidae	16.	*Platycephalus indicus* (Linnaeus, 1758)
Serranidae	17.	*Epinephelus malabaricus* (Bloch and Schneider, 1801) (syn. *Epinephelus tauvina* (nec Forsskal)
Theraponidae	18.	*Authistes puta* (Cuvier, 1829)
	19.	*Pelates quadrilineatus* (Bloch, 1790)
Apogonidae	20.	*Apogonichthyoides nigripinnis* (Cuvier,1828)
Sillaginidae	21.	*Sillago sihama* (Forsskal, 1779)
Rachycentridae	22.	*Rachycentron canadum* (Linnaeus, 1758)
Carangidae	23.	*Atule djeddaba* (Forsskal, 1775)
Leiognathidae	24.	*Leiognathus klunzingeri* (Steindachner, 1898)
Lutjanidae	25.	*Lutjanus argentimaculatus* (Forsskal, 1775)
Mullidae	26.	*Upeneus asymmetricus* (Lachner, 1954)
	27.	*Upeneus moluccensis* (Bleeker, 1855)
Haemulidae	28.	*Pomadasys stridens* (Forsskal, 1775)
Sparidae	29.	*Crenidens crenidens* (Forsskal, 1775)
Pempheridae	30.	*Pempheris vanicolensis* (Cuvier, 1931)
Mugilidae	31.	*Liza carinata* (Ehrenberg, in: Cuv. @ Val., 1836)
Sphyraenidae	32.	*Sphyraena chrysotaenia* (Klunzinger, 1884)
Gobiidae	33.	*Oxyurichthys papuensis* (Valenciennes,1837)
Siganidae	34.	*Siganus luridus* (Rüppell, 1828)
	35.	*Siganus rivulatus* (Forsskal, 1775)
Scombridae	36.	*Rastrelliger kanagurta* (Russell, 1803)
	37.	*Scomberomorus commerson* (Lacepêde, 1800)
Callionymidae	38.	*Callionymus* (Calliurichthys) filamentosus (Linnaeus, 1758)
Cynoglossidae	39.	*Cynoglossus sinusarabici* (Chabanaud, 1931)
Monacanthidae	40.	*Stephanolepis diaspros* (Fraser-Brunner, 1940)
Tetraodontidae	41.	*Lagocephalus spadiceus* (Richardson, 1844)

number of Red Sea fishes in the Levant within the next 50 years, bringing the total number to about 80. Although impressive in proportion to the total of about 310 species known from the Levant, the number constitutes only 8% of the amount of species recorded in the Red Sea (Dor, in press). The high rate of colonization stresses the uniqueness of the Levant as a biogeographical unit characterized by the phenomenon of Lessepsian immigration (Por, 1978). The reasons for such a successful colonization have been discussed elsewhere (Ben-Tuvia, 1973; Por, 1978; Ben-Tuvia, 1983a).

IMMIGRATION OF ATLANTIC SPECIES

The occurrence of Atlantic species not previously recorded in the Mediterranean Sea is not necessarily evidence of recent immigration. Such new records are usually based on only one or few specimens, indicating their rarity in the Mediterranean Sea. These are more likely to be old inhabitants of the Mediterranean Sea that have not come to the attention of a competent taxonomist until now.

There is no case among fishes that an Atlantic species suspected to be a recent immigrant in the Mediterranean Sea increases in numbers as occurs with the Suez Canal colonizers.

THE ORIGIN OF IMMIGRANTS

Among the immigrants 30 species (73.2%) are of wide Indo-Pacific distribution ranging from northern Australia and southern Japan to eastern coast of Africa and the Red Sea; 7 species (17.0%) are known from the western Indian Ocean only; 4 species (9.8%) are circumtropical. None of the colonizers belong to any of the endemic Red Sea species.

The typical coral reef families have no representatives among the Suez Canal colonizers. None of the Red Sea pomacentrids, chaetodontids, scarids, labrids, acanthurids and pomacanthids was found in the Levant. The following factors might be responsible:

1. These families, known to be associated in tropics with coral reefs, cannot colonize Levant prior to the settlement of the reef building coelenterates.
2. Both corals and coral reef fishes are thermophilic and cannot survive the low winter temperatures of the Mediterranean Sea.
3. Lack of suitable habitat prevents coral reef fishes from entering the Suez Canal.
4. Coral fishes are mostly bottom spawners with non-existing or short-lived pelagic stages; they therefore lack sufficient opportunities of drifting with currents to the northern end of the Suez Canal.

It is of interest to stress that the Suez Canal immigration added to the Mediterranean Sea 13 new families and 34 new genera (Table 1). New taxonomic entities have better opportunities of succeeding in a new environment by utilizing under-exploited ecological niches.

The largest number of immigrants, 4 species, belong to Clupeidae (including Dussumierinae), thus doubling the total number of the species of the family in the Levant. Among them Dussumieria acuta and Herklotsichthys punctatus are common and contribute to the inshore-pelagic fishery mostly in the Gaza and El Arish regions. Theraponidae are interesting since of the total of 3 species known in the Red Sea, two are immigrants. Several families represented by numerous species in the Red Sea such as Gobiidae with 84 species, Apogonidae with 39 species and Serranidae with 33 species (Dor, in press) are represented by one immigrant only.

The Suez Canal colonizers of the Levant arise from the populations of the Gulf of Suez, the most northern section of the Red Sea. There is no precise inventory of the Gulf of Suez fishes in comparison to other parts of the Red Sea but some observations (Ben-Tuvia and Grofit, 1972; Botros, 1971) indicate that the fauna is impoverished. The shallow Gulf contains good trawling grounds of muddy bottom in depths of 20-50 m. The main stream of colonizers come from this demersal stock, being able to find a similar habitat in the Mediterranean Levant.

ECOLOGICAL IMPORTANCE AND FISHERIES

The large amount of Red Sea fishes in the Levant, in terms of numbers of species or in terms of biomass, implies that the immigration must have an immense impact on the ecological balance of the region. However, there are no data that enable us to analyse the complicated ecological relationship among the fish communities in various habitats.

The only quantitative data available are those from routine catch statistics collected in marine fisheries by the Israel Department of Fisheries. A large proportion of the Suez Canal immigrants (45%) now belong to the common fishes of Levant. Among them at least 7 species, namely Dussumieria acuta, Saurida undosquamis, Atule djeddaba, Upeneus asymmetricus, Upeneus moluccensis, Sphyraena chrysotaenia, Siganus rivulatus and Scomberomorus commerson are of commercial importance.

Although we must assume "a priori" that the colonizers compete with some of the native species there is no evidence of a drastic change in abundance of any of the Mediterranean commercial fish that could be attributed to a new competitor. Seldom is enough information available about life-history cycles and ecological relationships to establish which are the competing species. In terrestrial ecology cases have been recorded of a successful colonizer exterminating a related species. No such case can be recorded in the Levant Basin. The native lizardfish Synodus saurus, always a rare fish in the Israeli coast, is still occasionally caught in spite of the massive invasion of the Red Sea lizardfish Saurida undosquamis. The appearance of two Indo-Pacific goatfishes Upeneus moluccensis and U. asymmetricus does not seem to influence the abundance of the native Mullus barbatus and M. surmuletus.

The ecological relationship between Saurida undosquamis and the Atlanto-Mediterranean hake Merluccius merluccius has been discussed by Ben-Yami and Glazer (1974). The Red Sea lizardfish was found for the first time along the coast of Israel in 1952 and started to appear in commercial quantities in 1955. It is caught on trawling grounds in depths of 20-100 m. It is a predator feeding almost exclusively on fish. In winter hake occupies the same trawling grounds but in summer is found usually in deeper waters between 80 and 250 m. About equal quantities of fish and crustaceans were found in the stomach contents of hake (Shmida, 1964).

Analysis of statistical data for the period 1950-1970 do not indicate a decline of catch of hake that could be attributed to the increasing catches of lizardfish (Ben-Yami and Glazer, 1974). Both catch per unit of effort data and the percentage of each fish in total trawl catches show year to year fluctuations, but no trend. According to the same authors lizardfish contributed about 20% of the trawl catches, about 5% more than hake. However in later years the quantities of hake decreased considerably. The average catch of hake for the years 1974-1982 amounted to 72.3 t or 7.8% of trawl catches while lizardfish constituted 21.2%. The decrease in the catches of hake in recent years is partially explained by less fishing in deeper waters, 100-250 m, where hake is abundant (E. Grofit, personal communication). According to Gorgy (1966) lizardfish accounted for about 50% of trawl catches in exploratory survey in 1959-1960 in El-Arish area (Sinai).

It is more difficult to estimate the catch of other immigrants since no separate statistics are available for the various statistical categories of mixed native and Red Sea fishes. It has been assumed on the basis of trawl catch composition analysis of the data for 1980-1982 (Golani, 1981-83) that Upeneus sp. constituted one third of goatfish (Mullidae) catches, Sphyraena chrysotaenia one third of sphyraenid catches and other Red Sea fish 2% of the total catches of trawl. Thus it has been estimated that the Red Sea immigrants constituted 33.6% of trawl fishery as compared to 21% in the earlier periods (Ben-Tuvia, 1973). By similar analysis it has been estimated for the same period of years that the catch of the Red Sea immigrants in the coastal fishery (trammel-net, gill-net, hook-and-line) constituted 12.4% and in the pelagic fishery (purse-seine) 12.1%. On the average the immigrants contributed 16.2% of the Mediterranean catch of Israel in the period 1980-1982.

Although no case has been recorded of a replacement of a native fish by a Red Sea colonizer there are observations indicating changes in abundance of species among the immigrants. Leiognathus klunzingeri was extremely common on trawling grounds in depths of 20-100 m in the early fifties but has become much less abundant since the sixties. The invasion of the piscivore lizardfish could be the cause.

Rabbitfishes are another example of changes in space and time. Siganus rivulatus is known to be an early immigrant found not only along the coast of Israel and Lebanon (George, 1972) but also Cyprus, Rhodes and Libya. S. luridus was found for the first time on the Mediterranean Coast of Israel in 1956 and later in the Aegean Sea (Ondrias, 1971) and even off Tunisia. Collections and observations taken during the research project: "Biota of the Red Sea and the Eastern Mediterranean, 1967-1972" (Por et al., 1972) confirm that S. luridus is now common in the coastal waters of the Islands of Cyprus and Rhodes while S. rivulatus could not be found.

NEW IMMIGRANTS

Among the recent records, three immigrants are of particular interest since they appeared almost immediately in large numbers. Sillago sihama (Forsskal) was found in 1976 off the coast of Lebanon and in 1977 off the Coast of Israel (Mouneimne, 1977; Ben-Tuvia, 1978). Further observations in Israel revealed that the fish is common on sandy bottoms in shallow waters, and contributes slightly to commercial catches of trawl and trammel net. It occurs in sizes of 90-140 mm total length. It is known to be of considerable commercial importance in the Indo-West Pacific area.

Pempheris vanicolensis Cuvier was found for the first time in the Mediterranean Sea in 1978, almost simultaneously off the Coast of Lebanon and Israel. Mouneimne (1979) reported 4 specimens under the name of P. molucca Cuvier which most probably represent the same species. In using the name P. vanicolensis I follow Randall (1983) who recently studied the Red Sea specimens. Schools of this fish were observed near rocks and artificial reefs in depths of 5 to 20 m. It is caught occasionally in gill net and trammel net. Size 90-140 mm (Ben-Tuvia and Golani, in preparation).

Oxyurichthys papuensis (Valenciennes), a gobiid fish of wide Indo-Pacific distribution, was found in large numbers on trawling grounds off Tel-Aviv and Ashdod in depths of 20-50 m in 1982 (Ben-Tuvia, 1983b). Preliminary observations seem to indicate that it might compete with the Atlanto-Mediterranean Gobius niger Linnaeus which is common on the same trawling grounds.

Single specimens of four additional Suez Canal immigrants were found in the Mediterranean Levant since 1976. Lutjanus argentimaculatus (Forsskal), 284 mm long, was reported from the coast of Lebanon (Mouneimné, 1979). One specimen of Rachycentron canadum (Linnaeus), 766 mm standard length was handlined in 1978 off Haifa. Muraenesox cinereus (Forsskal), 1200 mm long, was caught off Tel-Aviv in 1980 (Golani and Ben-Tuvia, 1982). Arius thalassinus (Rüppell), 152 mm (SL) was caught in Haifa Bay on 12 October, 1983 (Ben-Tuvia and Golani, in preparation). This species has not as yet, been recorded either from the Suez Canal or from the Gulf of Suez and Gulf of Aqaba.

SUMMARY

The number of Red Sea immigrant fishes recorded in the eastern Mediterranean reached 41 species, amounting to 13.2% of the fish fauna of the Levant. The majority of the immigrants belong to Indo-Pacific species of wide distribution. None of them belongs to the endemic Red Sea species. The ecological impact of the Lessepsian colonizers is demonstrated by their contribution to the fishery, which during the year 1980-1982 was estimated to amount to about 16% of the Mediterranean catches of Israel. The immigrants are most abundant among demersal fishes, inhabiting sandy and muddy bottoms. Seven new immigrants were found between 1976 and 1983; Sillago sihama, Pempheris vanico ensis and Oxyurichthys papuensis are now common. Lutjanus argentimaculatus, Rachycentron canadum, Muraenesox cinereus and Tachysurus thalassinus were found as single specimens.

ACKNOWLEDGEMENTS

I wish to thank Messrs S. Pisanty and M. Tom for providing specimens of some of the Red Sea immigrants, Mr. D. Golani for assistance in processing catch composition data and Professor D.F. Por for critically reading the manuscript. This study was partially supported by the Fauna Palestina Committee of the Israel Academy of Sciences and Humanities.

REFERENCES

Ben-Tuvia, A., 1973, Man-made changes in the eastern Mediterranean Sea and their effect on the fishery resources, Mar. Biol., 19:197.

Ben-Tuvia, A., 1978, Immigration of fishes through the Suez Canal, Fish.Bull., 76:249.

Ben-Tuvia, A., 1983a, The Mediterranean Sea, B. Biological Aspects. in: "Estuaries and Enclosed Seas," B.N. Ketchum, ed., Elsevier, Amsterdam

Ben-Tuvia, A., 1983b, An Indo-Pacific goby, Oxyurichthys papuensis in the eastern Mediterranean, Israel J. Zool.,32:1.

Ben-Tuvia, A., and Golani, D., in preparation, Additions to the fish fauna of the Mediterranean coast of Israel.

Ben-Tuvia, A., and Grofit, E., 1973, Exploratory trawling in the Gulf of Suez, November 1972, Fishing and Fish-Breeding (in Hebrew with English Abstract).

Ben-Yami, M., and Glaser, T., 1974, The invasion of Saurida undosquamis (Richardson) into the Levant Basin - An example of biological effect of interoceanic canals, Fish. Bul., U.S., 72:359.

Botros, G.A., 1971, Fishes of the Red Sea, in: Oceanogr. Mar. Biol. Ann. Rev., H. Barnes, ed., 9:221.

Dor, M., (in press), "Check lists of the fishes of the Red Sea (Clofres)," The Israel Academy of Sciences and Humanities, Jerusalem, Israel.

Garrick, J.A.F., 1982, "Sharks of the genus Carcharhinus," NOAA Technical Rep., NMFS Circul. 445.

George, C.J., 1972, Notes on the breeding and movements of the rabbitfishes Siganus rivulatus (Forsskal) and S. luridus (Rüppell), in the coastal waters of the Lebanon, Annali Museo Civico Storia Natur. Genova, 79:32.

Gorgy, S., 1966, Les pêcheries et le milieu marin dans le secteur méditerranéen de la R.A.U., Rev. Trav. Inst. Pech. marit., 30:25.

Golani, D. (ed.), 1981-83, "Israeli fisheries in figures: 1980-1982," Department of Fisheries, Ministry of Agriculture, Tel-Aviv, Israel.

Golani, D. and Ben-Tuvia, A., 1982, First records of the Indo-Pacific daggertooth pike-conger, Muraenesox cinereus, in the eastern Mediterranean and in the Gulf of Elat (Gulf of Aqaba), Israel J. Zool., 31:31.

Mouneimné, N., 1977, Liste des poissons de la côte du Liban. (Mediterranêe orientale), Cybium, 1:37.

Mouneimné, M., 1979, Poissons nouveaux pour les côtes libanaises (Mediterranêe orientale), Cybium, 6:105.

Ondrias, J., 1971, A list of the fresh and sea water fishes of Greece, Prakt. Inst. Oceanogr. Fish Res., 10a:23.

Por, F.D., 1978, "Lessepsian Migration. The influx of Red-Sea biota into the Mediterranean by way of the Suez Canal," Ecological Studies, V. 2, Springer-Verlag, Heidelberg.

Por, F.D., Steinitz, H., Farber, J., and Aron, W., 1972, The biota of the Red Sea and the eastern Mediterranean (1967-1972): A survey of the marine life of Israel and surroundings, Israel J. Zool.,21:459.

Shmida, A., 1964, Feeding habits of the Mediterranean and Red Sea fishes. Thesis for Reali Gymnasium, Haifa (in Hebrew).

Tillier, J.B., 1902, Le Canal de Suez et sa faune ichthyologique, Mém Soc. Zool. Fr., 15:279.
Randall, E.J., 1983, "Red Sea Reef fishes," Immel Publ., London.

MEDITERRANEAN MARINE ECOSYSTEMS: ESTABLISHMENT OF ZOOPLANKTONIC COMMUNITIES IN TRANSITIONAL AND PARTLY ISOLATED AREAS*

V. Kiortsis

Zoological Laboratory and Museum, University of Athens
Panepistimiopolis, GR-157 71, Athens, Greece

INTRODUCTION

Marine life is, by no means, simple. The planktonic ecosystems make no exception to this rule. Their complexity in nature defies description and inferences from field and/or laboratory studies give only a blurred picture of the reality.

We usually complain of the roughness of our sampling methods and of the sometimes 10fold errors in our estimates. Therefore, the present trend to replace the guesses of a scientists' mind by fashionable simulation models and computing devices, seems legitimate. However, a large-scale synthesis of multidisciplinary data, of various hypotheses and theories, yet cannot be done by the computer alone.

The hydrogeological history of the Mediterranean - if not the biological one - is rather well known, at least in its broad lines (see other reports of this meeting). Various theories were formulated, but all are based on a series of well established events: Mediterranean waters originated in and evolved from the Tethys and Paratethys; there was, in the past, intermittent communication with the Atlantic through Gibraltar, isolation from the Indo-Pacific, junction with the Black Sea, sinking of the Aegean, the Adriatic, the West Mediterranean; important climatic changes occurred during the glacial and interglacial periods, with wide variation of the sea level.

---

*Dedicated to Prof. Marie-Louise Furnestin, "grande dame" of Mediterranean planktonology, a tribute of appreciation and esteem by the author.

The ecological conditions prevailing in present-day Mediterranean are:

A negative water balance, due to excessive evaporation; temperature stratification; lack of significant tides; high salinity; temperate to sub-tropical climatic gradient, from north to south; a west-east gradient in nutrients deficiency and decreased longitudinal species diversity with regional particularism, if not true endemism, etc.

With those elements at hand one has to consider the different biota. The choice of the zooplankton may seem odd for the description of geographically distinct communities. The stability and bidimensional character of benthos, as compared with the moving, tridimensional one of plankton, makes this exercise a difficult task. However, to quote M.L. Furnestin (1979): "even in the otherwise rather uniform Mediterranean real planktonic units exist, subordinate to hydrological barriers; the effect of marine circulation, which ensures faunistic exchanges between adjacent areas, sometimes result in variation of the species transported".

Consequently, the origin, maintenance and survival of planktonic elements in the Mediterranean and the establishment of planktonic communities could be attributed to historic geomorphological and hydrological circumstances and the prevailing ecological conditions.

MEDITERRANEAN REGIONS

Present-day Mediterranean belongs to the Atlanto-Mediterranean Province (Ekman, 1953). It forms a distinct subregion, stretching from Gibraltar - that cannot be recognized as a strict faunistic boundary - to the easternmost coast; the Black Sea is left apart, belonging to the Sarmatic subregion (Klausewitz, 1969).

The main transitional areas, connecting the Mediterranean with adjacent seas, are:

1. The Alboran Sea, which communicates through Gibraltar with the Ibero-moroccan region of the Atlantic (Maurin, 1968).

2. The North Aegean Sea: in that case the communication channel with the Black Sea being the Straits (Dardanelles, Bosphorus and the Sea of Marmara).

3. The South-East Mediterranean: that "subtropical saline cul-de-sac" (Por, 1978), which connects with the upper part of the Red Sea (Gulf of Suez) through a man-made communication channel: the Suez Canal.

Subdivisions of the Mediterranean: traditionally divided, mainly on topographical grounds, into a Western and an Eastern Basin, the Mediterranean region is often subdivided into the following subregions:

I. Western Basin

1. Alboran (Spain-Morocco)
2. Catalunian (Spain mainland-Balearic islands)
3. Balearic (Spain-Algeria-Balearic I.)
4. Gulf of Lion (S. France)
5. Ligurian-Provençal (S. France-Italy-Corsica)
6. N. Tyrrhenian
7. S. Tyrrhenian

(6 and 7 separated approximately by 40°N)

Eventually there is also a central area, W. of Sardinia.

II. Eastern Basin

1. Adriatic (N. of Otrando strait) — Upper Adriatic; Lower Adriatic
2. Ionian (Italy-Greece)
3. Sicilian-Libyan (from Tunisia to approximately 20°E)
4. Aegean N.
5. Aegean S.

(4 and 5: Greece-Turkey, separated approximately by 38°N)

6. Levantine (no clear-cut western limits).

TRANSITIONAL AREAS

The boundaries of the above mentioned subregions constitute also transitional areas of secondary importance, as compared with the 3 major ones.

Oceanographers and marine biologists have stressed the importance of water circulation in planktonic horizontal migration, distribution, diversity and abundance. In the Mediterranean this was done in two ways:

(a) As a result of the negative water balance of this Sea, the quantitative importance of the Atlantic hyposaline current, entering the Mediterranean through Gibraltar and, to much less extent, Black Sea water entering in the Aegean through the Straits, amply surpass the counter-currents in opposite directions. Following the fate of those water masses all along the Mediterranean, was and still is a favorite exercise of physical oceanographers. The surface circulation pattern, described in the beginning of this century by Nielsen (1912), still holds true in its broad lines. More recent studies convincingly explain also the fate of intermediate and deep waters (Cruzado, 1983).

(b) The zooplanktonologists have tried to contribute to our

knowledge of water mass circulation and its biogeographic implications, by using various subspecies, species and groups as indicators. But, in characterizing water masses and in following their fate in space and time, biological indicators are, at most, subsidiary; they cannot compete with the simplicity and accuracy of the temperature/salinity gradients.

Nevertheless, by mere definition, ecosystems are essentially biological; planktonic ecosystems and communities are still characterized by faunistic composition and standing stock measurements. Species diversity and biomass variation would remain key factors in describing planktonic ecosystems. For sure, any correlation of those biotic factors with hydrological and climatic parameters, is of value. But does these correlations alone represent the necessary and sufficient conditions for the description of a planktonic community ?

Let me give an example on the ambiguity of these correlations.

The Atlantic inflowing currents entering the Mediterranean are important both from the hydrological and the biogeographical point of view. Most authors agree upon the general circulation pattern of the North Atlantic Water (N.A.W.) mass, especially in the W. Mediterranean Basin. In reaching the Sicilian and Balearic channels the N.A.W. mixes up with resident waters, in the proportion of 70% to 30% respectively. The Atlantic influence can be traced down to the end of the Levantine subregion, but, in the meantime the N.A.W. loses its identity in transiting the Ionian Sea (Crusado, 1983). Therefore, speaking of a N.A.W. community, eastward of the 22°E meridian (S o/oo>38), appears as an oceanographic heresy.

Many planktonic organisms transported by the Atlantic current "have the quality of hydrological indicators" (Furnestin, 1979). There is evidence as to the marking of Atlantic influence by Chaetognatha, Pteropoda and some other groups. On the other hand, species previously considered as indicators of Atlantic waters, are now known to be indigenous to the Mediterranean (Hure and Scotto di Carlo, 1968). According to the last author "the question of plankton indicators of Atlantic water still remains very much open to debate".

The apparent divergence of views between M.L. Furnestin and Scotto di Carlo and Ianora (1983) on the most important transitional area of the Mediterranean,that is the Alboran Sea (A.S.), is of a rather "quantitative" nature, the former author emphasizing the Atlantic origin of zooplanktonic species diversity and abundance in the Alboran Sea (A.S.) - which acts both as a communication channel and a barrier between the Mediterranean and the Ocean; the latter stressing that "the A.S. may not have a distinctive species composition, when compared to other Mediterranean subregions but it probably differs notably in plankton production" being "one of the most productive areas in the Mediterranean". Scotto di Carlo and Ianora (1983) attribute those characteristics of A.S. to the transient upwelling along its N.W. coast, due ultimately to the Atlantic current itself.

The analogy of the Alboran Sea with another major transitional area, namely the North Aegean Sea (N.A.S.), is striking but "the flow" is reversed. In both cases, the Mediterranean acts as a recipient of usually colder waters with lower salinity. It receives also immigrants of Atlantic and Pontic origin and eventually exports its own "endemic" forms. However, the final result goes the opposite way: the Atlantic ocean, with its wealth of production and specific diversity, imprinted, long ago, its faunal characteristics on the whole Mediterranean; even today and in spite of the Gibraltar barrier, this ocean plays an important role in the Western Basin; the "atlantization" of the Mediterranean is an established fact. On the other hand, the shrinkage of the Sarmatic into present-day Black Sea and the peculiar oceanographic conditions prevailing there, limit the actual pontic influences on the Mediterranean, to a few, ecologically insignificant, planktonic immigrants (Demir, 1959; Bogdanova, 1973; Moraitou-Apostolopoulou, 1976). The reverse trend, the so-called "mediterraneanization" of the Black Sea (Bacescu, this book) is an increasingly important process, in which to the natural causes man-made alterations are added. It seems obvious that, whatever the hydrological situation, it is always the ecologically superior region which affects the inferior one. In that case the Atlantic affects the Western Mediterranean and the Eastern Mediterranean the Black Sea. Transitional areas, as the Alboran and the N. Aegean, witness this, almost unidirectional, process.

PARTLY ISOLATED AREAS

When considering planktonic communities at smaller scale, e.g. in a gulf or bay, partly isolated from the open sea, one thinks that the composition of the fauna and the restricted species diversity would be determined, to a large extent, by local geographic, hydrological and ecological conditions.

These, ecologically sensitive areas are often under an intense, man-made environmental pressure: overfishing, urban and industrial pollution, tourism, heavy maritime traffic, oil leakage, etc. As a result, the initial structure of the ecosystem is profoundly altered. Many such areas exist in Greece and some are thoroughly investigated (Saronikos gulf, North Evoïkos gulf, Amvrakikos gulf). There are similar examples from all around the Mediterranean. Due to easy access and vicinity of marine stations and laboratories the planktonic communities are well described and subject to quantification and modelling. Next step, which has yet to be taken, is to consider these communities as <u>distinctive units</u>, whose main characteristics would be not a trivial response to temporarily prevailing ecological conditions - identical or comparable with those of other places in the Mediterranean - but the result of an evolution and differentiation from the biogeographical subregion, to which the smaller area belongs. In other words: does the ecology of a partly isolated area make up the local zooplanktonic community out of a common mediterranean faunal stock or has biogeography created in the past a distinctive unit there, merely reacting to environmental challenges of the present ? The ambiguity of relations between Biogeography and Ecology are based on such dilemmas. A planktonologist, whose preferences lie definitely on the biogeographical rather than the ecological side, writes: "Those considerations.... bear witness to the fact that the ecological and zoogeographical fields are related and that it is often difficult to define their limits" (Furnestin, 1979).

ENDEMISM

We actually know that the faunistic composition comes from a variety of sources and that, in the past, a number of endemics evolved from ancestral, autochthonous and heterochthonous forms, according to local conditions.

The endemism in the Mediterranean is still a subject of controversy (Tortonese, 1978). In particular, there is conflicting evidence as to the level - generic, specific or subspecific - on which the endemism is expressed. Naturally, the answer varies with the zoological group considered. Evolutionary isolation mechanisms, known to play a major role in speciation among terrestrial, benthic and nectonic animals (fishes), are difficult to demonstrate in zooplankters. However, physiological, reproductive and ecological barriers do exist in the continuum of marine waters. Molecular genetic studies of marine zooplanktonic species, up to now limited to lagunar animals like Artemia, are needed to provide us with convincing answers to this problem.

When considering the establishment of zooplanktonic communities in transitional or in partly isolated mediterranean areas, it is not the scarcity of data but the lack, as yet, of a comprehensive theory, that characterizes the field. General descriptions of the planktonic ecosystem, in both the Western (Gaudy, this book) and the Eastern Basin (Moraitou-Apostolopoulou, this book) are available. Comparative studies of some planktonic groups suggest that in semi-isolated coastal areas the zooplanktonic communities are more the product of prevailing ecological conditions than of biogeographic history (Cladocera: Moraitou-Apostolopoulou and Kiortsis, 1973; 1977; 1982; Kiortsis and Moraitou-Apostolopoulou, 1975; Mysidacea: Hadjakis, 1982; Siphonophora: Malouchou-Grimba, 1984; references limited to recent work in Greek waters). The situation may be quite different in major and secondary transitional areas (Rampal, 1975; Pavlova, 1966; Kiortsis 1974; Ghirardelli and Rottini, 1970) but no conclusive evidence is yet at hand.

In the context of this brief report more questions arise than responses are given. For instance, why is it not the dominant species of a transitional region, which preferentially migrates into new grounds, but others ? Among the N. Aegean Copepods, the predominant Temora stylifera is not invading the Black Sea, as do 21 other species of secondary importance. Instead of the most frequent epipelagic Siphonophore of the Aegean, Bassia bassensis, it is the less common Eudoxoides spiralis, which reaches also the Black Sea. It is true that describing an ecological niche in marine planktonic community - a niche to become vacant and subsequently occupied by a newcomer - is not an easy task.

In the Mediterranean, as in other semi-closed seas, natural and man-made communication channels allow the continuation of faunal exchanges. More questions arise: what is the extent and biological significance of the extraneous contributions to the zooplanktonic Mediterranean fauna ? Is the gene flow of species already established in the Mediterranean but of Atlantic, Pontic or Erythrean origin, important to the actual genetic constitution of Mediterranean populations ? Does the encounter of the gene pools of indigenous and exogenous populations have any significant effect ? How the challenge of increasing pollution of the Mediterranean coastal waters would be met by its evolving zooplanktonic populations ? Although, we hope to get soon answers to those questions, through experimental work on population genetics and environmental toxicology, a clear understanding of the taxonomic status of the planktonic populations in transitional and partly isolated areas, is a prerequisite to such studies, especially at the biochemical and molecular level.

At this point, I would like to emphasize the theoretical importance of getting convincing answers to these questions. There are also practical implications, especially for the protection of the marine environment and the stabilization and upgrading of sensitive ecosystems. The Mediterranean Sea is ecologically threatened; only an in depth knowledge of the actual dynamics and evolutionary prospects of its communities would give us the guidelines for its ultimate preservation.

REFERENCES

Băcescu, M. (this book). The effects of the geological and physicochemical factors on the distribution of marine plants and animals in the Mediterranean.

Bogdanova, A.K., 1973, Hydrological conditions of penetration of

Mediterranean Sea organisms into the Black Sea, Rapp. Comm. int. Mer Medit., 21:507.
Cruzado, A., 1983, Water masses and general circulation, in: "Quantitative analysis and simulation of Mediterranean coastal ecosystems: the Gulf of Naples, a case study", UNESCO Reports in Marine Science, No 20.
Demir, M., 1959, Pontellidae and Parapontellidae (Copepoda) from the Southern Black, Marmara and North East Aegean Sea, Publ. Hydrobiol. Res. Inst. Univ. Istanbul, Ser. B, IV:17.
Ekman, S., 1953, "Zoogeography of the Sea", Sidwick & Jackson, London.
Furnestin, M.L., 1979, Aspects of the zoogeography of the Mediterranean plankton, in: "Zoogeography and diversity in plankton," S. van der Spoel and A.C. Pierrot-Bults, eds., Arnold, London.
Gaudy, R., (this book), Features and peculiarities of zooplankton communities from the Western Mediterranean.
Ghirardelli, E., and Rottini, L., 1979, Chétognathes récoltés dans la mer Egée occidentale et la mer Ionienne orientale, Rapp. Comm. int. Mer Medit., 25/26:153.
Hatzakis, A., 1982, Les Mysidacés des mers grecques (morphotaxonomie, écogéographie, bionomie) Thèse Doctorat, Univ. Athènes (in greek).
Hure, J., and Scotto di Carlo, B., 1968, Comparazione tra lo zooplancton del Golfo di Napoli e dell'Adriatico meridionale presso Dubrovnik. I. Copepoda, Publ. Staz. Zool. Napoli, 36:21.
Kiortsis, V., 1974, Quelques considérations sur l'écologie du zooplancton en mer Egée. Répartition géographique et variations saisonières, Rapp. Comm. int. Mer Medit., 22:139.
Klausewitz, W., 1969, Remarks on the zoogeographical situation of the Mediterranean and the Red Sea, Ann. Mus. Sci. nat. Genova, 77:323.
Malouchou-Grimba, V., 1984, The Siphonophore Hydrozoans of the greek seas (morphotaxonomy, eco-geography, histology) Doktorate thesis, Univ. Athens (in greek).
Maurin, C., 1968, Ecologie ichthyologique des fonds chalutables atlantiques (de la baie ibéromarocaine à la Mauritanie) et de la Mediterranée occidentale, Rev. Trav. Inst. Pêches marit., 32(1):147.
Moraitou-Apostolopoulou, M., 1976, Influence de la Mer Noire sur la composition de la faune planctonique (Copépodes) de la Mer Egée, Acta Adriat., 18(16):271.
Moraitou-Apostolopoulou M., (this book), The zooplankton communities of the Eastern Mediterranean (Levantine Basin, Aegean Sea); influence of man-made factors.
Nielsen, J.N., 1912, Hydrography of the Mediterranean and adjacent waters, Rep.Dan.Oceanogr. Exped. 1908-1910, Medit. adj. Seas, Hydrography, 2(2):77.
Pavlova, E.V., 1966, Composition et repartition du zooplancton en mer Egee, in: "Recherches sur le plancton des mers du sud", Ed. Nauka (in russian).

Por, F.D., 1978, "Lessepsian migration" Ecological studies, v. 23, Springer Verlag, New York.

Rampal, J., 1975, Les Thécosomes. Systématique et évolution. Ecologie et biogéographie méditerranéennes. Thèse Univ. Provence, Marseille.

Scotto di Carlo, B., and Ianora, A., 1983, Standing stocks and species composition of Mediterranean zooplankton, in: "Quantitative analysis and simulation of Mediterranean coastal ecosystems: The Gulf of Naples, a case study," UNESCO Reports in Marine Science, No. 20.

Tortonese, E., 1978, How is to be interpreted a "Mediterranean" species ? Thalassographica, 2(1):9.

CONTRIBUTORS

Prof. Carlos Almaça, Laboratorio de Zoologia, Faculdade de Ciências, Lisboa, Portugal

Dr. Mihai Băcescu, Musée d'Histoire Naturelle "Grigore Antipa", Bucuresti, Romania

Prof. Bruno Battaglia, Istituto Biologia Animale, Università di Padova, Italia

Dr. Gérard Bellan, Station Marine d'Endoume, Marseille, France

Dr. Denise Bellan-Santini, Station Marine d'Endoume, Marseille, France

Dr. Adam Ben-Tuvia, Department of Zoology, The Hebrew University of Jerusalem, Jerusalem, Israel

Prof. Francesco Cinelli, Istituto di Zoologia e Anatomia Comparata, Università di Pisa, Pisa, Italia

Dr. Marta Estrada, Instituto de Investigaciones Pesquera, Barcelona, Spain

Dr. Giancarlo Fava, Istituto Biologia del Mare C.N.R., Venezia, Italia

Dr. Gaston Fredj, Laboratoire d'Océanographie Biologique, Nice, France

Dr. Raymond Gaudy, Centre d'Océanologie de Marseille, Faculté des Sciences de Luminy, Marseille, France

Dr. Josep Maria Gili, Departament d'Ecologia, Universitat de Barcelona, Barcelona, Spain

Dr. Lydia Ignatiades, Nuclear Research Center "Demokritos", Athens, Greece

Prof. Vassili Kiortsis, Zoological Laboratory, University of Athens, Athens, Greece

Dr. A. Kocatas, Department of General Zoology, Ege University, Bornova, Izmir, Turkey

Prof. Ramon Margalef, Department of Ecology, University of Barcelona, Spain

Prof. Maria Moraitou-Apostolopoulou, Zoological Laboratory, University of Athens, Athens, Greece

Dr. Artemis Nikolaidou, Zoological Laboratory, University of Athens, Athens, Greece

Dr. G. Høpner Petersen, Zoological Museum, University of Copenhagen, Copenhagen, Denmark

Dr. Trevor Platt, Marine Ecology Laboratory, Bedford Institute of Oceanography, Dartmouth, Canada

Prof. Joandomènec Ros, Departamento de Ecologia, Universidad de Murcia, Murcia, Spain

Prof. Michele Sarà, Istituto di Zoologia, Università di Genova, Genova, Italia

Prof. Enrico Tortonese, Istituto Zooprofilattico, Genova, Italia

Dr. Nikos Tsimenidis, Institute of Oceanographic and Fishery Research, Athens, Greece

Dr. George Verriopoulos, Zoological Laboratory, University of Athens, Athens, Greece

Dr. Kostas Yannopoulos, Zoological Laboratory, University of Athens, Athens, Greece

Mrs Argyro Zenetou, Zoological Laboratory, University of Athens, Athens, Greece

SPECIES INDEX

SUBJECT INDEX

GPSR Compliance

The European Union's (EU) General Product Safety Regulation (GPSR) is a set of rules that requires consumer products to be safe and our obligations to ensure this.

If you have any concerns about our products, you can contact us on ProductSafety@springernature.com

In case Publisher is established outside the EU, the EU authorized representative is:

Springer Nature Customer Service Center GmbH
Europaplatz 3
69115 Heidelberg, Germany

Batch number: 10755805

Printed by Printforce, the Netherlands